ALLE · ZEIT · WACH
1842

I. A. Tarchevsky · G. N. Marchenko

Cellulose: Biosynthesis and Structure

Translated from the Russian by L.V. Backinowski and M. A. Chlenov

With 170 Figures and 53 Tables

Springer-Verlag Berlin Heidelberg New York
London Paris Tokyo Hong Kong Barcelona

Professor Dr. I. A. Tarchevsky
Professor Dr. G. N. Marchenko

Institute of Biology, USSR Academy of Sciences.
P.O. Box 30, Kazan 420084, USSR

Translated from the Russian by
L. V. Backinowski and M. A. Chlenov

ISBN 3-540-52247-6 Springer-Verlag Berlin Heidelberg New York
ISBN 0-387-52247-6 Springer-Verlag New York Berlin Heidelberg

Typesetting: Macmillan India Ltd., Bangalore, India. Printing: Mercedes, Berlin.
Bookbinding: Lüderitz & Bauer, Berlin.
2151/3020-543210 – Printed on acid-free paper

Introduction

Cellulose constitutes the most abundant organic material of the biosphere and is widely used in national economies. Despite the tremendous importance of cellulose, elucidation of the mechanism of its synthesis and formation of molecular and supermolecular structures has attracted much less attention than other biopolymers, viz. proteins and nucleic acids. Many textbooks and specialized monograhs on cellulose point to the scarcity of our knowledge of the mechanism of biosynthesis of this polymer. Reference may be made to METZLER [1], the author of the three-volume edition 'Biochemistry. Chemical reactions in Living Cell', who states, "In spite of the evident significance of the problem, details of synthesis of cellulose and chitin, built of regular polymeric chains with β-1,4-linkages, are practically unknown."

At the same time, considerable experimental material has accumulated to date which allows us to work out a concept on regularities of alteration in cell wall composition, of main stages in biosynthesis of cellulose, of in-cell localization of the sites responsible for biosynthesis of various cell wall polysaccharides. Certain regularities are revealed concerning the regulation of synthesis of cellulose and other polysaccharides, orientation of cellulose microfibrils on the plasmalemma surface. Attempts are made to understand structural and functional peculiarities of cellulose-synthetase incorporated in the plasmalemma. Experiments are carried out with cellulose-synthetase preparations isolated from cells.

Some questions remain to be solved. They are the regularities of association of molecular chains of cellulose into microfibrils, synchronization of synthesis of different polysaccharides and mechanisms of their interaction leading to a complex and rather labile cell wall structure which is the basis of a plant "skeleton".

Much attention is paid to the studies of the fine structure of cellulose depending on species and growth conditions of plant, to changes in cellulose structure during its isolation from raw materials and modification aimed at the preparation of industrially-important products.

The combination, in one book, of two different, i.e. at first glance, trends in cellulose studies, viz. biological and physico-chemical, seems to be quite justified. Experts in analysis of cellulose structure have come to the conclusion on its relationship to conditions of biosynthesis, on the 'memory effect' of peculiarities of biogenesis. Specialists who study cellulose biosynthesis, especially its later stages, when formation of microfibrils occurs, display growing interest in structural characteristics of cellulose due to the practical importance of this property. In addition, current knowledge allows one to suggest that specific characteristics of the biosynthesis of cell wall polysaccharides largely predetermine their chemical structure and their physico-mechanical properties. However, up till now, these two interrelated aspects of cellulose physicochemistry have usually been discussed separately.

Taking into account the exceptional importance of cellulose as a potentially efficient, versatile, and essentially inexhaustible source of raw material for the synthesis of a broad spectrum of organic products which are at present mainly of petrochemical origin, it seems quite imperative and urgent to combine together the experimental data on the biosynthesis of cellulose and the formation of its structure in the broad sense of the word. That is why the authors hope, by bringing together, in a single book, sections on the biosynthesis and the structure of cellulose, to attract the attention of both biologists who study cellulose as a specific polymer and encourage them to promote this process.

The biosynthetic section is written by I. A. Tarchevsky and the one on structure by G. N. Marchenko. The authors have included many results obtained in their laboratories.

Our sincere thanks are due to academician A. L. Kursanov for permanent support of our investigations and for valuable advice in the preparation of this book. We are thankful to staff members who participated both in performing experiments and in book design, especially to V. V. Lozovaya Ph. D. (Biology), V. V. Bulidorov Ph. D. (Chemistry) and Skvortsova.

Kazan, July 1990

I. A. Tarchevsky
G. N. Marchenko

Contents

PART II Cellulose Structure

Abbreviations

ADP	adenosine diphosphate
ADPG	adenosine diphosphate glucose
Ara	arabinose
ATP	adenosine triphosphate
ATPase	adenosine triphosphatase
BD	Brounian dynamics
C	cellulose
CI	cellulose I
CII	cellulose II
CTN	cellulose trinitrate
DNA	deoxyribonucleic acid
DP	degree of polymerization
EBR	electric birefringence
ER	endoplasmic reticulum
FB	birefringence
Gal	galactose
GDPG	guanosine diphosphate glucose
GTP	guanosine triphosphat
Hyp	hydroxyproline
IR	infrared
Lys	lysine
MC	mercerized cellulose
MD	molecular dynamics
mRNA	messenger RNA
MW	molecular weights
MWD	molecular weights distribution
NADP	nicotineamide adenine dinucleotide phosphate
NC	nitrocellulose
NCC	nitrocellulose cotton
NDPG	nicotineamide adenine dinucleotide glucose
NDPS	nicotineamide adenine dinucleotide sugar
NLC-1	nitrocellulose bleached normal
NLC-2	nitrocellulose short staple
NMR	nuclear magnetic resonance
NTP	nicotine triphosphate
PEG	polyethyleneglycol
PMR	paramagnetic resonance
PPi	pyrophosphate
RC	regenerated cellulose
RNA	ribonucleic acid

Ser	serine
UDP	uridine diphosphate
UDPG	uridine diphosphate glucose
UTP	uridine triphosphate
UV	uridine netravisible
Xyl	xylose

Introduction

With the present-day level of progress of industrial potential in developed countries, its further advance largely depends on the prospects of solving resource problems. It should be emphasized that the primary carbonaceous raw materials consumed by industry are mainly oil and natural gas. The proportion of other carbonaceous materials (coal, shale, peat) does not exceed 5% [2].

In 1980, the world production of coal, oil, and natural gas amounted to 9.3×10^9 tons, of which only approx. 4% (330×10^6 tons) was converted into products of basic organic synthesis [2] and the major part being consumed for energy production. The waste of combustion products of various fuels is an essential consequence and its influence on the ecological situation is quite pronounced. The overall industrial exhaust of carbon dioxide alone increased by 4% per year during the decade 1970–1980, amounting to approx. 20×10^6 tons in 1980 [2, 3] that is approx. 10% of the total amount of CO_2 consumed in the biosphere for photosynthesis [3, 4]. Thus it is possible to state that, in fact, the biogeochemical turnover of carbon depends largely on anthropogenic factors.

It should be stressed that the sources of carbonaceous raw materials have been created during the extremely long period of evolution of Earth. Taking into account the continuously increasing rate of their consumption, it is possible to state that the resources are being irreversibly exhausted.

This necessitates the exploration of new approaches to the solution of the problem of carbonaceous resources which are promising with regard to possible overcoming of the aforementioned unfavourable trends. One approach is the elaboration of new, progressive technological processes. Photosynthesis is the largest chemical process which proceeds under natural conditions. It is this process that satisfies human requirements in food and oxygen and already serves as a source of raw material for some branches of industry.

Total annual production of biosynthetic products ranges according to different estimates, from 100 to 400×10^9 tons dry biomass [2, 4, 5] whereas the proportion of practically employable, the so-called economical harvest amounts to only 800–900 million tons.

Starting materials for photosynthesis are mainly water and carbon dioxide. Energy supply does not present any problem since photosynthesis consumes about 0.03% solar energy [6]. What is essential is that even in future no problem of environmental pollution arises and the only 'sideproduct' of photosynthetic activity is oxygen.

The basic characteristics of photosynthesis in nature as a biochemical process is its continuous operation, relatively short-term 'industrial' cycle, and essential inexhaustibility of starting materials and energy consumed. It should be pointed out, however, that the actual concentration of carbon dioxide in atmosphere is much below the optimum for photosynthesis [7]. Nevertheless, there exists, in principle, a possibility of solving the problem of finding sources of renewable raw material based on the utilization of structural polysaccharides formed during photosynthesis. Judging from the most conservative estimates the annually renewable plant biomass exceeds, by reserves of potential energy, all kinds of energy resources by a factor of 4 to 6 [2]. Biomass consists of direct products of photosynthesis (which are formed principally in chlorophyll-containing cells upon their illumination) and products of secondary synthesis which are formed in nonphotosynthetic cells from incoming transport products. Cellulose is synthesized from the intermediate and transport products of photosynthesis.

Of especial interest is the fact that annual production of cellulose amounts to 50 to 100×10^9 tons [2]. It is the more so important, in contrast to the other part of the biological harvest of aquatic and terrestrial plants, that it is a product that is not extensively used as a food or industrial raw material, although for a long time cellulose has been the basis for large-capacity industrial production of a broad spectrum of goods such as cotton fabrics and paper, artificial fibers and plastics, explosives and gun powders, emulgators and lacquers, etc.

Thus, the problem of renewable organic raw material can in principle be solved by natural photosynthesis and its intensification supplemented with processing of the corresponding products (carbohydrates, proteins, cellulose, etc.) into products presently available through petrochemical processes, viz. ethylenic, aromatic, and dienic hydrocarbons, alcohols, acids, etc.

Productivity of photosynthesis, like of any physiologo-biochemical processes, can be increased by optimization of external parameters such as temperature, water supply, mineral nutrition, concentration of CO_2, intensity of illumination. It is known that, under optimal conditions, the intensity of photosynthesis can be increased by a factor of 2 to 5 when compared with that under normal conditions. The other route for increasing the productivity of photosynthesis is the alteration of a plant genotype.

Without dwelling on the mechanism of photosynthesis, it is necessary to indicate that the most substantiated scheme for photosynthetic reduction of CO_2 carbohydrates is that by Calvin (Calvin cycle) who has shown that the primary act of photosynthesis is the addition of CO_2 molecule to ribulose-1,5-bisphosphate and the primary stable product of CO_2 fixation is a three-carbon molecule, phosphoglyceric acid. It is this acid, that is further converted into the final products of photosynthesis—sugars, polysaccharides, lipids, and proteins.

Studies of photosynthesis in sugar cane, maize, and other plants, performed between 1954 and 1966, revealed another route of primary fixation of CO_2 which precedes the Calvin cycle. This route starts with the addition of CO_2 to phosphoenolpyruvate, giving rise to a very labile four-carbon compound, oxaloacetate and the first stable products of CO_2 fixation are mainly malate or aspartate. They undergo oxidation or deamination to give pyruvate and CO_2 which is then involved into the Calvin cycle.

Thus, higher plants can be subdivided into two groups depending on the mechanism of photosynthesis, viz. C_3- and C_4-plants. The former group comprises almost all the plants of moderate climates and algae, arboraceous species and certain tropical plants. The C_4-group consists of sugar cane, maize, sorghum, millet, etc. It is essential that under the conditions of hot, dry climates the C_4-plants exhibit a higher intensity of photosynthesis (60–80 mg CO_2) than the C_3-plants (15–40 mg CO_2). And accordingly, the biological harvest of C_3-plants can reach 25 tons dry mass per hectare per year whereas in C_4-plants this figure can be as high as 55 tons. Under conditions of moderate climate, however, productivity of C_3-plants exceeds that of C_4-plants [8].

The exhaustion of resources of fossil organic raw materials makes it pertinent to forward wider involvement of plants into economic industrial turnover as a question of necessity. It is even more important that only 2000–3000 of the known 250,000 species of annual plants find a variety of applications, and as a rule this is not a production of biomass for subsequent chemical processing [6].

Since the major part of products of photosynthesis is unsuitable or is not employed for nutrition purposes it is quite natural to direct it for processing into traditional products of organic synthesis, including petrochemical one.

Primarily, wastes are concerned which are left after crop harvesting and which are potential sources of cellulose material,—straw, chaff, stems and husk of sunflower and cotton, and so on, as well as those in forestry and the woodworking industry. From tentative estimates, agricultural and forestry wastes in the U.S.S.R. amount to 600 million tons per year—that is equivalent from the calorific value to 100 million tons oil [2]. Prospects exist for cultivation of rapidly growing and highly productive plant species.

Of obvious interest is chemical processing of the main products of photosynthesis, namely, carbohydrates. They can be subjected to hydrogenation and hydrogenolysis, etherification, oxidation, etc., to yield products which are considered currently as purely petrochemical ones. Widely developed is the enzymatic transformation of carbohydrates into ethanol which is further dehydrated into ethylene, the most abundant petrochemical product. Fig. 1.1 summarizes possible transformations of organic raw material in question. Naturally, this figure does not reflect all the possible variants of chemical intermediates which can be derived from cellulose as the starting raw material. Nevertheless, it follows with certainty that even now, in the framework of the practically accepted technology on this basis, the production of a broad spectrum of chemicals is possible including large-scale ones: ethylene, butadiene, acetaldehyde, etc. currently produced by the petrochemical industry.

This allows us to formulate a thesis that a plant biomass including cellulose, which was considered until recently as a raw material which can be used along with other organic raw materials, may become, in the not too distant future, a predominant raw material source for the industrial chemistry.

This is undoubtedly one of fundamental problems of present-day science. Naturally, its solution is impossible without both detailed insight into the mechanism of cell wall formation, cellulose as its basic component and the progress in studies on structure of intact cellulose, its changes upon extraction

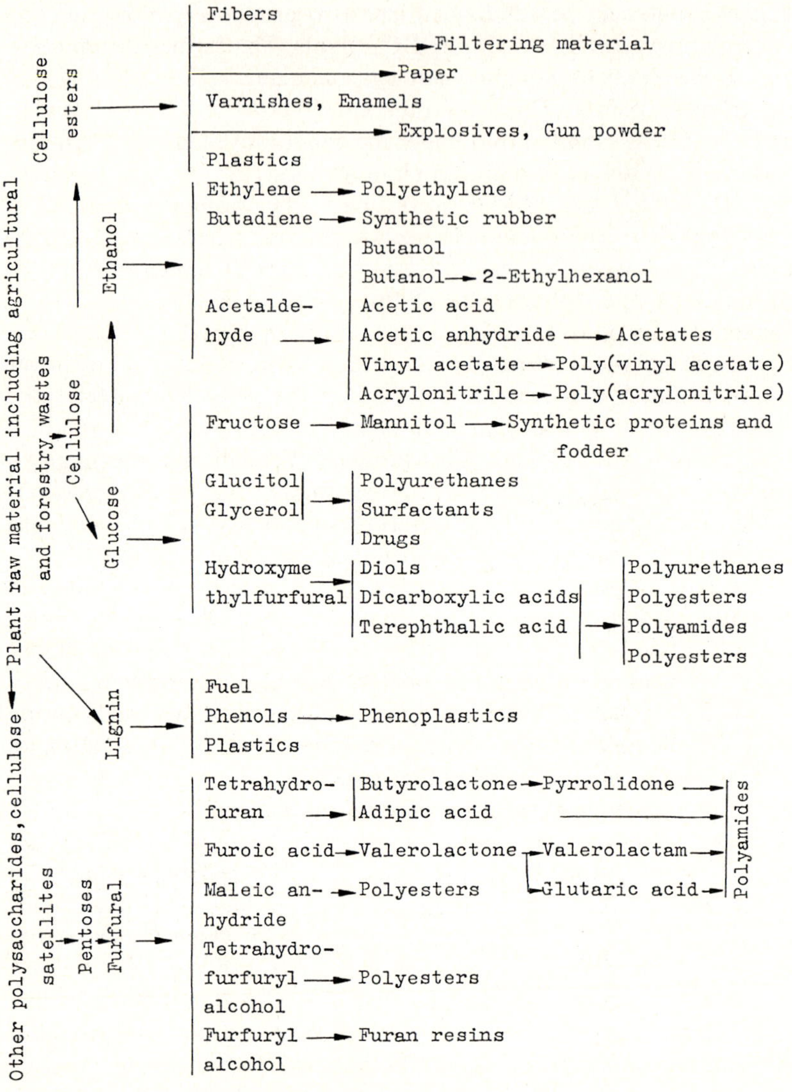

Fig. 1.1. Processing of structural polysaccharides

from cell walls and controlled modification for the purpose of preparing industrially important products.

In recent years, more and more attention has been paid by chemists to chitin as a potentially useful raw material for the chemical industry.

It should be mentioned that specific aspects of biosynthesis and structure of polysaccharides, cellulose in particular, have been extensively reviewed, though they have not been considered in total. This book should, to a certain extent, fill the gap. In preparing this monograph the authors thought it impossible to give a comprehensive bibliography and confined themselves to the most important literature sources. Their choice is probably subjective to some extent since it reflects the authors' concept of the questions under consideration.

Part I

Biosynthesis of Cellulose

1 Cell Wall Composition

A plant cell differs from an animal one in many respects, one of the main ones being the presence of a rather rigid cell wall which functions as a skeleton and protects the cell contents against the action of osmosis, the disruptive action of external mechanical forces, and the penetration of pathogens (bacteria etc.) into the cell.

An exception is represented by several algae with naked cells (protoplasts) which are therefore unable to maintain a definite shape, they exist permanently in an amoeba-like state. These cells are most often at the gamete or zoospore stage [9]. The enormous diversity and unique appearance of unicellular algae are determined by the specific structure of their cell walls, this being one of the main systematic features.

Tens of specialized tissues and cells of higher plants are characterized by large diversity of shape, structure, and chemical composition of cell walls. The weight of algal cell walls usually amounts to 13–15% cell weight, in certain cases it constitutes up to 40%. In higher plants these values can also vary considerably depending on species and tissue, cell development stage and climatic conditions.

Cell walls are differentiated into primary and secondary ones. The former is typical for dividing and rapidly growing cells. Their thickness remains almost constant though their total surface area is increased severalfold during cell growth. When it is stopped it is accompanied with appearance and subsequent thickening of a secondary cell wall which moves the thin primary cell wall towards the cell surface. Primary and secondary cell walls differ considerably in chemical composition and structure. It is by secondary cell walls that the industrially important properties of plants (mature wood, cotton and flax fibers, etc.) are determined.

The largest variation in cell wall composition is observed in algae. This refers to the ratio of different polysaccharides, proteins and inorganic components (iron in *Volvocales*, calcium in *Characeae*, silicon in diatoms). According to Sedova [10] "Nature decided to conduct a vast experiment with algae to select from numerous polysaccharides the one which suits the best as a cell wall material". Noteworthy, a whole range of these "trial" polysaccharides, e.g. agar and carrageenans of red algae, laminarin and alginate of brown algae, are used as raw materials in the chemical and food industries and for cosmetics [8].

Cell wall polysaccharides differ from each other in a set of parameters. Cellulose and callose are homopolysaccharides built of glucose residues only. Hemicelluloses and pectic substances are heteropolysaccharides whose constituents are hexoses, pentoses, and uronic acids (Fig. 1.2). Unlike cellulose and

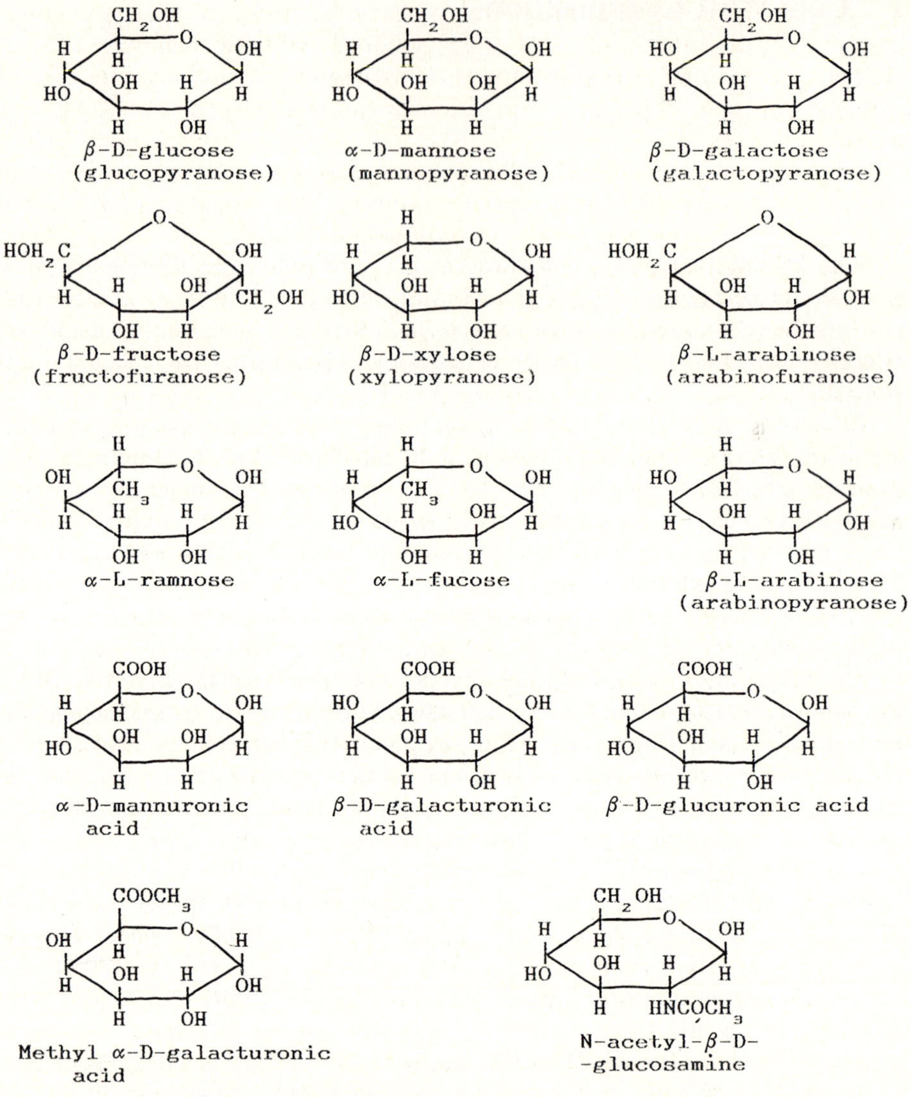

Fig. 1.2. Chemical structure of different monosaccharides

callose which are linear polymers, the majority of other cell wall polysaccharides are branched. They also differ in the number of the component monomers which ranges from thousands of glucose units in cellulose to hundreds and tens of monosaccharides and their derivatives in other polysaccharide molecules.

Differences in primary structures determine specific features of secondary and tertiary structures. Molecular chains of cellulose can form an ordered supermolecular structure, the so-called microfibrils.

It should be noted that the architectonics of cell walls can be very complex and diversified. Interesting data were obtained with the help of scanning electron microscopes on peculiarities of structure (or architecture) of cell surfaces, primarily, of pollen grains with inherent, complicated sets of apertures and excrescences.

An insight into the peculiarities of mechanisms of synthesis of cell wall components may probably be gained from studies of the formation of the shell of microspores and of pollen grains derived therefrom.

The cell wall of parent microspores consists mainly of callose which is pierced with plasmodesmas. The callose-containing cell wall of microspores derived from the parent cells is free from plasmodesma. The process of conversion of microsporous cell walls into a shell (sporoderma) involves formation of primexyne composed of cellulose microfibrils beneath the callose layer subject to hydrolysis. It is at this stage that formation of a unique, species-specific architecture of the shell of pollen grains begins [11]. The shell architecture is determined, in particular, by the outline of apertures, thin or perforated shell areas which serve for the exit of pollen tubes.

The structure of cell walls of conducting tissues is also very complex. Tracheids of the xylem of *Gymnospermae* possess ring-shaped, helical, and ladder nodes of cell walls, cellular elements of vasculi of xylemma of phanerogamous plants possess multiform perforations (pores and cross-beams).

The mechanism which determines the complex outline of cell walls of tracheids and xylem vasculi is also insufficiently clear. In xylem cells deposition of the secondary cell wall does not proceed continuously as it does with cotton fiber, rather it stratifies from the inside of the primary one as circles or helices. The secondary cell wall can contain cavities approaching the primary one, the so-called bordered pores. In the sieve platelets of phloem sieve tubes inside the primary cell wall there are callose-covered apertures (pores).

The sieve tubes of phloem are characterised by sieve plates of various complexity and porous regions of primary cell walls pierced with protoplasm cords from one sieve segment to another.

The architecture of cell walls of conducting vasculi of plants is formed as a result of both directed movement, within plasmalemma, of cellulose synthetase complexes and enzymatic digestion, dissolution of definite sections of the to-be-formed cell walls.

1.1 Cellulose

Cellulose is the highest-molecular-weight polysaccharide, the β-(1-4)-D-glucan of general formula $(C_6H_{10}O_5)_n$ where degree of polymerization (n) is species-dependent. For the cellulose of cotton seed-hairs the n-value is 2,000–6,000 for the primary cell wall and 13,000–14,000 for the secondary cell wall; for the cellulose from *Acetobacter xylinum* and *Acanthamoeba castellani* it ranges from

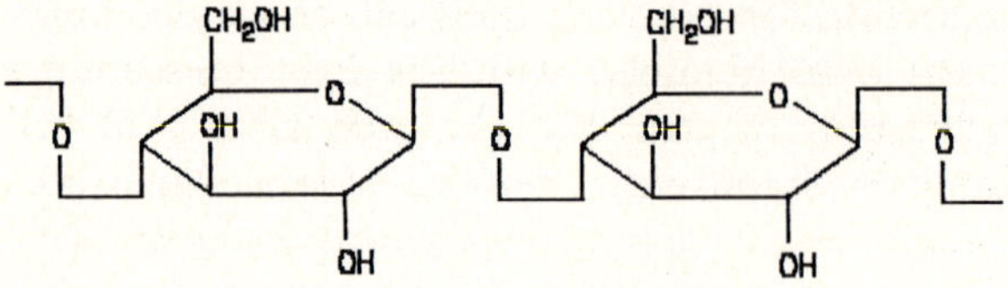

Fig. 1.3. Chemical structure of cellobiose unit

2,000 to 5,700 and from 2,000 to 6,000 respectively. The cellulose of the alga *Valonia* has the highest degree of polymerization (26,500), the *n*-values for cellulose from bast-fiber and wood from angiosperms and gymnosperms are 9,500, 8,200 and 8,400 respectively [12]. Despite being built of only glucose residues, cellulose can be regarded as a polymer of a β-1-4-linked disaccharide (cellobiose) units. This disaccharide is detected in fact in cellulose hydrolysate (Fig. 1.3).

Cellulose is almost absent from the middle lamella which partitions the maternal cell during formation of the daughter cells, it amounts to several per cent in the cell wall which is produced by isolated protoplasts. Its content is not very high (20–30% dry mass) in the primary cell wall, however, it constitutes the bulk of the dry material of the secondary cell wall. In several specialized cells, e.g. cotton-seed hairs, its content exceeds 90–95%, especially in secondary wall regions formed at the end of cotton fiber maturing.

The molecular chains of cellulose form, in cell walls, the supermolecular structures, microfibrils, comprising 60 to 70 cellulose threads.

Cellulose belongs to polymers that are very difficult to dissolve and this is determined by compact packing (mainly crystalline, with single amorphous region) of cellulose chains in microfibrils. Adjacent chains are linked not by covalent but by multiple hydrogen bonds and this is why cellulose is insoluble in water, dilute acids and alkalies, and in organic solvents.

1.2 Hemicelluloses

The term hemicelluloses refers to heteropolysaccharides that are associated with cellulose and are built of different hexoses, pentoses, and derivatives thereof, their D.P. ranging from 50 to 200. They are much more easily soluble than cellulose and this is due to the "looser" packing of component molecules, thereby being more accessible to solvents.

The majority of hemicelluloses are water-insoluble and only some of them can form aqueous, colloidal solution. Some hemicelluloses can be dissolved in dilute alkalies and precipitated by the addition of acids and ethanol. Hydrolysis of hemicelluloses into component sugars is usually carried out with 2 M hydrochloric acid at 100 °C.

Hemicelluloses can be subdivided into several groups on the basis of monosaccharide composition and ratio: 1) arabinans; 2) xylans (homo- and heteropolysaccharides including arabinoxylans, glucuronoxylans, arabinoglucuronoxylans, etc.); 3) glucans (homopolysaccharides including β-(1-3)-D-glucan callose, and heteropolysaccharides such as arabinoglucans); 4) galactans (both homo- and heteropolysaccharides, e.g. arabinogalactans); 5) mannans (homo- and heteropolysaccharides represented by glucomannans, galactomannans, and glucogalactomannans); 6) fructosans and glucofructosans; 7) galacturonans, and 8) mannuronans.

The main hemicellulosic component of monocotyledonous and dicotyledonous plants is represented by xylans [13] with a backbone built of β-(1-4)-linked D-xylose residues bearing 0–2 branches of single 4-*O*-methyl-D-glucuronic acid residues. Xylans of dicotyledonous plants contain, as minor components, glucose, arabinose, galactose, rhamnose, glucuronic and galacturonic acids. Xylans of monocotyledonous plants may have single arabinose branching attached to 0–3 of 'central' xylose or glucuronic acid residue attached to 0–2. There are data on the existence of nonbranched, linear homoxylan.

Another essential component of hemicelluloses is represented by xyloglucans. They are composed of glucose (its content is not less than 45%), xylose (30–36%), galactose (approximately 10%), fucose (6–7%), and arabinose [13]. Its backbone is built of β-(1-4)-linked D-glucose residues bearing mono-, di-, or trisaccharide side-chains. Shown below is one of regions of xyloglucan of D.P. of approx. 50 (Fig. 1.4).

It is of note that the content of xyloglucans in cell walls of monocotyledons (2%) is by one order of magnitude lower than in dicotyledons (20%).

Callose is a wound-induced, linear β-(1-3)-D-glucan. Its name comes from callus tissues formed on mechanically damaged parts of plants where it largely accumulates. Callose formation is also promoted by pathogenic microorganism infection or under stress environmental conditions.

Normally callose is the major polysaccharide in a cell plate at its early development stages and in primary cell wall of pollen tubes [14].

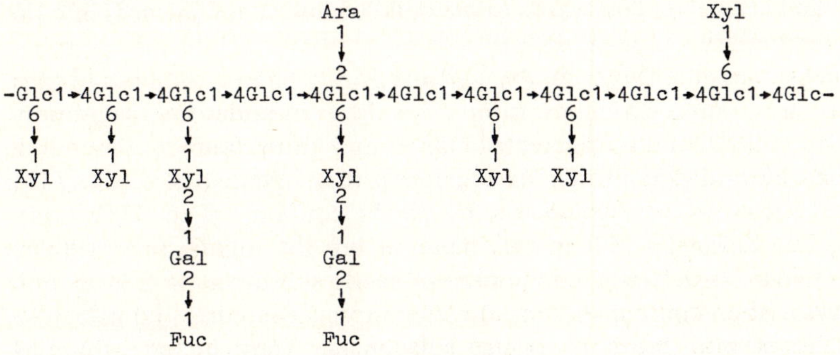

Fig. 1.4. General structure of the xyloglucans unit

The degree of polymerization of callose is approx. 600. Two linear β-(1-3)-glucans of this D.P. and identical primary structures but different in quaternary structures were shown to be synthesized by callose synthetase under conditions differing in concentrations of Ca^{2+}, Mg^{2+}, and β-glucoside [15].

The major hemicelluloses of dicotyledonous plants are neutral xyloglucans and acidic arabinoxylans, in monocotyledonous plants they are acidic arabinoxylans and neutral β-(1-3, 1-4)-glucans [16]. The glucans of barley (40 MDalton) can be digested by proteases to give polysaccharides of smaller molecular mass (1 MDalton). A conclusion was made on this basis that about 40 1-MDa-glucan molecules are attached to a common protein core.

1.3 Pectic Substances

Some pectic substances are soluble in warm water, the rest may be solubilized (pectin) upon boiling in dilute acids, oxalic acid in particular, or ammonium oxalate, or alkaline EDTA solution [17]. Pectin can be precipitated with ethanol from its aqueous solution to form a gel.

Given below is an example of fractionation of matrix polysaccharides based on the sequential extraction of cell wall material [18].

1) 0.05 M EDTA for 6 h at 20 °C, twice;
2) 0.05 M Na_2CO_3 plus 0.02 M $NaBH_4$ for 16 h at 1 °C;
3) 0.05 M Na_2CO_3 for 3 h at 20 °C;
4) 0.5 M KOH plus 0.01 M $NaBH_4$ for 2 h at 20 °C;
5) M KOH for 2 h at 20 °C, and
6) 4 M KOH for 2 h at 20 °C.

The first four fractions are mainly pectic substances while the last two are mainly hemicelluloses.

Pectic substances extractable from meristematic and parenchymatous tissues comprise 15–30% of the cell wall material while this quantity is only 0.5–1.5% for heavily lignified tissues. Lignification begins in the primary cell wall region and is extended outwards to the middle lamella and inwards into the developing secondary cell wall [18].

Pectic substances are rather labile, which seriously hampers studies of their primary structure. Their extraction from cell walls even under mild conditions can result in artefacts, that is fragments of larger and more complex compounds of polysaccharide and glycoprotein nature, rather than in native molecules [18].

Primary cell walls of dicotyledonous plants contain up to 35% pectic substances, but especially rich in this material are the middle (intercellular) plates. The most characteristic component of pectic substances is galacturonic acid, much less abundant constituents are rhamnose, arabinose, and galactose.

There are several types of pectic substances. They include rhamnogalacturonan I, homogalacturonan, arabinans, galactans, arabinogalactans, rhamnogalacturonan II, apiogalacturonan [13]. The backbone of the rhamnogalacturonan I is built of galacturonic acid and rhamnose residues to which arabinose and galactose units are attached, the ratio of these monosaccharides being approx. 2:1:1.5:1.5 and the degree of polymerization of about 2.000. Homogalacturonan is a linear, nonbranched acidic polysaccharide, carboxylic groups being both free and methylated and regions being present of low- and high-methoxy content. Degree of polymerization is equal to 25 or somewhat more. Arabinans are highly branched homopolysaccharides with alternating linear and branched regions. Two individual arabinans of D.P. 34 and 100 were isolated. Galactans belong to one more type of homopolysaccharides with β-(1-4)-linkages (the main linkage type) and β-(1-6)-linkages (approx. 4%). The degree of polymerization varies from 33 to 50. Arabinogalactans possess a galactan backbone with arabinobiose attached to O-3. Rhamnogalacturonan II is a highly branched molecule (D.P. 25–30) with galacturonic acid as the major backbone-building sugar (α-1-4-linkages) and rhamnose insertions; galactose, arabinose, 2-*O*-methylxylose, 2-*O*-methyl-fucose, apiose, and fucose are the components of side chains. Apiogalacturonans are characterized by the presence of apiose disaccharide side chains attached to O-2 or O-3 of the main galacturonan chain.

Rhamnogalacturonans are the major constituents of pectic substances. Galacturonic acid residues are partially methyl esterified. The polysaccharide backbone is built of α-(1-4)-D-galacturonic acid chains interspersed with (1-2)-linked L-rhamnopyranose residues. Side chains consist of D-galactose, L-arabinose, D-xylose, D-glucose and less frequently of D-mannose, L-fucose, D-glucuronic acid and such rare sugars as apiose, 2-*O*-methyl-D-xylose and 2-*O*-methyl-L-fucose, the methylated sugars occupying terminal positions of the side chains.

Cultured sycamore cells produce rhamnogalacturonan 1 built of D-galacturonic acid and rhamnose (300 residues each). About half of rhamnosyl residues are substituted with side chains at O-4.

The so-called rhamnogalacturonan II of cultured sycamore cells possesses similar basic polysaccharide chain composition. Short side chains of one to three sugar units are attached to O-4 of the L-rhamnose residues and to O-3 of some galacturonic acid residues. The backbone of this polysaccharide is built of blocks containing 25 galacturonic acid residues separated by L-rhamnose residues. Polygalacturonic blocks of adjacent pectin molecules interact with Ca^{2+} to form an egg-box-like structure as can be seen from the diagram (Fig. 1.5).

The following conclusions were made on pectin structures: 1. Pectins of the middle lamellae possess a rhamnogalacturonan backbone with regions of non-esterified galacturonic acid residues separated by highly esterified and branched regions which carry oligosaccharide substituents. The ratio of non-esterified to esterified regions is approx. 1:3. Neutral pectic polymers are probably not present in the middle lamellae. 2. The pectic polymers of primary cell walls have a relatively higher proportion of oligosaccharide chains and these chains are much longer than those of the pectins of the middle lamellae. (In the latter

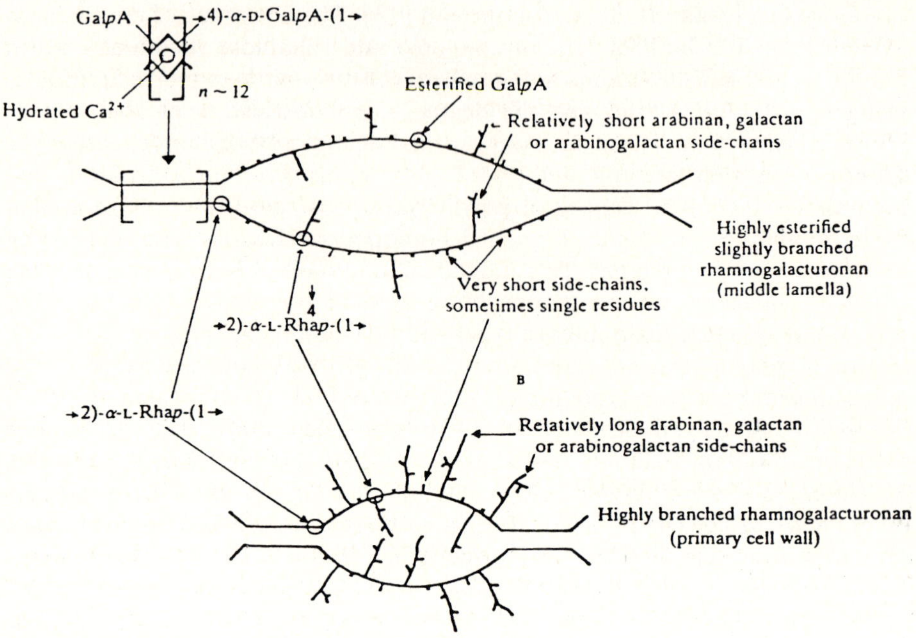

Fig. 1.5. Diagrammatic representation of some structural features of pectins from the middle lamellae (*A*) and primary cell walls (*B*)

the homopolygalacturonan regions are shorter.) In the flexible regions a fair proportion of the galacturonic acid residues is methyl esterified and some of them are ester-linked to hydroxyl groups of sugar residues of hemicellulosic origin or even to hydroxyl groups of polyphenolic compounds which are the part of some hemicellulosic complexes.

Calcium salts of pectins, the so-called pectinates, produce cohesion of the middle lamella due to ionic rather than covalent bonds. Ca^{2+} ions form stable complexes with blocks comprising 15–20 continuous galacturonic residues. Under the action of citrate, solubilization of pectic substances of the middle lamellae of potatoes occurs and neutral (N) and acidic (A) components can be isolated. The former was subdivided into N1- and N2-fractions. N1 is rich in proteins and is seemingly of intracellular origin. It may contain lectins and arabinogalactan proteins. N2 is very rich in mannose and is probably of enzymatic origin. Substitution pattern of sugar components in A-fraction is, in decreasing order, as follows: 4-substituted galacturonic acid, 4-substituted galactose, 5-substituted arabinose, terminal arabinofuranose, terminal galactopyranose, 2-substituted rhamnopyranose, 2,4-disubstituted rhamnose, and terminal *xylo*pyranose. Prolonged heating of A affords a "neutral" pectic polysaccharide containing 75% galactose and 15% arabinose residues. Fractionation of hot-water-soluble pectic substances of potatoes gives a "galactan" which contained 88% galactose, 5% arabinose, and 3% rhamnose residues [18].

Our knowledge of the functions of oligo- and polysaccharides is being extended as studies progress. The two well-known major functions are the reserve function, the classical reserve polysaccharides being starch and glycogen, and the structural function. Several matrix polysaccharides of cell walls of seed tissues are reserve ones, *xylo*glucans of tamarind resting seeds and pectic galactans of lupine seed cotyledons, for example. Much evidence has accumulated showing their information (signal) functions of the hormonal and recognition type.

1.4 Cell Wall Proteins

Analysis of aminoacid composition of cell-wall glycoproteins of higher and lower plants has shown them to be rich in hydroxyproline and to resemble structural proteins of animal origin. Those who studied the hydroxyproline-rich protein of cell walls have ascribed to it the leading role in the extension ability of cells and named it therefore 'extensin' [19]. Indirect proof of its possible involvement in cell extension can be obtained from extension of *Chlamydomonas* cells whose walls are composed of glycoprotein with the protein portion rich in hydroxyproline.

Proteins isolated from plant cell walls are representative arabinogalactan-glycoproteins. The extensin molecule has a protein backbone as a rigid rod [20], it is rich in hydroxyproline and poor in methionine [21, 22].

Extensin is a basic and poorly soluble glycoprotein of primary cell walls. The molecular mass of its protein portion is 40,000 Da, it is also rich in serine, lysine, and tyrosine. The ratio of protein to carbohydrate portions is 1:1.

The majority of hydroxyproline residues of extensin of dicotyledons is glycosylated with oligosaccharide chains built of 3–4 arabinofuranosyl units, the extent of protein glycosylation in monocotyledons is smaller (for a review see Kristen, [23]). It is of note that deglycosylation of extension results in loss of its helical conformation [24].

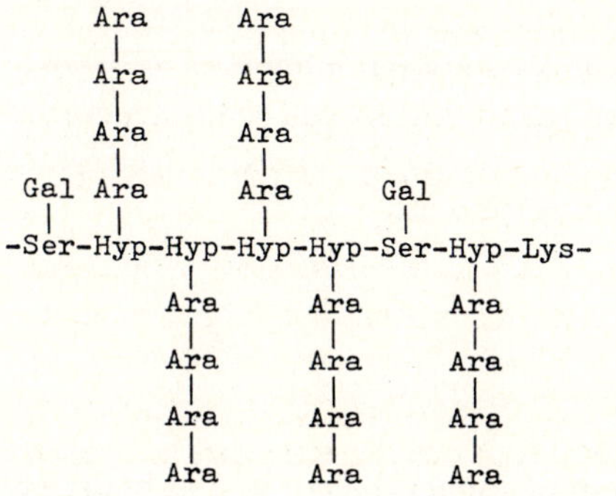

Fig. 1.6. General structure of glycoproteins unit

The glycan chains of extensin are composed mainly of arabinose, galactose content is lower and even less than that of xylose and glucose [25]. Total mass portion of the glycan part is about 45% [22, 26] or, according to some other estimates [27] even 74%. The galactose side-chains are attached covalently to serine [20] and oligoarabinoside ones—to hydroxyproline moieties [28].

Shown below is a partial structure of glycoprotein of cell walls of dicotyledonous plants with characteristic arabinotetraose side-chains [13] (Fig. 1.6).

On the basis of kinetic studies with [^{14}C]-proline as the extensin precursor Lamport and coworkers [29, 30] have shown that the cell wall contains three hydroxyproline-rich proteins. One of them as being more 'mature' contains many tetra-, tri-, and mono-arabinosyl residues. Of the two others, present in minute quantities but exhibiting high specific radioactivity of the component hydroxyproline, one is almost totally nonglycosylated and the second is characterized by the presence of arabinose moieties. Most likely they reflect different stages of 'maturing' of the main cell wall protein.

The glycoprotein of the cell walls of monocotyledonous plants differs from that of dicotyledons described above in that 65–75% of hydroxyproline residues of the polypeptide backbone are not glycosylated and the rest bear mainly tri- and, to lesser extent, tetra-, di- and monoarabinosyl side-chains.

Arabinogalactan-containing proteins are present not only in the primary cell wall but also in the middle lamellae and differ from extensin in better solubility in water and much lower protein percentage, approx. 7% [23]. The protein moiety is rich in hydroxyproline, serine, alanine, and glycine and the carbohydrate moiety contains galactose, arabinopyranose and arabinofuranose, rhamnose, mannose, galacturonic and glucuronic acids. Carbohydrate chains with a β-D-galactosyl-hydroxyproline linkage region are somewhat branched. Lectins can also belong to cell wall components.

Primary cell walls contain not only 'structural' proteins but enzymes as well, mainly of glycoprotein nature. They include phosphatases, glycosyl hydrolases, glycosyl transferases, endoglycanases and isozymes of peroxidase.

Diversity of linkages between different components of cell walls (primary cell walls and of middle lamellae) which is but to a minor extent reflected in a scheme suggested by Fry [16] gives an impression of a very complex three-dimensional structure of this essential cell compartment.

Quite obviously the diversity of cross-links between polymers is largely favoured by multiple modes of endogenous regulation of cell wall structures. This involves changes in secretion rate of ions, primarily H^+ and Ca^{2+}, enzymes which modify polymeric structure, various effectors, etc.

1.5 Lignin

Lignin is produced by oxidative condensation of coniferyl alcohol and hydroxycinnamic acid derivatives. Its molecular mass exceeds 10,000, it cannot be solubilized even with 70% sulphuric acid, and in forms a complex, three-

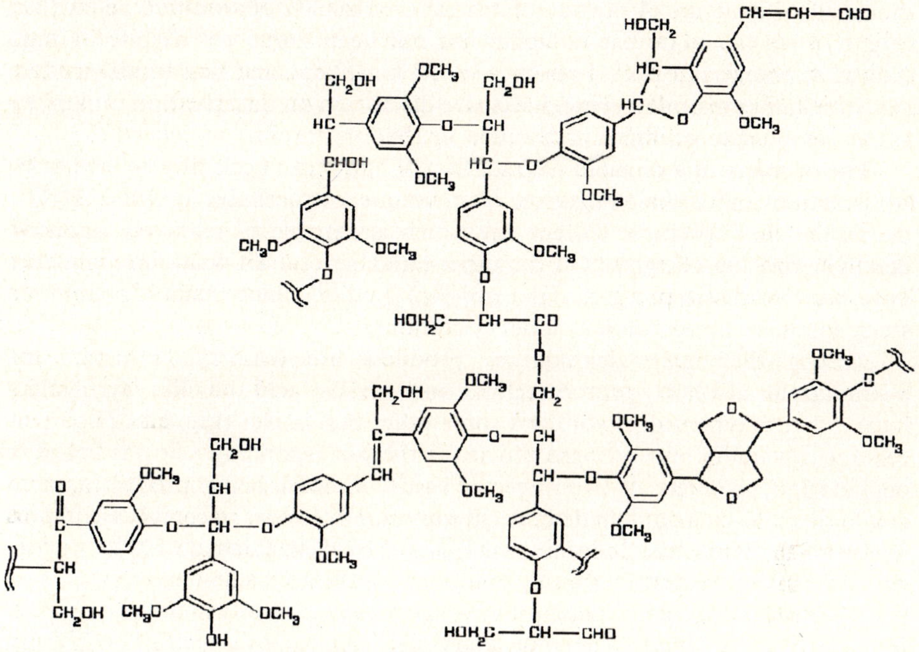

Fig. 1.7. Chemical structure of lignin macromolecule portion

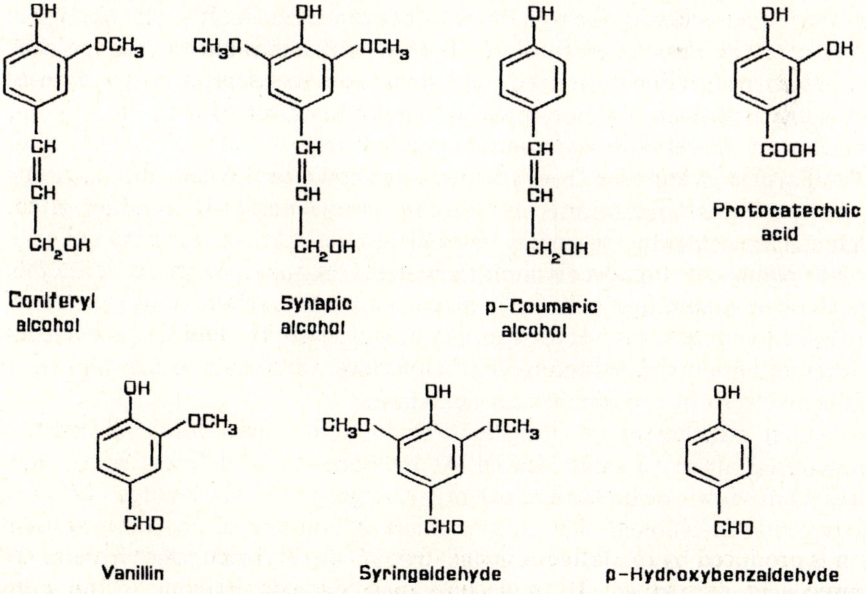

Fig. 1.8. Chemical structure of different lignin degradation products

dimensional structure. A portion of its macromolecule is reproduced below [31] (Fig. 1.7). An adequate use of lignin has not been found yet despite its high industrial potential, colossal reserves in the biosphere, and waste side-production of cellulose manufacturing. Extensive discussion on this question took place at the symposium on lignin in Sweden in 1986.

The problem of a directed degradation of lignin can hopefully be facilitated by isolation and characterization of a unique extracellular enzyme (H_2O_2-dependent diarylpropane oxygenase) which is capable of relatively intensive depolymerization of lignin. On the other hand, several artificial lignomimetics have been synthesized, e.g. arylglycerol-β-aryl ether, which make it possible to study ligninase action under model conditions.

Among the lignin degradation products are coniferyl, synapic, and *p*-coumarylic alcohols, protocatechuic and vanillic acid, vanillin, syringaldehyde, *p*-hydroxybenzaldehyde, and some other derivatives (Fig. 1.8). High percentage of vanillin and syringaldehyde (or the corresponding acids) is found in degradation products of arboraceous lignin, vanillin being predominant in coniferous and syringaldehyde in deciduous plants. Leaves of cereals are almost free from the latter, they contain much *p*-hydroxybenzaldehyde [32].

1.6 Interactions Between Cell Wall Polymers

Known to date are only some general features of the interaction between cell-wall polysaccharides and glycoproteins [13, 33, 34].

It has already been mentioned that cellulose molecules are associated with each other in microfibrils due to numerous hydrogen bonds. It seems likely that the surface of cellulosic microfibrils is 'covered' with a monolayer of xyloglucan which is hydrogen-bonded. Indeed, xyloglucan can be extracted with 8 M urea which splits hydrogen but not covalent bonds. Removal of urea leads to the restoration of the cellulose-xyloglucan initial structure.

The purified xyloglucan from the pea was shown to prevent ribbon formation from cellulose microfibrils in *Acetobacter xylinum* [14]. Binding of the xyloglucan molecules by the surface of cellulose microfibrils in primary cell walls of higher plants can supposedly limit their size. It is quite possible that another hemicellulosic material of cell walls, glucuronoarabinoxylan, is also noncovalently linked with cellulose. Hydrogen bonding was shown to be responsible for the interaction of glucoarabinoxylan molecules with each other and with xyloglucans to form supermolecular aggregates.

Covalent attachment of xyloglucan chains and pectic polysaccharides is demonstrated albeit to small degree. At the same time, different pectic substances were shown to be both covalently (the preponderant linkage type) and noncovalently interlinked. The former exists in bonding of neutral arabinans and galactans to acidic rhamnogalacturonan I, of homogalacturonans to rhamnogalacturonans I and II. The linkage via calcium ions is the most significant of the latter bonding type. Calcium is capable of chelating carboxylic groups of four galacturonic residues in two adjacent polysaccharide chains. A suggestion on covalent linkages between hydroxyproline-rich proteins and cell-wall polysaccharides has not been yet convincingly proved [13]. Shown in Fig. 1.9 is a tentative scheme of molecular organization of a plant cell wall with possible interlocation of the so-called 'structural' polymers [33].

In a review on cross-links between matrix polysaccharides of a growing (primary) cell wall the following linkage types are considered [16] (Fig. 1.10).

I. Covalent linkages.

1. Glycosidic [e.g. Arabinogalactan (Gal-rha) Rhamnogalacturonan I]
2. Ester, via uronic acid residues [e.g. Pectin (uronoyl-Glc) cellulose]
3. Ester, via feruloyl residue, etc. [e.g. Feruloyl (Ara) pectin]
4. Phenolic coupling
   a) ether type [e.g. Extensin (Tyr-Tyr) extensin]
   b) biphenyl type [e.g. Pectin (Fer-Fer) pectin]
5. Ether [e.g. Polysaccharide (ether) feruloyl-polysaccharide]
6. Disulphide (Cystine)

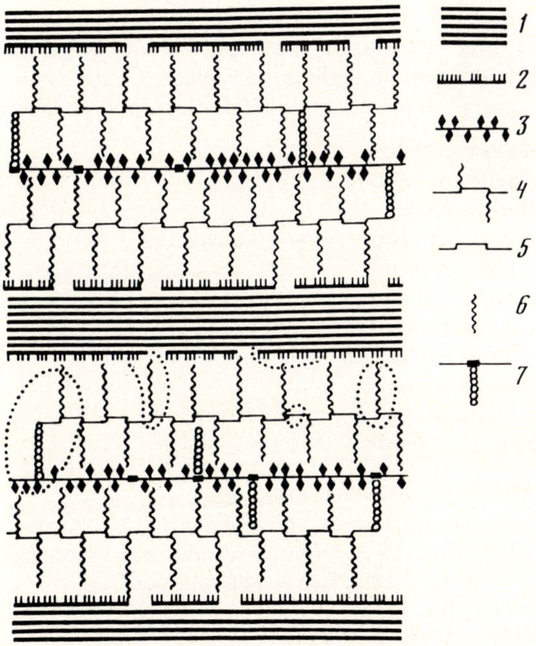

Fig. 1.9. Tentative structure of plant cell wall [33].

1 elemental cellulose fibril, *2* xyloglucan, *3* cell-wall protein, *4* pectic polysaccharides, *5* basic chain of rhamnogalacturonan, *6* arabinogalactan chain, *7* arabinogalactan attached to serine of the cell-wall protein

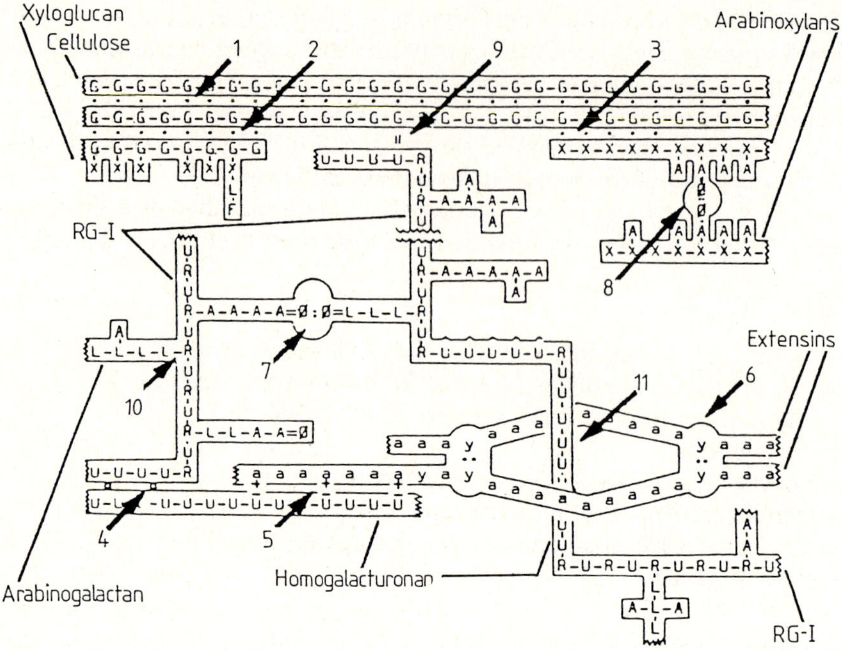

Fig. 1.10. Representative primary structures and possible cross-links of wall polymers [16]. This is not a model of the plant cell wall, and no significance is placed on the chain-length, orientation, conformation, or spacing of the molecules. The diagram illustrates:

(.) Hydrogen-bonds	1. cellulose–cellulose
	2. xyloglucan–cellulose
	3. xylan–cellulose
(○) Calcium bridges	4. homogalacturonan–homogalacturonan
(±) Other ionic bonds	5. extensin–pectin
(:) Coupled phenols	6. extensin–extensin
	7. pectin–pectin
	8. arabinoxylan–arabinoxylan
(=) Ester bonds	9. pectin–cellulose
(–) Glycosidic bonds	10. arabinogalactan–rhamnogalacturonan
(◇) Entanglement	11. pectin-in–extensin

Abbreviations used in the structures are:

A = Arabinose, *B* = Fucose, *G* = Glucose, *L* = galactose, *R* = Rhamnose, *U* = Galacturonic acid, $\bar{U}$ = Galacturonic acid methyl ester: *a* = Amino acid other than tyrosine; *y* = Tyrosine; *y*:*y* = *Iso*dityrosine, ∅ = Ferulic acid; ∅:∅ = Diferulic acid

Linkage of the type I.4a is produced with the participation of a cell-wall peroxidase. This gives rise to *iso*dityrosine fragments from two tyrosine residues of the neighbouring extensin molecules (Fig. 1.11). There is an evidence that one more tyrosine residue can be attached to *iso*dityrosine to form a trimer.

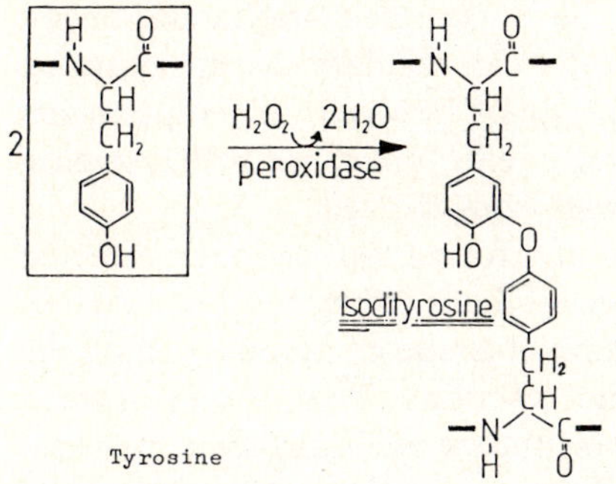

Fig. 1.11. The scheme of isodityrosine unit formation

In pectic substances the role of a cross-linking unit is played by cinnamic acid derivatives, ferulate and *p*-coumarate, which esterify pectins (about 10 feruloyl residues per 1000 sugar units in a pectin molecule) [16]. Mild hydrolysis of pectin affords, inter alia, two feruloylated disaccharides, 4-*O*-(6-*O*-feruloyl-β-D-galactopyranosyl)-D and 3-*O*-(3-*O*-feruloyl-α-L-arabinopyranosyl)-L-arabinose. Shown below is a scheme of oxidative coupling of two ferulate residues to produce a diferulate cross-link between pectin molecules (type I.4b linkage) (Fig. 1.12). Pectic substances with diferulate cross-links become resistant to hydrolysis.

II. Noncovalent linkages

1. Hydrogen bonds [e.g. Hemicellulose · cellulose]
2. Ionic bonds [e.g. Extensin · pectin]
3. Calcium bridges [e.g. Pectin · (Ca^{2+}) · pectin]
4. Hydrophobic interactions [e.g. pectin gelling (H-bonds also involved)]
5. van der Waals bonds [widely distributed]
6. Lectin bonds [e.g. Lectin · polysaccharide]

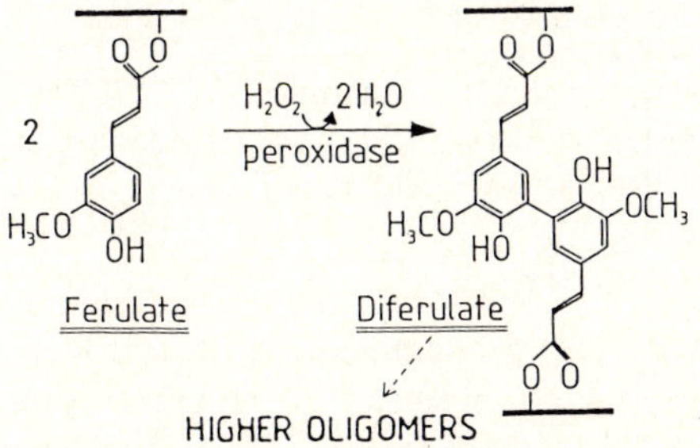

Fig. 1.12. Oxidative coupling of ferulate residues to produce diferulate cross-links

The above data point to cell wall to be a complex, multicomponent, regulated compartment of a plant cell. The latter can regulate this structure by changing (via scission or creation) various links between its components through the intermediacy of proton or ion level, or effectors of cell-wall enzymes, or the transport of the enzymes to the cell wall from the cell.

Effective regulation implies the existence of information feedback directed from the cell wall inside the cell. It can be expected that some general regularities will here be displayed typical for animal- and plant-tissue metabolism-catabolic regulation of intensity of synthetic, anabolic reactions. This is seen from a general scheme where synthesis of biopolymers (nucleic acids, proteins, polysaccharides) and lipids is regulated by intermediate, signal degradation products (Fig. 1.13). It was shown in several works that oligonucleotides, oligopeptides, oligosaccharides, and oxygenated derivatives of polyunsaturated fatty acids can act indeed as regulators or hormones. Under stationary conditions of metabol-

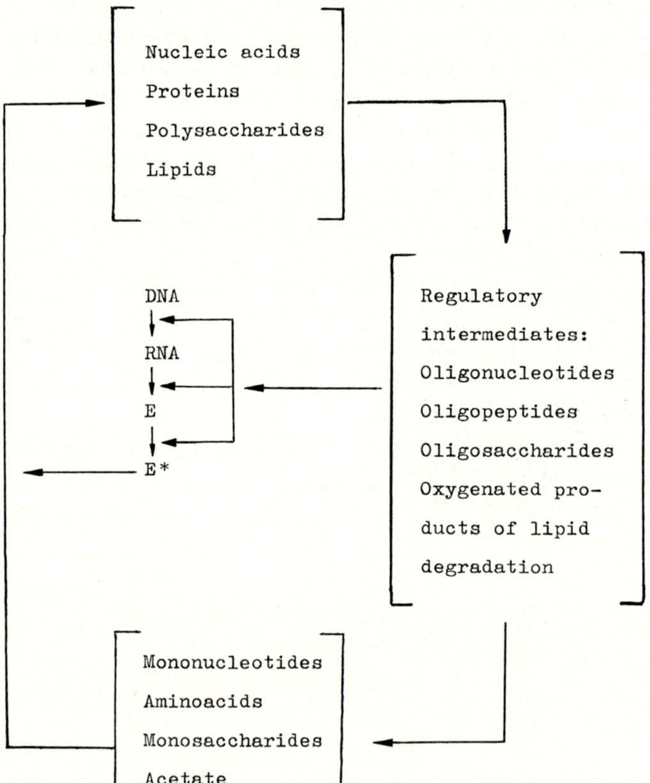

Fig. 1.13. The scheme of the biosynthesis reactions regulation by intermediates of catabolic processes

ism the regulatory function of the catabolism intermediates, which are permanently formed, is unnoticeable but the feedback is certainly realized in maintenance of the synthetic processes at a definite level.

Regulatory mechanisms usually become apparent upon rapid and essential changes in environmental conditions, e.g. under stress action of pathogenic organisms, mechanical damage of tissues. These factors result in appearance of 'alarmogens' (the alarm hormones) [35] which trigger mechanisms securing the organism's survival.

Studies of plant cell response to invasion by fungal or bacterial pathogens and stress factors performed by Albersheim and coworkers resulted in the discovery that regulatory molecules responsible, for example, for the production of antibiotics in the host cells are oligosaccharides which represent fragments of cell wall polysaccharides of a pathogenic fungus or a host plant (for a review see Albersheim, Darvill [36]).

In the case of fungal attack, a host-cell enzyme came into action to liberate, from the fungal cell wall, an active, branched heptaglucoside (Fig. 1.14).

The other mechanism is involved in the formation of regulatory oligosaccharides upon bacterial attack. They secrete an enzyme which splits the host-cell-wall pectins to produce an active, linear oligogalacturonide.

One more mechanism for the formation of active oligosaccharides consists in the liberation from a damaged plant cell of an enzyme which splits structural polysaccharides of neighbouring cells of the same plant into fragments.

A nonsaccharide derived from the sycamore-cell-wall xyloglucan was shown to possess high 'antiauxin' activity.

All the aforementioned fragments of cell-wall matrix polysaccharides were named 'oligosaccharines' [36]. This term is used to unite 'a new class of regulatory molecules' which can participate in the realization of a protective function, mediatory function when phytohormones act on cells, and a function of plant growth and morphogenesis regulation.

The number of biologically active oligosaccharides, newly found in different plants, is being continuously increased [16, 37, 38, 39].

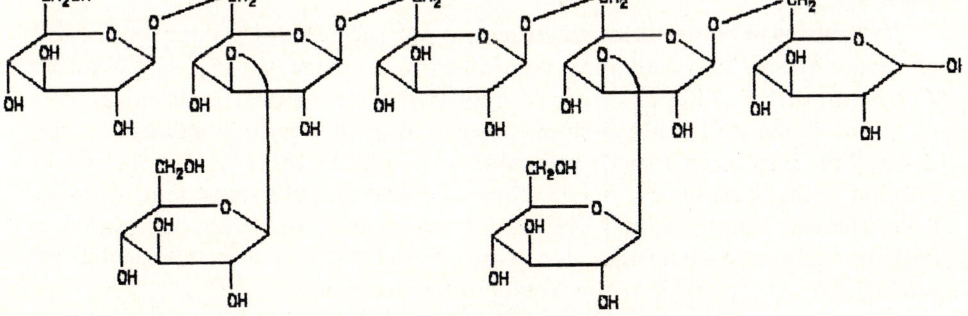

Fig. 1.14. General structure of the heptaglucoside—"oligosaccharine"

1.7 Changes in Content of Cell Wall Components

During cell development, upon passage of primary into secondary cell wall and its forming the composition of polysaccharides changes regularly [40, 41, 42, 43]. Meinert and Delmer (loc. cit.) have shown that percentage of proteins, hemicelluloses, and pectic substances in the developing cotton-fiber cell wall drops while that of cellulose increases (Table 1.1).

Formation of secondary, strongly reinforced cell walls in arboreal plants is connected with the deposition of hemicelluloses and especially cellulose [40, 44]. Lignification of cell walls consists, to a large extent, in the accumulation of lignin and impregnation of polysaccharides [45].

The composition of cell walls of different tissues of higher plants is discussed in a review [18].

Non-lignified tissues of growing cells of dicotyledons contain the following components (in decreasing order): pectic substances, cellulose, hemicelluloses (e.g. xyloglucans), proteoglucans, glycoproteins, phenolics; the predominant hemicelluloses of cambial tissues are acidic xylans. The same composition is characteristic of non-lignified non-endospermic pea cotyledon, lupine cotyledons are rich in pectic (arabino) galactans.

The major components of cell walls (mainly secondary) of lignified tissues (as in xylem elements and sclerenchyma fibers) are cellulose, hemicelluloses (acidic xylans), lignin, very small amounts of pectic substances and glycoproteins.

Constituents of non-lignified cell walls of growing cells and parenchymatous tissues of monocotyledons are the same as for dicotyledons with the exception that the content of proteoglycans, glycoproteins, and phenolics is smaller. Non-lignified wheat endosperm contains hemicelluloses (mainly arabinoxylans), very small amounts of cellulose, some phenolics and proteins.

Lignified tissues of stem and stalk fibers contain cellulose, hemicelluloses (mainly acidic arabinoxylans), lignin, and small amounts of proteoglycan complexes. Wood fibers of gymnosperms (e.g. Jack pine) contain cellulose, hemicelluloses (mainly galactoglucomannans and arabinoglucuronoxylans), lignin, and small amounts of proteoglycan complexes. Selected data are listed in Table 1.2 [18].

Differences in composition of structural polysaccharides in connection with particular species and ecological conditions were reviewed [46]. Angiosperms and gymnosperms differ mainly in composition of matrix polysaccharides of the latter are in fact galactoglucomannans, i.e. they always contain galactose while those of the former are usually galactose-free. In many cases it is arabinogalactan that is the basic matrix polysaccharide in gymnosperms and that is absent from angiospermous plants. The major xylan in gymnospermous plants is *arabino*-4-*O*-methylglucuronoxylan while in angiosperms it is represented by 4-*O*-methylglucuronoxylan or arabinoglucuronoxylan.

Tissues of monocots contain much more xylans than do dicots. In general, xylans predominate in cereals as compared to other herbaceous plants. Dicots

Table 1.1. Changes in composition of developing cotton-fiber cell wall expressed as 1.10^{-10} g per mm of fiber length

Component	Age of hairs, days after anthesis										
	5	8	10	12	14	16	18	20	22	25	29
Cellulose	1.9	4.2	10.8	14.4	23.0	49.9	170	245	362	594	1363
Neutral sugars	6.9	10.5	1.3	12.0	15.6	22.5	27.5	29.9	37.4	46.2	58.0
Uronic acids	2.1	6.7	10.2	12.6	14.7	14.4	6.9	6.0	8.3		
Protein	8.4	10.1	10.4	13.2	23.0	33.6	16.1	12.0			
Ash		1.3	1.4	1.2	10.1	17.6	9.2	6.0			
Cell wall as a whole	21	42	47	60	92	160	230	299	416	660	1450

Table 1.2. Carbohydrate composition of cell walls of some plant tissues (expressed as g anhydrosugar per 100 g cell wall material)

Sugars	Dicotyledons		Monocotyledons	
	Apple parenchyma (non-lignified tissues)	Pea seed hulls (lignified tissues)	Grass mesophyll (non-lignified tissues)	Wheat seed hulls (lignified tissues)
6-Deoxyhexose	2.0	1.6	1.0	0.3
Arabinose	12.3	2.7	13.9	4.6
Xylose	3.3	11.8	17.6	21.7
Mannose	4.3	0.3	—	0.6
Galactose	5.7	1.1	4.0	1.6
Glucose	22.7	57.0	57.2	32.6
Uronic acids	32.8	15.4	6.5	6.0

contain small amounts of gluco- and galactoglucomannans which are totally absent from monocots. Among the polysaccharide constituents of the former (and not of the latter) was found fucose albeit in low concentration. It is likely that galactose is a more important 'building material' for matrix polysaccharides in dicots than in monocots.

Leaves of cereals contain more hemicelluloses and cellulose as compared to leaves of dicotyledons, difference in polysaccharide content in stems is less pronounced. Usually the content of pectic substances in dicots exceeds several-fold than in monocots.

Leaves of tropical cereals (*Panicoideae* and *Eragrostideae*) are richer in cellulose than those in moderate climate (*Festuciformis*) [46]. The ratio of structural polysaccharides is also different. The *Festuciformis* cereals always contain more cellulose than hemicelluloses, while in *Panicoideae* in the level of hemicelluloses in equal or even higher than that of cellulose.

The monosaccharide composition of hemicellulose of tropical cereals differs from that of *Festuciformis*, glucose being its main component in the former and xylose in the latter. It should be mentioned, however, that these differences are conditioned primarily by climatic factors. Indeed, growing of *Panicoideae* and *Festuciformis* cereals in greenhouse under identical conditions leads to the reduction, but not to the disappearance, of changes in cellulose-to-hemicellulose ratio. It can be concluded, however, that levels of cellulose and matrix polysaccharides and their changes are genetically determined. In this connection it is of interest to note that the level of soluble carbohydrates in tetraploid form of rye grass is higher than that in diploid one, while the level of cellulose is the same for both.

The seasonal changes in content of structural polysaccharides is the consequence of two reasons: the alteration of the ratio of the mass of leaves to that of stems in favour of the latter and ontogenetic changes in polysaccharide level in each organ. Increase in hemicellulose and especially cellulose percentage in moderate-climate plants is higher than in tropical ones. Thus, for example, the dried mature rye grass leaves contained 21–22% cellulose against 8–10% in developing leaves.

The highest productivity of cellulose (per hectare) is exhibited by forests, followed by savanna, and, finally, cultivated areas. Annual, terrestrial production of cellulose considerably exceeds the aquatic one.

1.8 Possible Mechanisms of Endogenous Regulation of Cellulose Biosynthesis

Various soil-climatic, technological, and other factors intervene in the polysaccharide synthesis indirectly, through a system of endogenous regulation of biosynthesis of cellulose, hemicelluloses, and pectic substances. Several mechanisms for this regulation exist which can as a rule be referred to but in general due to scarcity of experimental data.

1.8.1 Genetic Regulation

This mechanism governs variations in modes of cellulose formation in different objects belonging to remote taxonomic groups, viz. Bacteria, algae, higher plants, as well as changes in the synthesis of polysaccharides during primary and secondary cell wall formation. Importance of this regulation can be judged from suppression of cellulose synthesis upon action of transcription and translation inhibitors. Phytohormones play an essential role in regulation of genetic activity, the most important of them being cytokinins [47, 48] as well as gibberellic acid and auxins.

This is the strategic route of regulation by which the intensity of different biosynthetic steps is augmented due to the additional production of enzymes.

The possibilities of elucidating the peculiarities of genetic regulation in high plants can be associated with cotton fiber.

The rate of cellulose deposition can increase more than 100-fold within a few days in cells of cotton fibers, associated with the onset of secondary wall formation. This could involve changes in the level of expression of genes coding for proteins required for cellulose synthesis [49]. The authors have shown that, in vitro, there are notable decreases in the level of translatable mRNAs that code for polypeptides of 25×10^3 M_r and 30×10^3 M_r, and a notable increase or appearance of two abundant mRNAs for polypeptides of 27×10^3 M_r and 32×10^3 M_r.

Experiments in vivo show a notable increase in the rate of various proteins synthesis, especially that of polypeptides of 39×10^3 M_r and 45×10^3 M_r.

The synthesis of the catalytic polypeptide of 18×10^3 M_r of cellulose synthase increases substantially at the onset of secondary wall cellulose synthesis in the cotton fiber [14].

1.8.2 Regulation of Enzymatic Activity

The second mechanism of regulation of synthesis of polysaccharides is the allosteric regulation of activity of enzymes already present in cells. This is realized through direct or indirect (mediated by secondary messengers) interaction of growth hormones with enzymes (cf. Chap. 7). Enzymes can also be activated by substrates and reaction intermediates. It should be noted that this regulation mechanism has seldom been investigated and it may be judged mainly by information gained from studies of carbon metabolism in photosynthesizing cells. This mechanism applies equally to the formation of structural polysaccharides in photosynthesizing cells from the products of CO_2 uptake and is most obviously manifested upon changes in illumination conditions. Transfer of plants from dark to light brings about activation of ribulose-1,5-bisphosphate carboxylase, triose phosphate dehydrogenase, fructose-1,6-bisphosphatase, sedoheptulose-1,7-bisphosphatase, and ribulose-5-phosphate kinase and inactivation of glucose-6-dehydrogenase [50, 51, 52].

Light-induced activation and inactivation of enzymes mentioned is accounted for by their interaction with photoreduced components of the electron-transfer chain. Light-induced activation of 3-phosphoglycerate kinase is caused by interaction with ATP whose level inside the cell is increased with illumination. Alteration in activity of the Calvin cycle enzymes can also be caused by a photo-induced increase in pH and magnesium level in stroma of chloroplasts.

Regulation of activity of enzymes of carbon photosynthetic metabolism can be exemplified by inhibition of ADPG-pyrophosphorylase, which is involved in starch biosynthesis, by elevated concentrations of 3-phosphoglycerate. Phosphorylated intermediates of photosynthesis and derivatives thereof can supposedly affect activity of NDPS-pyrophosphorylases which participate in biosynthesis of structural polysaccharides, e.g. UDPG- and GDPG-pyrophosphorylases. Most probably, this mechanism is also realized through interaction of enzymes of carbohydrate metabolism with certain growth hormones (cf. Chap. 5) or through changes in concentrations of the starting substrates or intermediates of the synthesis of structural polysaccharides in photosynthesizing and nonphotosynthesizing cells.

According to some data [49] cellobiose is a potent activator of glucan synthetases. Since cellobiose is not the intermediate of carbohydrate metabolism the question arises on what is the effector of cellulose synthesis that is similar to cellobiose.

1.8.3 Membrane Regulation

This regulation mechanism governs both the transport of substrates and intermediates of cellulose synthesis across plasmalemma and intercellular membrane systems. It controls, in addition, the intensity of movement of cellulose synthetase complexes along the membrane, as well as in all probability, the activity of glucan synthetases. The regulating effect of the plasmalemma is determined by transmembrane ion gradients and the magnitude of transmembrane potential differences. Active synthesis of cellulose in intact cells occurs under conditions of high transmembrane potential, relatively high levels of magnesium at the inner surface and of calcium at the outer surface of the plasmalemma. The membrane regulation of transport of compounds taking part in the synthesis of structural polysaccharides provides for both regulation of transport across membranes and participation of flows of membrane vesicles in the delivery, from the cell to cell wall, of polysaccharides and glycoproteins in various stages of formation. One more type of regulation which is similar to that under consideration is of the orientation type. This influences the arrangement of the newly formed cellulose microfibrils of a cell wall and is determined by the direction of protein microtubules beneath the plasmalemma.

Membrane regulation of synthesis of structural polysaccharides can be exemplified by the synthesis of cellulose and callose in damaged cells compared to intact ones [49]. Intensity of callose synthesis in intact cells is too low compared to that of cellulose whereas rupture of the plasmalemma results in considerable activation of callose synthesis and inhibition of cellulose synthesis. This is accounted for in the following manner. After rupture of the plasmalemma the magnesium ions, UDPG, and other nucleotides escape the cell while calcium ions enter the cell. This results in an increase in activity of β-1,3-glucan synthetase and of callose production while β-1,4-glucan synthetase is inhibited.

1.8.4 Substrate Regulation

One of the important results obtained in the author's and some other laboratories is the disclosure of considerable undersaturation with substrates for synthesis of structural polysaccharides. This is not surprising, at least because of 'attenuation' of a plant's photosynthetic apparatus for a much higher concentration of CO_2 than that existing now. It is the low level of CO_2 in the atmosphere (0.03%) that is the main limiting factor of photosynthesis in the majority of terrestrial plants. This is proved by the fact that considerable increase in the CO_2 content results in multiple enhancement of the intensity of photosynthesis, this being due to higher throughout of all the channels of carbon metabolism including metabolic pathways from CO_2 to structural polysaccharides. In nonphotosynthesizing organs the increase in the level of substrates for cellulose synthesis can occur as a result of activation of input of assimilates to these organs. Various practical ways of employing this regulatory mechanism for increasing raw-cotton crops is discussed in Chap. 5.

1.8.5 Energetic Regulation

To this regulatory mechanism can be attributed alterations in active transport of substrates and intermediates in the synthesis of structural polysaccharides as well as in the intensity of their synthesis that is connected with the uptake of ATP or other nucleoside triphosphates. This type of regulation becomes especially manifested when some specific factors, e.g. drought, low temperatures, diminished illumination, bring about suppression of the main ATP-producing processes, viz. photophosphorylation or oxidative phosphorylation (cf. Chap. 6 for detailed discussion). ATP can influence the synthesis of structural polysaccharides via indirect routes, which involve activation of general cell metabolism including the synthesis of enzymes of carbohydrate metabolism.

It thus can be concluded that there are numerous pathways for the regulation of synthesis of structural polysaccharides and hence practical approaches which can be used in agriculture, forestry, and biotechnology.

2 Objects Useful for the Study of Synthesis of Cell Wall Polysaccharides

Studies on peculiarities of biosynthesis of cellulose and other structural polysaccharides are being carried out with various objects and the most popular are cotton seed hairs and isolated protoplasts.

Synthesis of cell-wall polysaccharides is rather often studied with classical fast-growing objects which are used in studies of growing processes, viz. epicotyls, hypocotyls, and coleoptiles of cereals. In certain cases, e.g. in analysis of regularities of synthesis of mucilage polysaccharides, plant roots serve as an object.

2.1 Cotton Fiber

Cotton fiber can be regarded as a fortunate subject for study on the following basis [53]:

1) Each hair is a single cell and all seed hairs develop and elongate synchronously:
2) The formation of primary and secondary cell walls is of rather long duration and is sharply time-segregated, thus both processes can be studied separately;
3) The developing hairs with their ovules can be detached from the maternal plant and cultured in artificial media, this enables one to study the influence of hormones and other exogenous additives on the biosynthesis of cellulose and other polysaccharides;
4) Cellulose deposits in the secondary cell wall of cotton hairs is the purest (the most concentrated) form of cellulose occurring in the plant kingdom;
5) The practical value of cotton fibers allows us to hope that the study of mechanisms of regulation of cellulose biosynthesis carried out with this subject will help in the search for effective ways to increase cotton fiber quality and the harvest.

2.2 Acetobacteria

For some time past studies of cellulose biosynthesis have been carried out and more frequently with bacteria of the genus *Acetobacter*, in particular, with *Acetobacter xylinum*. Cellulose of this microorganism forms an extracellular 'ribbon' attached to the cell membrane, allowing the cells to be kept closer to the liquid phase surface and so receive a better supply of oxygen. There is no continuous layer of cellulose surrounding the cell. Development of *A. xylinum* is accompanied by the formation of a cotton-like layer composed of ribbon-shaped cellulose structures. *A. xylinum* is considered to be an organism valuable for large-scale production of uniform cellulose by biotechnical methods. It can be used for the preparation of cellulose derivatives of highest purity for some special purposes.

Other organisms useful for the study of cellulose biosynthesis were reviewed recently by Delmer [12]. Few studies have been conducted with algae despite the cell wall structure of many of them being well described. Tissues of arboraceous plants with well-developed cellulosic cell walls contain much lignin and are therefore not used in studies of this kind.

2.3 Isolated Protoplasts

Protoplasts were isolated mechanically for the first time in 1892 from *Stratoides aleides* by slicing up cells plasmolyzed in 0.4–0.7 M sucrose whereupon protoplasts sedimented [54]. Since 1960, enzymatic methods for the isolation of protoplasts have become popular, they involve lysis of cell-wall polysaccharides by a mixture of enzymes [55]. Enzymes frequently used are pectinase and cellulase. The former disturbs intercell contacts in tissues resulting in cell detachment (maceration of tissue) while the latter destroys cellulose and 'liberates' water-soluble hemicelluloses. The protoplasts devoid of cell walls acquire a spherical form and following washing out of the enzymes, new cell walls are produced. This was first observed with tomato [56] and tobacco [57] protoplasts.

The so-isolated protoplasts are cultured under sterile conditions to prevent utilization of nutrients by extraneous microorganisms and invasion of the latter into the protoplasts, the absence of cell walls facilitating penetration of infection into a cell.

At present, protoplasts can be isolated from any organ of plants, from root to pollen, and plant species from which protoplasts have been isolated amount to over 200. Methods of isolation and culturing of protoplasts are dealt with in detail in several reviews [54, 58, 59]. The culture medium contains an osmoticum, capable of maintaining the osmotic pressure at the level produced by 0.5 M, and, in addition plant hormones, vitamins, mineral salts, amino acids, and sugars which are the starting substrates for the synthesis of polysaccharides. In the case of protoplasts isolated from a chlorophyllous tissue, the organic substances necessary for cell wall construction can be produced by photosynthesis.

Regeneration of cell walls can be judged from the transition of ball-shaped protoplasts into ellipsoid ones, from plasmolysis of cells, whose walls are being formed, when placed into a hypertonic medium (the protoplast subjected to plasmolysis can be recognized by the appearance of a "halo" around it due to light scattering by a newly formed cell wall), from staining of cell-wall polysaccharides with specific dyes [60]. Information on cellulose biosynthesis by isolated protoplasts during cell wall production can be obtained from electron microscopy data.

It was supposed earlier that enzyme-mediated removal of cell walls is followed by a rather long period prior to wall regeneration. Duration of this lag period is thought to be determined both by objective reasons (i.e. by the status of the isolated protoplasts) and by now efficient are the methods for culturing the isolated protoplasts from which cellulases and pectinases have been washed off and for establishing the start of new cell wall formation.

Today it has been established by means of grating electron microscopy that the appearance of short cellulose microfibrils on the naked surface of plasmalemma begins several minutes after washing off the cell-destroying enzymes [61]. The use of labelled substrates for the synthesis of cellulose and other polysaccharides supports these results obtained by electron microscopy (Fig. 1.15). Incorporation of [^{14}C] or ^{3}H into newly formed cell-wall polysaccharides is very rapid (Figs 1.16 and 1.17). Difficulties in identifying the beginning of this process stem from the fact that the amounts of the polysaccharides produced are very small and they can hardly be detected upon analysis. To overcome this the substrates used are of as high specific (and total) radioactivity as possible and, secondly, nonlabelled polysaccharides are added to the fixed protoplasts as carriers prior to analysis of cell walls whose formation has just begun.

Fig. 1.15. Cellulose microfibrils appear on the surface of the protoplast after 10 minutes of culture

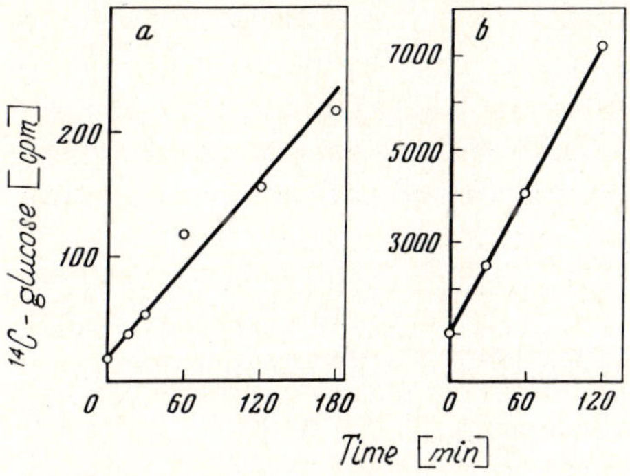

Fig. 1.16a, b. Short-term incorporation of [^{14}C]-glucose into cellulose (**a**) and alcohol-sedimented, noncellulosic polysaccharides (**b**) by soybean protoplasts [62] (ordinate, [^{14}C]-glucose, cpm; abscissa, 'Time, min')

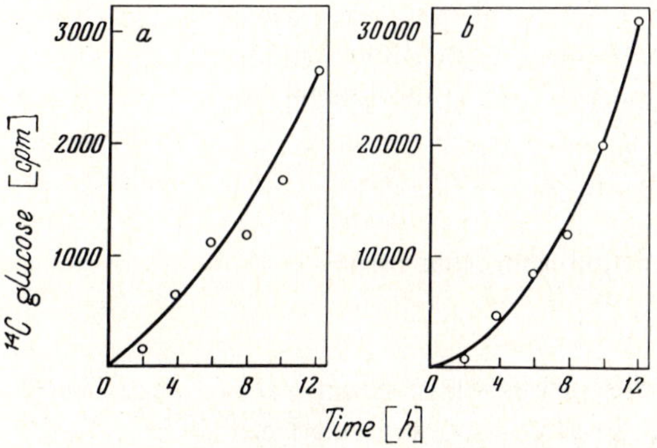

Fig. 1.17a, b. Incorporation of [^{14}C]-glucose into cellulose (**a**) and other polysaccharides (**b**) by soybean protoplasts vs time. (ordinate, '[^{14}C]-glucose, cpm', abscissa, 'Time, hours')

2.4 Methods of Extraction of Cell Wall Polysaccharides

Isolation of cellulose is usually based on dissolution and removal of all other tissue components. One of the popular methods for determination of cellulose content consists of treatment of a sample of defatted, homogenized plant tissue with a mixture of concentrated acetic and nitric acids at 100 °C whereby all high-polymeric substances except cellulose are solubilized (hydrolysed) [17]. The acid solution is separated, the residue is washed with water to remove the hydrolysis products and cellulose content is then determined from percentage of glucose liberated from the residue upon subsequent hydrolysis with sulphuric acid.

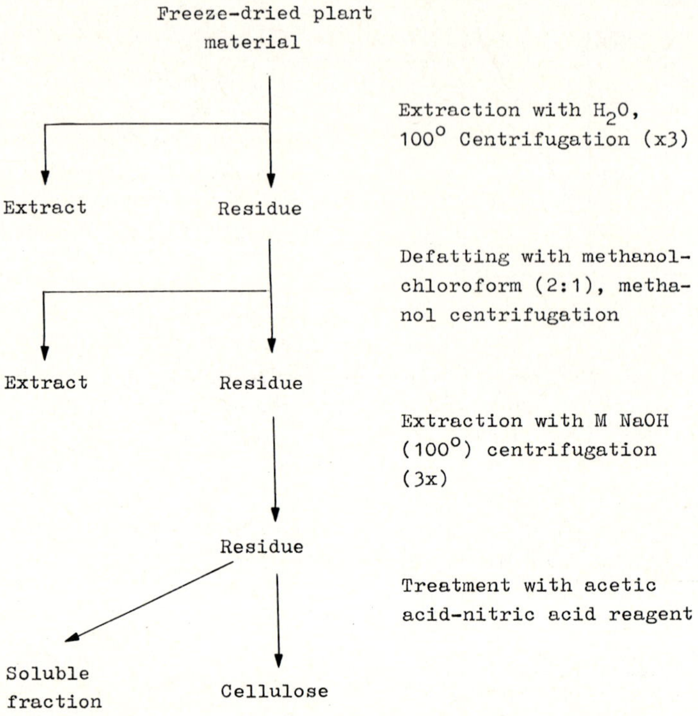

Fig. 1.18. The scheme of cellulose extraction from plant tissue

In studies of the mechanism of the formation of structural polysaccharides, isolation of cellulose from plant material is now performed mostly by alkaline extraction [63] the cellulose remaining insoluble, and acid extraction [64] cellulose being insoluble in acetic acid-nitric acid reagent [65] (Fig. 1.18).

Dissolution of β-(1-3)-glucan was shown not to proceed completely upon treatment with the hot alkali and the cellulose fraction always contained some β-(1-3)-glucan [66–69]. Treatment with acetic acid–nitric acid reagent does allow one to obtain pure cellulose noncontaminated with other cell-wall components. The polysaccharide fractions are further analysed using hydrolysis with acid or specific enzymes. Quantitation of β-(1-3)-glucan involves digestion of cell walls with *endo*-β-(1-3)-glucanase [70–72]. Galacturonic acid can be detected following digestion of cell-wall fractions with *Aspergillus niger* pectinase [42, 43]. Qualitatively the hydrolysis products can be analysed by means of paper, thin-layer, gas-liquid chromatography, and HPLC.

3 Sequence of Reactions of Biosynthesis of Cellulose and Other Structural Polysaccharides

3.1 Production of Nucleoside Diphosphate Glucose in Photosynthetic Cells

Structural polysaccharides are not regarded usually, by those engaged in studies of photosynthesis, as products of carbon dioxide assimilation and therefore their formation from CO_2 has not been considered [73, 74].

At the same time, all or at least a portion of structural polysaccharides (hemicelluloses and pectic substances) do not differ, in principle, with respect to mode and site of formation in cells, from sucrose which is the generally accepted final product of photosynthesis. This allows one to consider CO_2 as the main substrate for synthesis of polysaccharides in photosynthetic cells. Following diffusion inside the cell, carbon dioxide is reduced, by enzymes present in stroma of chloroplasts to, the level of carbohydrates. Ribulose-1,5-bisphosphate is the acceptor of CO_2 which under catalytic action of ribulosebisphosphate carboxylase, is added to $C_{(2)}$ of the pentulose. Subsequent hydrolysis of the thus formed six-carbon-atom molecule affords two molecules of 3-phosphoglycerate.

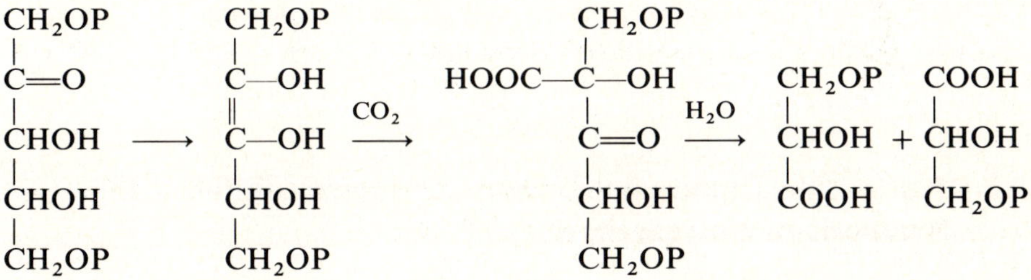

The latter is converted, through mediation of 3-phosphoglycerate kinase into glycerate-1,3-bisphosphate which is the necessary reactive intermediate for subsequent reduction.

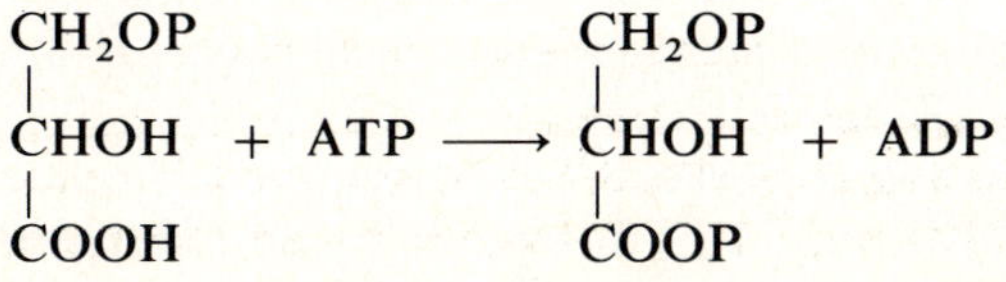

Glycerate-1,3-bisphosphate is reduced with NADP-dependent triosephosphate dehydrogenase to the level of carbohydrates.

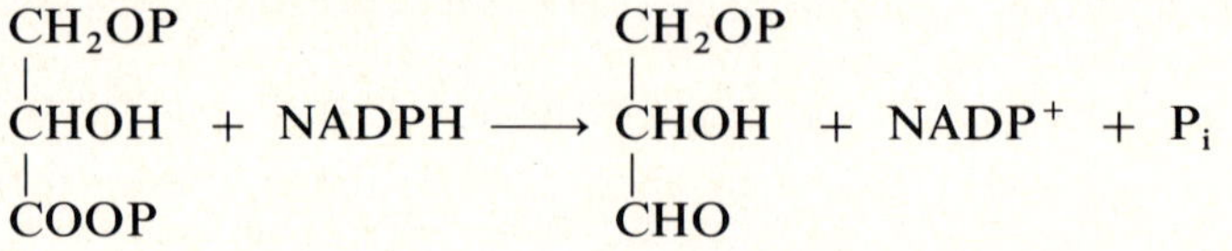

The two thus produced glyceraldehyde 3-phosphate is isomerized into dihydroxyacetone phosphate and these two three-carbon-atom molecules are condensed in the aldolase-catalyzed reaction to produce fructose-1,6-bisphosphate.

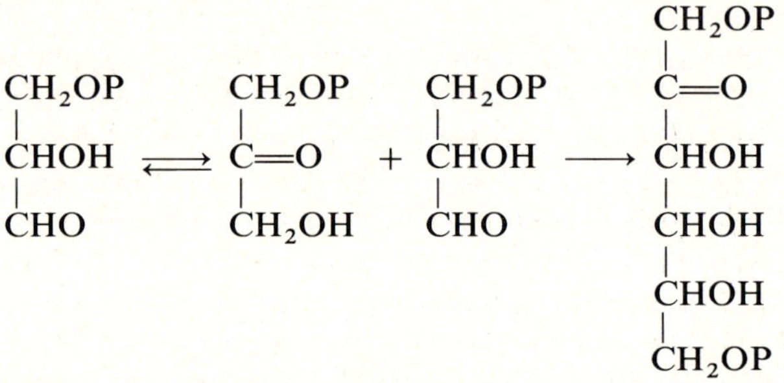

From this point the interconversion of hexose phosphates begins and which is terminated with formation of nucleoside diphosphate glucose (NDPG) as follows: dephosphorylation of fructose-1,6-bisphosphate, isomerization of fructose-6-phosphate into glucose-6-phosphate and of the latter into glucose-1-phosphate, whose reaction with a nucleoside triphosphate affords the NDPG.

$$\text{Fructose-1,6-P}_2 \rightarrow \text{Fructose-6-P} \rightarrow \text{Glucose-6-P} \rightarrow \text{Glucose-1-P} \xrightarrow[-\text{PP}_i]{\text{NTP}} \text{NDPG}$$

3.2 Nucleoside Diphosphate Sugars as Intermediates in the Synthesis of Polysaccharides

The role of nucleotide diphosphate sugars (NDPS) in anabolism of carbohydrates has been established by L. F. Leloir who was awarded the Nobel Prize for these studies. Different NDPG's were shown to serve as glucosyl donors in the synthesis of different glucans, ADPG for starch and UDPG for glycogen ('animal starch'). Hemicelluloses and pectic substances are synthesized with participation of NDPS derivatives of corresponding neutral sugars and uronic acids.

As was suggested by Hassid and coworkers [75] biosynthesis of cellulose in cotton-seed hairs requires GDPG only. This viewpoint, based on experiments with enzymes isolated from cells, did not find total support. Other authors have shown that UDPG is a glucosyl donor for cellulose [76–78]. This controversy was settled by Delmer and coworkers [53, 79] who studied peculiarities of cellulose biosynthesis in developing cotton fibers. Exogenous, [^{14}C]-glucose-labelled UDPG and GDPG were used as substrates and the following conclusions were drawn. During the period of rapid fiber elongation and intensive primary-wall synthesis (7 → 14 days after anthesis) the low-molecular-weight cellulose was synthesized from GDPG, incorporation of glucose from UDPG was very low. Conversion to cellulose with a high degree of polymerization characteristic of secondary cell wall was accompanied with a change in the glucosyl donor: incorporation of glucose from GDPG is sharply decreased and UDPG becomes the main substrate (Fig. 1.19).

The author's data also proved that the change (albeit incomplete) of substrate occurs when elongation of cell wall of cotton-seed hairs is replaced by thickening. Utilization of glucose from GDPG for the synthesis of primary-cell-wall cellulose of cotton-seed hairs on the 8th day after anthesis was estimated as 343×10^3 cpm/g dry fiber/hour as compared with 7×10^3 for glucose from UDPG. During formation of the secondary cell wall (23 days after anthesis) these values were 28×10^3 and 170×10^3 respectively.

Chemical analysis of cotton fibers [80] has confirmed again that UDPG is especially important at the stage of secondary cell wall synthesis. GDPG could not be detected in substantial amounts in seed hairs on days 13 to 24 after anthesis whereas UDPG level increased from 0.2 to 2.1 μmole per boll. The main flow of carbon from exogenous [^{14}C]-glucose into cellulose is mediated by UDPG. Up to the moment of secondary cell wall formation (on day 17 after

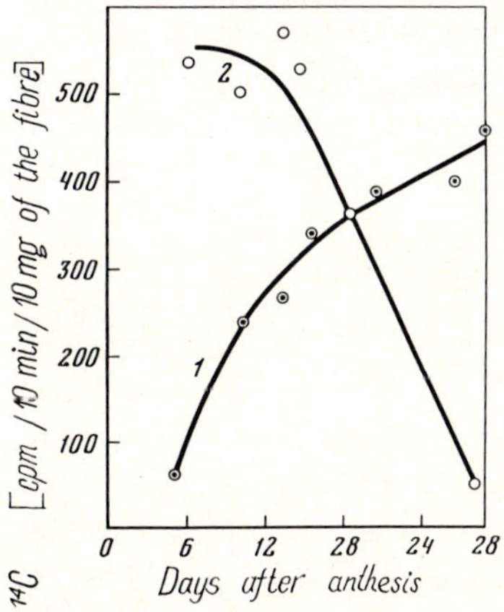

Fig. 1.19. Incorporation of ^{14}C from UDP-[^{14}C]G (1) and GDP-[^{14}C]G (2) into cellulose as a function of cotton-fiber age [53]

(ordinate, cpm/10 min/10 mg fiber)
(abscissa, days after anthesis)

anthesis) when intensive synthesis of cellulose took place the content of UDPG was 1.2 μmole per boll, molar concentration of UDPG on the 24th day after anthesis was 5 mM.

Computer-assisted analysis of kinetic data on carbon flux from [^{14}C]-glucose via glucose phosphate and UDPG into cellulose and callose-like glucan allowed one to conclude that the concentration of UDPG is quite sufficient to enable the rates required for the synthesis of cellulose, sucrose, glycolipid, and β-(1-3)-glucan.

This callose-like glucan is synthesized most intensively in response to injury to cells and tissues. Therefore in studies on peculiarities of synthesis of structural polysaccharides it should be taken into account that the reaction to injury which accompanies preparation of a sample for an experiment (cutting of tissues, removal of cell walls, etc.) can result in a shift of direction of synthesis, from UDPG, of cellulose and callose in favour of the latter [81] while usually callose synthetase, which is localized in plasmalemma, exhibits low activity. Some data indicate this enzyme to be activated by proteases.

In nonphotosynthetic plant tissues the main starting substrate for the synthesis of structural polysaccharides is sucrose which is the principal transport product of photosynthesis. Sucrose incoming from photosynthetic organs can be converted within cells into glucose and fructose by invertase localized in plasmalemma. Glucose is first phosphorylated, then incorporated in UDPG and participates in the synthesis of structural polysaccharides. This route requires consumption of two moles of ATP, for glucose kinase and phosphorylase reactions, and this energetic difficulty can be overcome by direct conversion of sucrose into UDPG:

$$\text{Sucrose} + \text{UDP} \rightarrow \text{UDPG} + \text{Fructose}$$

This reaction is catalyzed by plasmalemma-linked sucrose synthetase [82]. UDPG was shown to be better substrate than glucose in in vivo and in vitro syntheses of cellulose in cotton fiber (Fig. 1.20). The rate of incorporation of

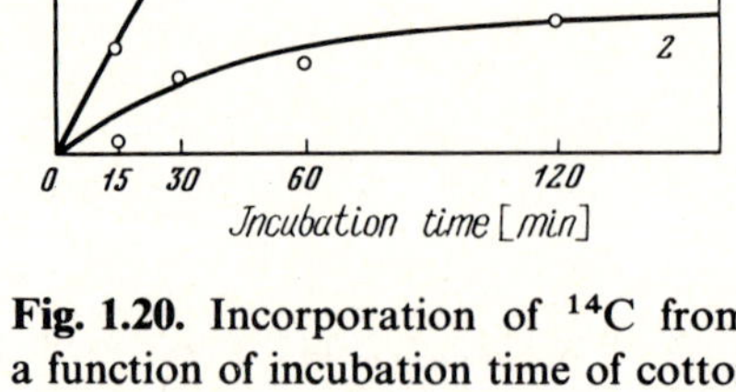

Fig. 1.20. Incorporation of ^{14}C from UDPG (1) and glucose (2) into cellulose as a function of incubation time of cotton-fiber (abscissa)

[^{14}C]-glucose from the labelled UDPG into secondary-cell-wall-cellulose increased during 60-min incubation of cotton-seed hairs in a solution of UDPG. Synthesis can be suspended by damage of isolated hairs or by inactivation of cellulose synthetase. If glucose label was not incorporated in hemicellulose in 30 min, it was found after 60 min in galactose and xylose of hemicellulose.

It is of interest that UDPG versus glucose utilization changes with cotton fiber maturing. Consumption was almost the same for both substrates (at their equal concentrations) during the first 5 days of hair-cell development, then the rate of UDPG consumption increased while that of glucose remained at the same level [76] (Fig. 1.21). Thus it can be concluded that it is the rate of glucose to UDPG conversion that limits the rate of synthesis of cellulose. Similar results were obtained with enzyme preparations from bean seedlings of various age. As to equal rates of glucose and UDPG consumption at early stages, it is supposed to be due to experimental errors: many noncellulosic polysaccharides (xylans and pectic substances) are produced in this period and they contaminate the cellulose fraction during extraction.

Pectic substances and hemicelluloses are synthesized from derivatives of UDPG and these transformations in general form have recently been discussed [83]. A fragment of this general scheme is represented below to illustrate the main pathways of UDPG utilization for production of other nucleotide diphosphate sugars (Fig. 1.22). Under the action of the corresponding epimerases UDPG is converted into UDP-galactose and UDP-mannose. Oxidation of hexose components of UDP-sugars affords UDP-glucuronate and UDP-galacturonate.

Methylation of uronic acids by methyl-transferases or decarboxylation to give UDP-xylose or UDP-arabinose complete the preliminary stage in synthesis of hemicelluloses. Carboxyl-methylated UDP-galacturonate is the starting monomer for the synthesis of pectic substances.

Matrix polysaccharides can also be synthesized by another metabolic pathway, via *myo*-inositol [84]. Tracer studies allowed to establish *myo*-inositol

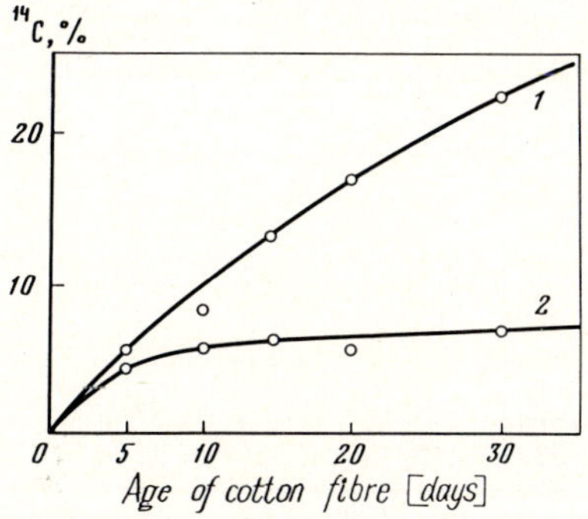

Fig. 1.21. Incorporation of ^{14}C from UDPG (1) and glucose (2) into cellulose as a function of cotton fiber age (abscissa)

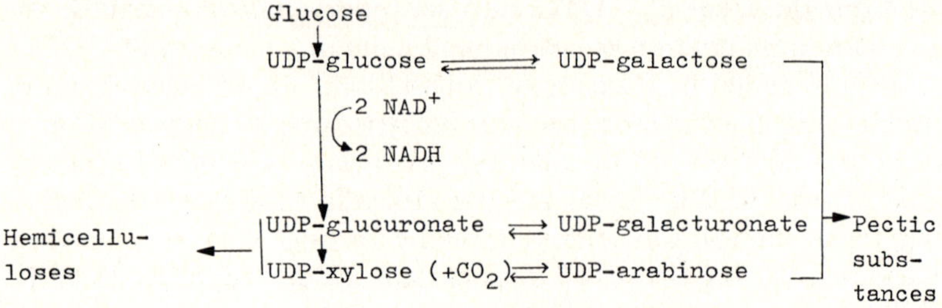

Fig. 1.22. The main reactions of UDPG utilization in the synthesis of hemicellulose and pectic substances

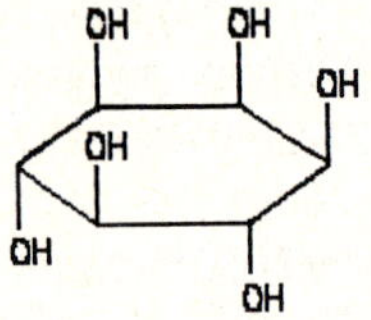

Fig. 1.23. Chemical structure of myo-inositol

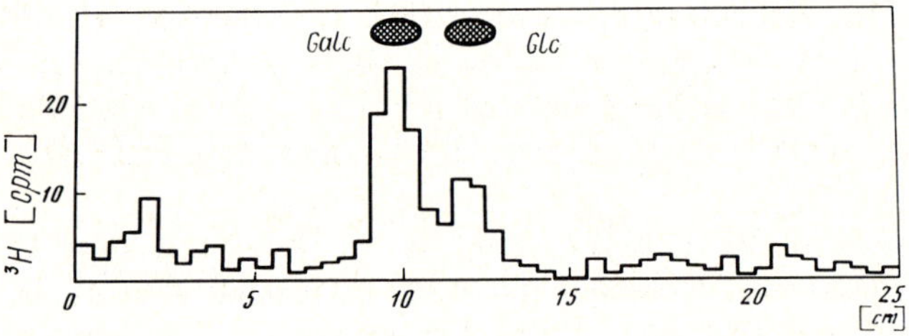

Fig. 1.24. Radioactivity of acidic sugar components in hydrolysate of alkalisoluble hemicellulose fraction isolated from cell walls of regenerated carrot protoplasts after 24 h culturing in [^{3}H]-myo-inositol [87]. (ordinate, radioactivity, cpm) (abscissa, mobility (in cm) upon paper chromatography. Gal A, galacturonic acid, GlcA, glucuronic acid)

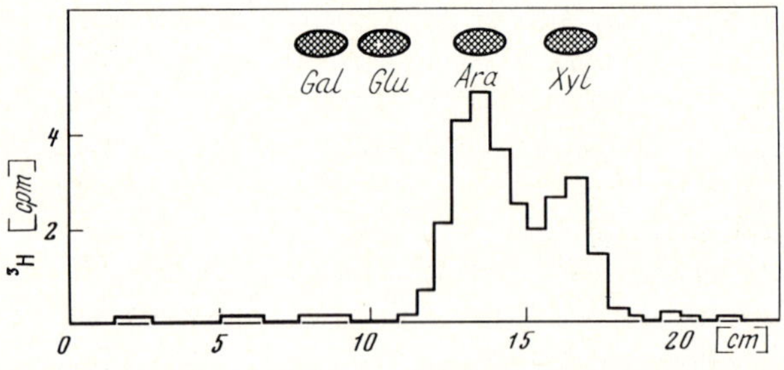

Fig. 1.25. Radioactivity of hydrolysate of neutral hemicellulose fraction isolated from carrot cell walls cultured for 6 h in [^{3}H]-myo-inositol [86]. For coordinates see Fig. 8. *Gal*, galactose; *Glc*, glucose; *Ara*, arabinose; *Xyl*, xylose

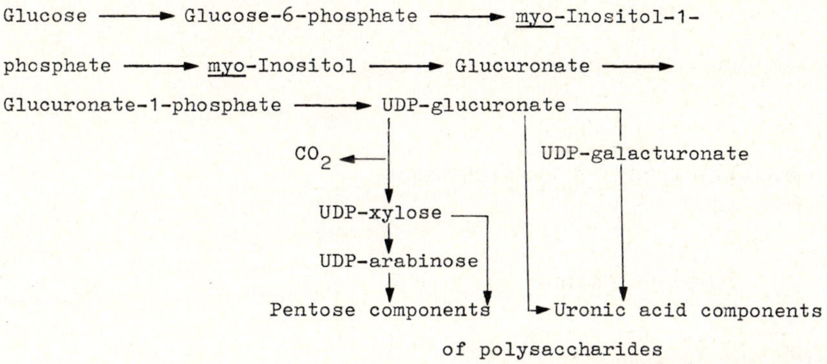

Fig. 1.26. The participation of myo-inositol in uronic acid and pentoses formation

to be the precursor of uronic acid and pentose components of matrix polysaccharides in sycamore [85] carrot [86–88] tobacco [89], and some other plants (Figs 1.24 and 1.25). The sequence of reactions of inositol oxidation pathway leading to uronic acids and pentoses is shown in Fig. 1.26. Nonconversion of [^{3}H]-*myo*-inositol into hexose components of cell-wall polysaccharides was observed [86].

3.3 On the Role of Membrane-bound Lipids and Proteins as Intermediates in the Synthesis of Matrix Polysaccharides

The concept of NDPS as the direct substrates involved in the synthesis of matrix polysaccharides has recently been subject to correction. It was suggested that lipids participate in this process as they do in the synthesis of cell-wall polysaccharides of gram-negative and gram-positive bacteria or in the synthesis of secreted glycoproteins in animals. The rationale for the use of lipids as intermediates in the synthesis of structural polysaccharides is the necessity in cross-membrane transport of NDPG. The latter is known to penetrate the membrane (Golgi apparatus, endoplasmic reticulum, plasmalemma) with difficulty and this may account for the role of membrane-embedded phospholipids which transfer sugar residues onto the polysaccharide chain being synthesized at the other side of membrane. Much attention was focused on the participation of a lipid intermediate in the synthesis of cellulose [90–92].

According to the modern concepts the transfer of glucose from UDPG to cellulose or of other sugars from respective NDPS's to pectic substances and hemicelluloses is mediated by lipids which are C_{90}- to C_{105}-isoprenoids (usually, C_{95}-C_{100}) with a saturated, terminal isoprene unit to which the phosphate group is attached [90, 91, 93, 94, 95] (Fig. 1.27). Lipids of this type are called dolichyl phosphates. The ability of dolichyl phosphates to participate, as

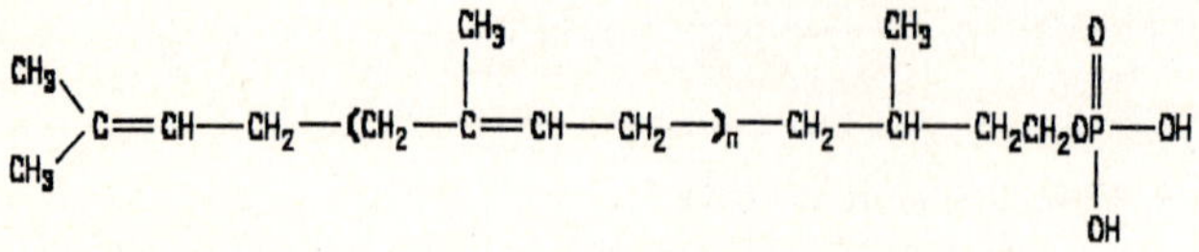

Fig. 1.27. Chemical structure of dolichyl phosphate

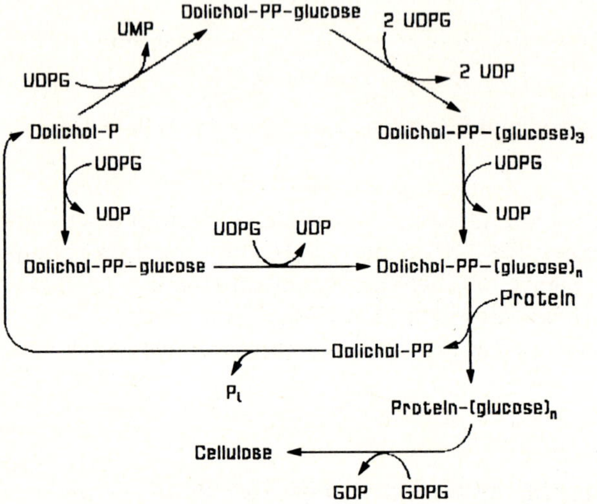

Fig. 1.28. The scheme of cellulose synthesis from UDPG, GDPG and dolichol

glycosyl-acceptors from NDPS, in the synthesis of polysaccharides was revealed with the help of enzymes which have been isolated from algae and higher plants and which were shown to catalyze the transfer of e.g. glucose from UDPG onto this lipid. The mitochondria is the site of dolichol synthesis from mevalonic acid [96]. Presented below is a tentative scheme of the reaction sequence in the synthesis of cellulose from UDPG and GDPG and phosphorylated dolichol [90] (Fig. 1.28). Some data indicate that sugar residues (mono- and oligosaccharides) are transferred from phospholipids onto membrane-bound proteins which themselves are the site of synthesis of polysaccharide chains [90, 91]. Following completion of synthesis the matrix polysaccharides probably become free of proteins unlike cellulose whose bonding with proteins is very strong.

Few data have been accumulated to date on the nature of a membrane-bound protein (molecular mass 13,000) which acts as the acceptor of lipid-linked oligosaccharides. It is similar to pea lectin β-subunit in electrophoretic mobility [97]. As to lectins, they have become attractive in recent years due to agglutinating ability. They are supposed to be the components of systems responsible for cell recognition and combination of cell surfaces in tissues [I].

The glycoprotein formed after glucose transfer seems to act as a primer for cellulose synthesis. It was found that coumarin, the well-known inhibitor of cellulose synthesis, inhibited the transfer of oligosaccharide chains from the lipid to the protein [91]. This is indicated by the distribution of the [^{14}C]-label from UDP-[^{14}C]-glucose between glycolipids, oligosaccharide-lipids, glycoproteins, and cellulose (Fig. 1.29).

Some authors have made the proposal that glucose residues may be transferred to cellulose without lipid intermediacy [12, 98, 99]. Studies by Helsper were carried out not with the whole cells but with membrane preparations isolated from petunia pollen tubes which exhibit cellulose-synthetase activity. Kinetic studies with [^{3}H, ^{14}C]-glucose-containing UDPG have to be taken into account and these show that the increase in specific radioactivity of polyprenyl-phosphate-glucose and sterylglucosides is slower than that of β-glucans.

It was suggested that a lipid intermediate participates in formation of β-glucan [100]. Experiments with inhibitors of glycosylation, viz. phospholipids, tunicamycin, and bacitracin, support this suggestion.

Some peculiarities of the kinetics of glucose incorporation into β-glucan in the presence of the above inhibitors, which necessitate care to be taken in evaluation of the possible role of glycolipids as intermediates in the synthesis of β-glucans, are rationalized by the existence of two metabolic pathways for utilization of the UDPG-derived [^{14}C]-glucose. The first operates at low UDPG level in cytoplasm and involves glycosyl transferase localized at the inner surface of the plasmalemma which performs a transfer of [^{14}C]-glucose

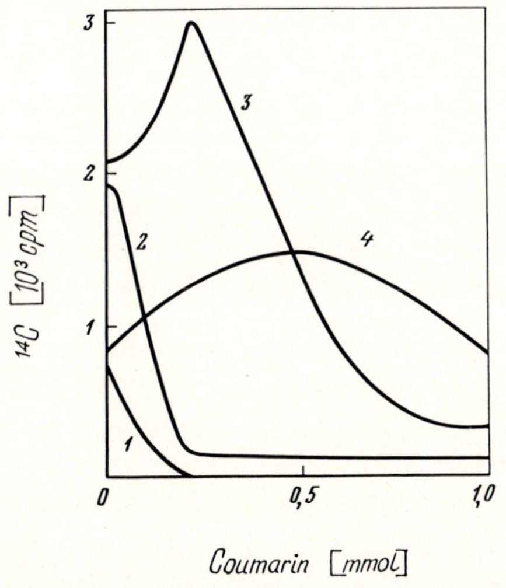

Fig. 1.29. Influence of coumarin on incorporation of label from UDP-[^{14}C]-glucose into cellulose (1), glycoproteins (2), oligosaccharide-lipids (3), and glycolipids (4). (ordinate, cpm × 10^{-3}) (abscissa, 'Coumarin, mM')

onto a lipid intermediate. Then the glucose is transferred onto β-glucan either directly or with participation of a glycoprotein.

At relatively high UDPG level at the outer surface of the plasmalemma, glucose can be utilized by a β-glucan synthetase directly, without recourse to a lipid intermediate. In the latter case the glycosylation inhibitors would not affect seriously the synthesis of the β-glucan.

3.4 Callose as a Possible Intermediate in the Synthesis of Cellulose

As far as intermediates of the biosynthesis of cellulose are concerned, one should take into account the suggestion forwarded rather long ago that callose is a precursor for cellulose, the former serving as a glycosyl donor for the latter [101]. This was not confirmed in one set of experiments [70], although Meier and coworkers have shown [102], using a brief [^{14}C]-sucrose- "gulp" technique with cotton-pod petioles, that the kinetics of a change in radioactivity of callose and cellulose follows the precursor-product relation. Incubation for 16 hours was accompanied with a decrease (not only relative but also absolute, and considerable one) in radioactivity of callose and increase in that of cellulose. In another set of experiments, where a shoot with a pod was exposed to $^{14}CO_2$ during photosynthesis for 2 hours and then the plants were transferred to the ordinary atmosphere, the general pattern of changes in radioactivity of callose and cellulose was similar to that observed with [^{14}C]-sucrose [102]. The only difference was in that radioactivity of callose dropped after one day (Fig. 1.30). Despite these facts the role of callose as a direct donor of glucose remains

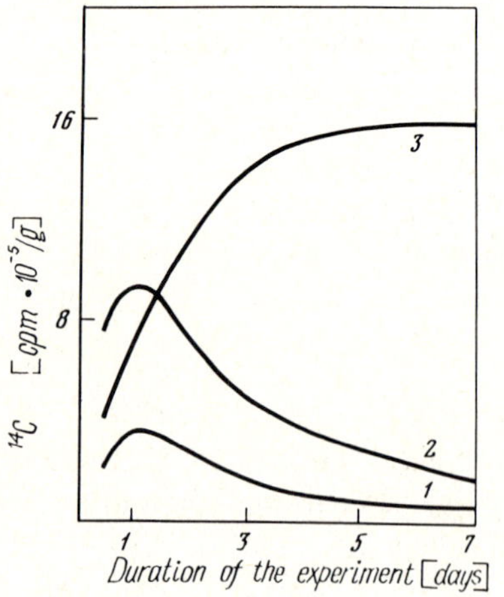

Fig. 1.30. Changes in radioactivity of callose (1), compounds soluble in 80% aqueous methanol (2), and cellulose (3) with time. (ordinate, cpm × 10^{-5}/h per one *cotton-seed* hairs) (abscissa, 'Duration of run in days')

questionable. This would imply the involvement of a transglycosylase in cellulose biosynthesis. Some data obtained with fluorescence microscopy (following Aniline Blue staining) indicate that during the process of a secondary cell wall formation there exists, on the inner surface, a thin layer of a substance reacting like callose [103].

Studies on specific action of glucanases and inhibitors of the β-glucan synthesis allowed one to support with confidence the hypothesis that callose is the possible intermediate in the synthesis of cellulose [102, 104]. Inhibition of the synthesis of callose was always accompanied with concomitant retardation of the cellulose synthesis whereas almost total inhibition of the latter affects the synthesis of callose only to a small extent.

Very intriguing data on the considerable decrease in intensity of the cellulose synthesis in cotton-seed hairs with [^{14}C]-sucrose as a precursor upon degradation of callose with exogenous exo-(1-3)-β-D-glucanase have been obtained. These data can serve as formal support for an idea of callose to be an intermediate in a chain of sucrose-to-cellulose transformation. They can also account for a relatively high [^{14}C]-callose/[^{14}C]-cellulose ratio (0.5–2.0) in short-term experiments with [^{14}C]-sucrose or $^{14}CO_2$ as a substrate compared to the ratio of the total content of these polysaccharides (below 0.02) in 36-days cotton-seed hairs.

Experiments with [^{14}C]-labelled precursors demonstrate that the rate of callose turnover is high and [^{14}C]-glucose derived therefrom is later accumulated in cellulose. There is no full confidence however that callose is the direct precursor for cellulose and that glucose interchange between them is performed in a transglycosylase-like reaction.

Possibly the short lifetime of the permanently produced and degraded callose is due to its specific function different from that to serve as a metabolic precursor of cellulose. This function can consist of ensuring an ordered stacking of newly formed cellulose microfibrils or a decrease in cell wall resistance against slipping of cellulose synthetase complexes. This can account for the fact that callose covers the plasmalemma as a continuous layer. This can also account for its relatively high turnover rate: having fulfilled its function in a lower cell wall layer, comparable in thickness to that of a definite (but unknown) number of cellulose microfibrils, callose disintegrates into glucose. In the case of a labelled substrate, the labelled glucose produced is involved in a metabolic flow which takes it away to be incorporated either irreversibly into cellulose or reversibly into callose.

In other words, instead of the former suggestion on functioning of (1-3) to (1-4)-β-glucan transglycosylase, another one is advanced for (1-3)-β-D-glucanase which functions within a secondary cell wall layer close to the outer surface of the plasmalemma. A question arises on the mechanism of callose degradation which is triggered at a specific distance from the surface of plasmalemma. Another question is why there exists a callose-free cellulose cover of a secondary cell wall. This seems to facilitate cell-directed alterations of cell-wall structure in response to an altered level of hormones, etc.

The relationship between syntheses of cellulose and callose was subject to analysis in a recent review [14]. One of the conclusions made, viz. considerable time-separation of biosynthesis of these two polysaccharides, is judged from the following data. The bulk of callose is deposited in developing cell platelets prior to intensive deposition of cellulose. It is callose that is mainly synthesized in primary cell walls of pollen tubes while in secondary ones it is cellulose. Cell damage upon mechanical or ecological stresses or upon action of pathogens results in dramatic retardation of cellulose synthesis whereas that of callose is accelerated.

The latter could be regarded as support for the role of callose as a cellulose precursor [102] provided the glycosyl transferase catalyzing callose-to-cellulose transformation is inhibited.

However, from comparison of properties of callose synthetase and cellulose synthetase an important conclusion could be made on their identity [14] (the well-known example of this kind refers to carboxylase and ribulose-1,5-bisphosphate oxygenase) and on the possibility of a controlled switch of synthetase complexes from β-(1-3)- to β-(1-4)-linkages in a glucan.

This suggestion is supported by the fact that both cellulose synthetase and callose synthetase activities are localized in plasmalemma and that enzymatic activities respond opposedly to endogenous and exogenous factors. Thus, callose synthetase is activated with increase in the level of UDPG as a substrate and of Ca^{2+} while cellulose synthetase is inhibited.

An important role in the switch of enzymatic activities from cellulose to callose synthesis is attributed to a regulatory polypeptide (18 kDa) [14]. It binds 2,6-dichlorobenzonitrile (DCB), the specific inhibitor of cellulose synthesis, and exist both in soluble state in cytoplasm and as a plasmalemma-bound species. This is

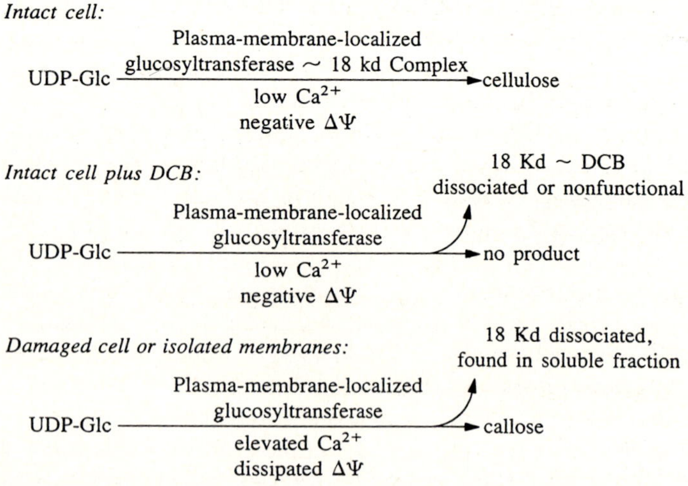

Fig. 1.31. The model of regulation of plasma-membrane-associated UDP-glucose: glucosyltransferase function

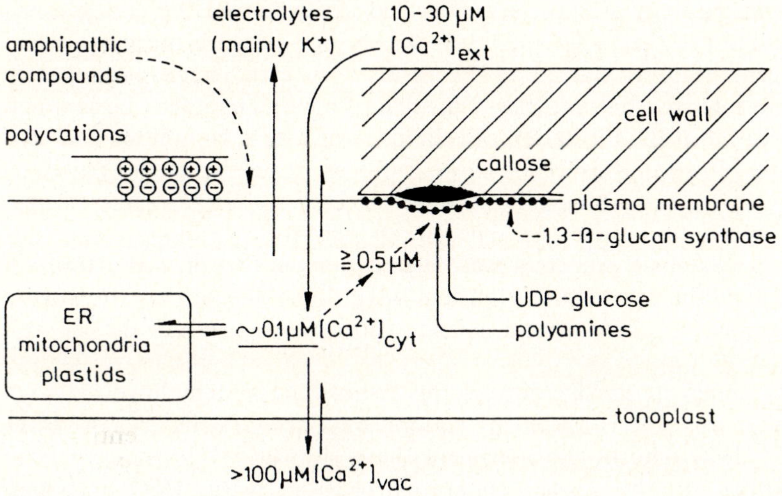

Fig. 1.32. Induction of callose deposition onto cell walls of suspension-cultured soybean cells. Polycations or certain amphipathic compounds at low concentrations perturb the membrane permeability to allow a limited leakage of electrolytes, K^+ representing the major cation. It is suggested that this is also accompanied by transient membrane depolarization and an increased influx of Ca^{2+}. The resulting elevated cytoplasmic Ca^{2+} can directly activate the 1,3-β-D-glucan synthase, which appears to be vectorially located in the plasma membrane and which in vitro is rendered more active and more sensitive to Ca^{2+} by polyamines. The events depicted side by side should be imagined to occur in the same surface region to explain the often localized callose deposition, whose restriction to the affected area may be promoted by the potent Ca^{2+}-buffering capacity of the cytoplasm

reflected in a regulation model for the function of UDPG:glucan glucosyl transferases (GT) localized in the plasmalemma (Fig. 1.31).

Chitosans (i.e. fragments of deacetylated chitin) and other polycations can elicit callose synthesis and deposition. Depending on the type and concentration of the elicitors used (analogous to nonspecific toxins produced from fungal hyphae), callose formation occurs with a lag time of less than 5–30 min. The primary alteration of plasma membrane by polycations (and other factors) and the stimulation of callose deposition are only indirectly linked [105] (Fig. 1.32).

There are some data that β-1,3- and β-1,4-glucan synthetases differ in cell localization in some lower plants [106] and this embraces the establishment of a precursor–product relationship between callose and cellulose [105].

3.5 Synthesis of Cell Wall Proteins

The question on the site of synthesis of cell-wall glycoproteins is of especial interest. It is known that it is proline which is incorporated into proteins during translation and its hydroxylation follows the peptide chain formation.

Electron-microscopy studies have revealed activation of the endoplasmic reticulum and not of the Golgi apparatus in certain isolated protoplasts where cell-wall regeneration is started [107]. Since it is with rough endoplasmic reticulum that intensive biosynthesis of proteins is conducted it may be assumed that it is involved in the formation of cellulose-synthetase complexes and cell-wall proteins. However this becomes doubtful as protein particles (rosettes) participating in cellulose biosynthesis were detected by electron microscopy, inside Golgi apparatus vesicles. And special experiments have finally shaken this assumption. An extensin precursor was detected for the first time in cytoplasm in 1964 [108] and the next step was the demonstration of the participation of the cell membrane systems in biosynthesis and export of extensin into the cell wall [109, 110].

Attempts aimed at localization of this membrane system have led to a suggestion that this role is played by endoplasmic network, its "rough" components [111, 112]. Thorough experiments have shown [^{14}C]-hydroxyproline to be localized in the Golgi apparatus together with UDP-arabinose transferase activity [113]. The latter is of especial importance since carbohydrate side-chains of extensin are represented mainly by arabinotetraose.

Experiments with tobacco cells have shown that labelled hydroxyproline from [^{14}C]-proline is first detected in dictyosomes of the Golgi apparatus, then in the microsomal fraction, and is accumulated in cell walls [114]. The use of [^{14}C]-arabinose allowed one to conclude that oligosaccharide side-chains of a glycoprotein are the main products synthesized in dictyosomes and that a degree of protein glycosylation and its solubility are somewhat higher for vesicles of the Golgi apparatus than those for dictyosomes [115].

4 Synthesis of Structural Polysaccharides: Localization of Reaction Sites

Cell wall formation has attracted much attention for a long time since it is a part of a more complex phenomenon, namely, plant cell division to give daughter cells. The most comprehensive review on this topic is the one published in 1968 [116].

4.1 On the Role of Golgi Apparatus in the Synthesis of Hemicellulose and Pectic Substances

Appearance of a cell wall between daughter cells is preceded by the emergence of the so-called cell platelet which is formed upon fusion of numerous vesicles which are accumulated in equatorial plane of a mother cell. These vesicles are separated from Golgi structures (cylinders) and directed to the expanding cell platelet which finally approaches the longitudinal wall of the mother cell and coalesces with it. The cell platelet contains more pectic substances than hemicellulose. It is first isotropic and then (prior to fusion with the mother cell wall) two anisotropic layers appear on both sides of the isotropic layer, the so-called median disk. These two layers constitute primary walls of daughter cells whose formation has just begun. Their birefringence is accounted for by the appearance of cellulose microfibrils. The membrane which surrounds vesicles of the Golgi apparatus integrates with the plasmalemma of daughter cells thereby increasing its dimensions and the contents of vesicles are "splashed" beyond the plasmalemma into the cell wall. A relatively high intensity of these processes as well as of biosynthesis of cellulose microfibrils seems to result in considerable cell wall thickening. This does not take place since following cell division extension of the daughter cells begins and of the cell walls which cover them.

Elongation of cells is accompanied with the continuous deposition of new "portions" of a cell-wall material. A question was discussed for some time on whether increase in cell-wall mass proceeds through insertion of new components between performed ones or by deposit of new layers onto inner cell wall surface adjoining plasmalemma. Using [^{14}C]-glucose as a substrate it was shown that the expanding cell walls become labelled throughout their thickness [117, 118] as if a mechanism operated whereby newly formed microfibrils were incorporated into the cell wall. It is much more probable that cell-wall labelling of this type is determined by easy penetration into the cell wall by matrix polysaccharides and not by cellulose. Thus the former mechanism is characteristic largely for pectic substances and hemicelluloses whereas the latter for cellulose microfibrils.

The role of the vesicles of the Golgi apparatus in the formation of a cell wall between daughter cells is clear while localization of sites of the synthesis of particular cell wall components, however, required many years of study and provoked animated discussion.

The Golgi apparatus is referred to as an excretory mechanism of animal and plant cells and it could be assumed that all the cell wall components including cellulose are synthesized in these organelles. Indeed, cellulosic plates were detected in vesicles of Golgi apparatus floating to the surface of the plasmalemma in the alga *Pleyrochrysis scheffelii* [119], although this is an exception rather than a rule. Analysis of distribution of labelled metabolites of exogenous [^{14}C]-sugars and NDPS among different organelles has shown that in all other plant species cellulose is synthesized on the cell surface, viz. on the plasmalemma or within its layer. Studies of the kinetics of labelled matrix polysaccharides allowed the establishment of the fact that the label appears at first in dictyosomes of the Golgi apparatus and then is transported by vesicles into the cell wall (Fig. 1.33). Intracellular transport and fusion of different membrane structures have led to a concept of intracellular membrane flows (Fig. 1.34).

The confusing fact was the presence of β-1,4-glucan synthetase activity in the Golgi apparatus. Cellulose itself is not synthesized in this organelle and it was concluded that synthesized there is the enzyme cellulose synthetase which is by some reason inactive in vivo until transferred into the plasmalemma, the site of cellulose synthesis [121]. However, experiments with cycloheximide, the inhibitor of protein synthesis, showed that β-1,4-glucan synthetase is not transported from dictyosomes of the Golgi apparatus and this enzyme plays its own

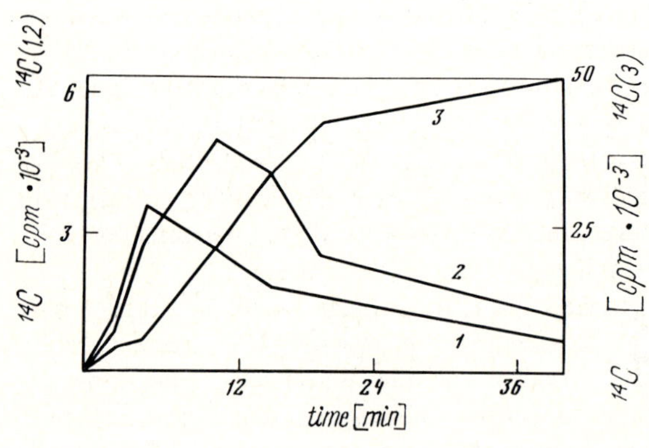

Fig. 1.33. Kinetics of incorporation of [^{14}C]-label from [^{14}C]-glucose into polymers of dictyosomes (1), secretory vesicles (2), and cell wall (3) of *Pisum* stem segments (Robinson, 1980). [^{14}C]-Glucose was consumed for first 6 min. (ordinate, cpm × 10^{-3}; left scale for (1) and (2) and right scale—for (3)). (abscissa, time in min)

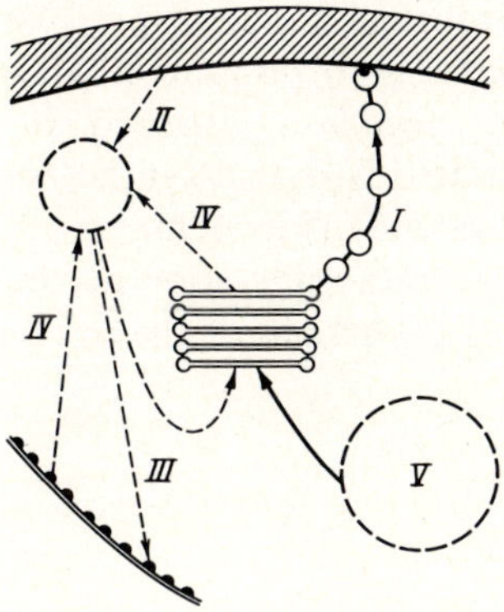

Fig. 1.34. Hypothetical scheme of membrane flows in elongating cells of higher plants [120]. I. Flow of secretory vesicles from dictyosomes of plasmalemma; II. Membrane subunits' reflux to pools localized in cytoplasma; III. Assembly of membranes of endoplasmic reticulum and dictyosomes from reserve subunits proceeding independently of protein synthesis and energetic regime of cells; IV. Liberation of membrane subunits from endoplasmic reticulum and dictyosomes and replenishment therewith of a pool of subunits in cytosol; V. Substrates, (intermediates) of synthesis of polysaccharides localized in cytosol

role in these intracellular formations [122]. This is shown to be reduced to the synthesis of the β-1,4-glucan backbone of hemicelluloses which are formed in the Golgi apparatus [123, 124].

4.2 Synthesis of Cellulose in the Plasmalemma and Properties of the Cellulose Synthetase Complexes

Localized in the plasmalemma, β-1,4-glucan synthetase participates in the synthesis of cellulose. Interestingly, the activity of cellulose synthetase in plasmalemma preparations isolated from cells is by an order of magnitude less than that in nondestroyed cells. It can thus be concluded that isolation is accompanied either with a violation of certain structural, molecular organisation, or with liberation of an inhibitor, or with a loss of synthetase-activating factor. In favour of the first suggestion is the demonstration of the fact that cellulose and callose synthetases require a definite transmembrane electric potential at the plasmalemma for high activity [125]. Damage of the plasmalemma which occurs upon isolation of membrane-bound enzyme preparations results in a loss of the intact membrane potential and, consequently, of a considerable portion of enzymatic activity.

Evidence for the role of membrane potential in the synthesis of cellulose from UDPG was obtained as follows. A potential difference was artificially induced at membranes of vesicles which are formed upon destruction of plasmalemma of cotton-seed hairs where formation of the secondary cell wall has already begun (16–20 days after anthesis). The membrane potential could be induced by placing vesicles in solutions with various combinations of cations and anions which differ in their velocities of penetration through membrane inside the vesicle. Thus, potassium ions could easily enter the vesicles in the presence of ionophore valinomycin. The inner surface of the membrane acquired positive charge while organic counter-ions of low mobility (2-morpholino-ethane-sulphonate, MES) made the outer surface of vesicles negatively charged. Increase in potential difference due to this system stimulated β-glucan synthetase activity, especially at concentration of potassium 50–60 mM (Fig. 1.35).

The use of ions varying in transmembrane penetratability allowed the induction of different potentials and the activation of the synthesis of glucans to differing extents (Fig. 1.36).

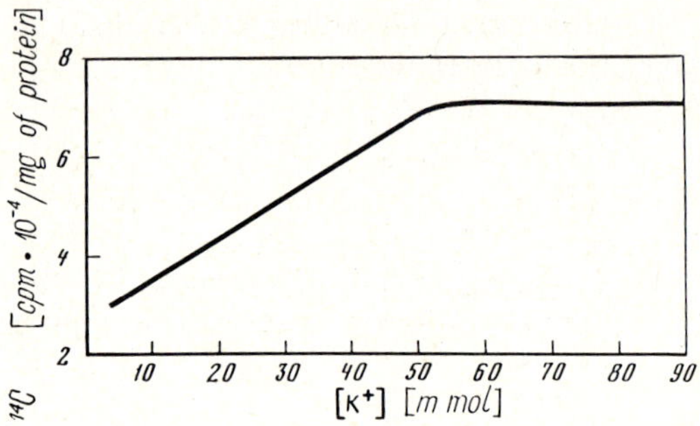

Fig. 1.35. Stimulation of synthesis of β-glucan from [^{14}C]-UDPG by membrane preparations (700–20,000 g) from developing cotton fibers by potassium. The curve refers to radioactivity of acetic acid–nitric acid–soluble fraction. (ordinate, cpm × 10^{-4} per mg protein)

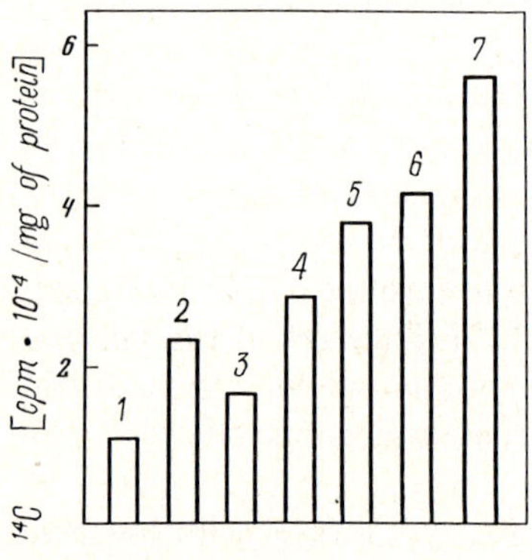

Fig. 1.36. Influence of anions on the synthesis of β-1,4-glucan from [^{14}C]-UDPG by membrane preparations (700–20,000 g) from developing cotton fibers. Vesicles were incubated for 5 min in the presence of potassium salt and valinomycin prior to incubation with [^{14}C]-UDPG for 15 min. Radioactivity given refers to that of acetic acid–nitric acid–soluble fraction. *1*—control; *2*—valinomycin alone; *3–7*—valinomycin in combination with KSCN, KNO_3, KCl, K_2SO_4, and potassium–MES respectively. (ordinate, cpm × 10^{-4} per mg protein)

Modification of membrane potential could be performed not only with model preparations of plasmalemma but with intact, actively metabolizing *Acetobacter xylinum* cells [126]. Variation of membrane potential $\Delta\psi$ was carried out by the aforementioned method, viz. by changing concentration of potassium ions in the presence of valinomycin (Fig. 1.37). Under conditions specified, the H^+-gradient was not considerably affected and its absolute value could be adjusted by varying pH.

It was found that the intensity of cellulose synthesis by *A. xylinum* cells depended on potassium concentration (Fig. 1.37); it almost stopped at 100 mM K^+ concentration which corresponded to the disappearance of ψ (Fig. 1.38).

It should be noted, however, that these two processes do not react equally to changes in potassium ion concentration.

Remarkably, almost total suppression of the synthesis of cellulose at $[K^+]=100$ mM was not detrimental to the synthesis of matrix (water-soluble) polysaccharides but rather stimulated it (Table 1.3). Intensity of synthesis of alkali-soluble matrix polysaccharides was also suppressed upon almost total disappearance of ψ at the plasmalemma, but less pronounced than that of

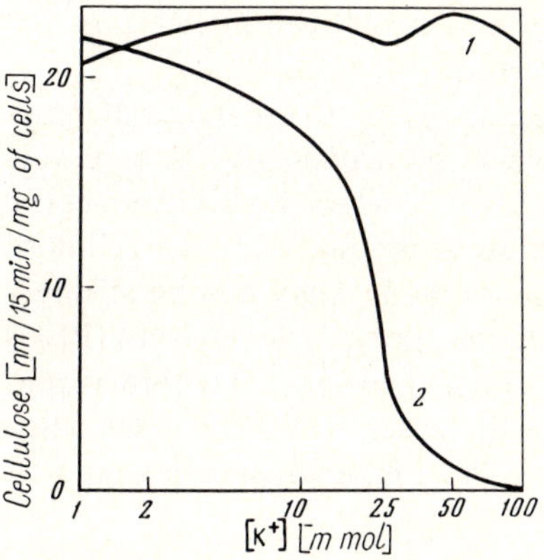

Fig. 1.37. Influence of valinomycin treatment on synthesis of cellulose in *A. xylinum* cells. *1*—control; *2*—valinomycin (5 μg per mg cells) (ordinate, Cellulose, nM/15 min/mg cells)

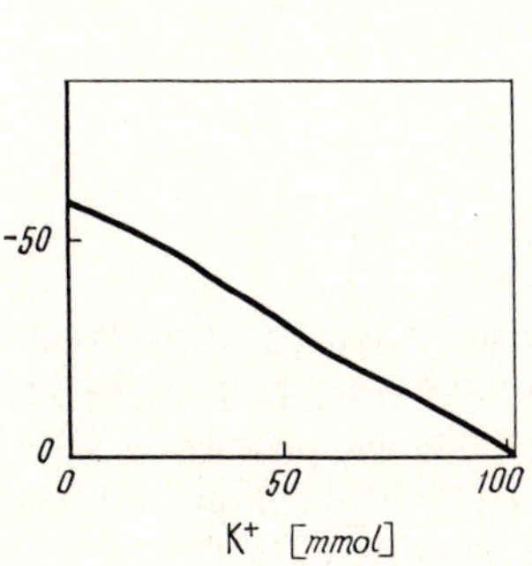

Fig. 1.38. Changes in membrane potential versus concentration of K^+ in *A. xylinum* pH 4.2 (ordinate, mV)

Table 1.3. Specific effect of valinomycin (5 $\mu g/g$ dry mass of cells) and potassium ions on intensity of synthesis of matrix polysaccharides and cellulose by *A. xylinum*

pH of incubation medium	Valinomycin plus salts added (100 mM)	Intensity of polysaccharide synthesis (nM/15 min · mg cell)		
		Water soluble	Alkali-soluble	Cellulose
6.0	Na_2SO_4	2.37	1.50	14.83
	K_2SO_4	4.01	0.27	0.50
5.3	Na_2SO_4	2.07	1.05	16.00
	K_2SO_4	2.24	0.38	1.13
4.5	Na_2SO_4	1.34	0.48	4.50
	K_2SO_4	1.11	0.12	0.01

cellulose. Replacement of potassium by sodium ions in the incubation medium restored (to a higher or lesser degree, depending on pH) the intensity of the synthesis of polysaccharides to that in control runs.

A sharp drop in the cellulose synthetase activity of the plasmalemma preparations upon isolation from cells could also be accounted for, as it was mentioned above, by a release of some specific factors, which possibly affect the activity of enzymes. In favour of this suggestion is the fact that the cellulose synthetase reaction in membrane preparations of *A. xylinum* can be strongly enhanced (Fig. 1.39) by guanosine triphosphate and polyethylene glycol (PEG) [127]. Production of cellulose, by the conventionally isolated membrane preparation, was 3 pmol/min/mg from UDPG, it increased to 9–30 pmoles when isolation and incubation were carried out in the presence of PEG, and at

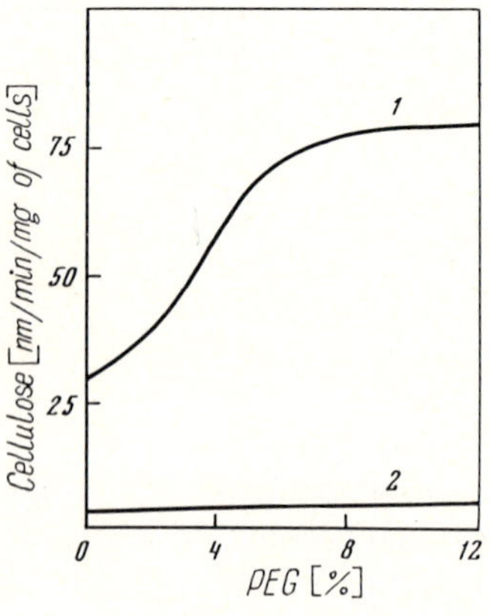

Fig. 1.39. Influence of GTP added (0.25 mM) on cellulose synthesis from UDPG (20 mM). Cell membranes of *A. xylinum* were prepared in the presence of PEG-4000 (20%). *1*—GTP; *2*—Control (ordinate, Cellulose, pmol/min/mg cells; abscissa, PEG, %)

saturating concentrations of GTP (0.15 mM) it reached 75 pmoles when concentration of UDPG was 0.02 mM and 600 pmoles (i.e. 40% of the in vivo productivity) for tenfold UDPG concentration. This was rationalized as follows (Fig. 1.40). The native cellulose synthetase (E) is bound in the plasmalemma with a specific protein activator (factor) P. Isolation of the membrane-bound enzyme using PEG-less buffers results in the removal of the protein factor from the enzyme and a considerable loss in its activity. Addition of PEG prevents the factor release and favours integrity of the E–P complex which is able to bind GTP with a transfer into more active form. Omission of PEG from the membrane preparation plus GTP leads to the shift of the equilibrium P–E–GTP$\rightleftharpoons$P + E + GTP to the right while addition of PEG shifts this equilibrium to the left, i.e. towards formation of the more active complex P–E–GTP.

Further development of these studies has led to the conclusion that the protein factor is in fact a specific guanylcyclase which catalyzes the condensation of two GTP molecules into bis-(3′→5′)-cyclic diguanylic acid (Fig. 1.41)

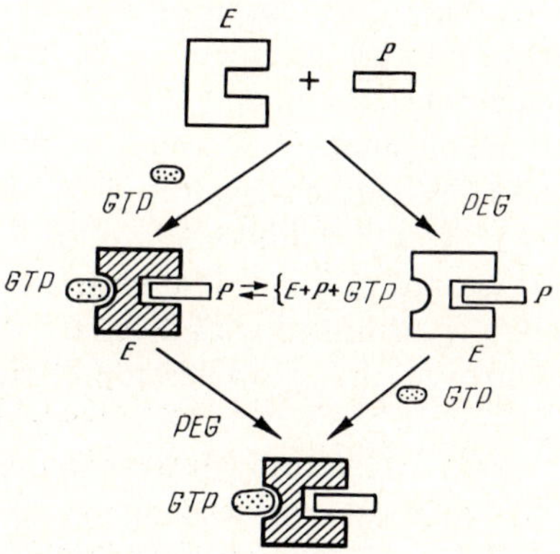

Fig. 1.40. Tentative scheme of interaction of cellulose synthetase (E), PEG, GTP, and a protein factor (P). The enzyme and protein factor in a buffer devoid of PEG are shown in the upper part. Enzymatic activity is low as is in the presence of PEG but without GTP (to the right)

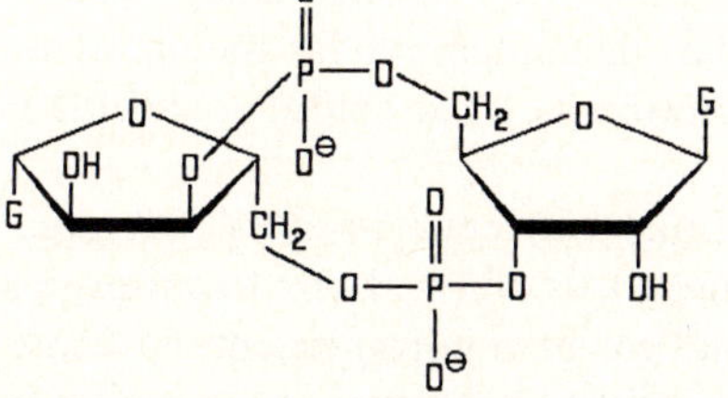

Fig. 1.41. Chemical structure of bis-(1′-5′)-cyclic diguanylic acid.

which, in turn, activates specifically cellulose synthetase [128]. It cannot be excluded that other bacteria possess a system of cellulose synthesis regulation analogous to that in *A. xylinum* which involves a combination of enzymes, regulatory proteins, and effectors [14]. This system also contains a membrane-bound enzyme which decomposes bis-(3′ → 5′)-cyclic diguanylic acid. The cyclo-dinucleotide regulation of cellulose synthesis is not apparently inherent in higher plants.

Studies of the synthesis of cellulose is complicated by the fact that the cell wall contains cellulose chains which are not separate but combined in microfibrils where chains are specifically packed. Most probably, this "multiple-wire" structure is synthesized by an aggregate of many glucan synthetase molecules, spaced in separate sites of the plasmalemma, cellulose synthetases being built of several subunits—each responsible for the synthesis of a single glucan chain [53]. This suggestion was subjected to experimental proof and it is now firmly established that cellulose microfibrils can be synthesized in algae and higher plants by mobile enzymic complexes embedded in the plasmalemma [129].

Attempts were undertaken to clarify the specificity of distribution, within the plasmalemma, of components of the cellulose synthetase system using a freeze-fracturing technique with scanning microscopy of fracture surfaces. Fracturing occurs between lipid layers inside the plasmalemma and microphotographs reveal the details (Fig. 1.42) of inner surfaces of the plasmalemma seen from the side of cytoplasma (C) and from cell wall (W). The W-surfaces exhibit globules [130–134] the so-called terminal complexes of synthesis of cellulose microfibrils. In certain cases, when the terminal globule is fractured in the course of replica preparation, it can be seen that the terminal complex is linked with a cellulosic microfibril tip [135, 136]. It is thus assumed proven that globules represent centres of formation of cellulose microfibrils [137, 138].

Such terminal complexes linked with growing tips of microfibrils were found in the algae *Oocystis* and Glaucocystis, in the cellulose-synthesizing microorganism *Acetobacter xylinum*, and in cells of various tissues of higher plants [139]. Each terminal complex is believed to give rise to one cellulose microfibril. "Rosettes" were also found, each built of six protein globules located at the protoplasmic C-side of the plasmalemma. Each globule in fern protonema 8–9 nm in diameter and a ring ("rosette") of these globules is 24 nm in diameter [140]. The highest density of rosettes (20 per square μm) is observed at the site of maximum cell wall growth and highest rate of microfibril formation (20 μm from the protonema tip), at a distance of 50 μm they almost disappear. In *Micrasterias* the rosettes are secreted into the plasmalemma by means of an endomembrane system consisting of the components of vesicular membranes [139, 141].

Elucidation of peculiarities of structural links between "rosettes", terminal complexes (blocks), and growing cellulose microfibrils allowed one to propose a hypothetical mechanism for microfibril formation in a complementary transmembrane structure of corn root cells: rosette–terminal complex–growing tip of microfibril [136].

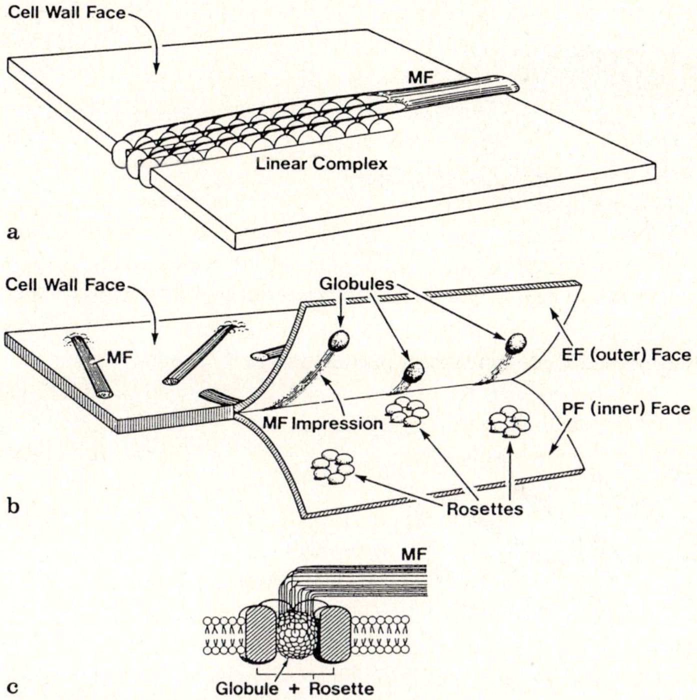

Fig. 1.42a–c. Stylized drawings of terminal complexes observed by freeze-fracture of plasma membranes of cellulosic algae and higher plants. **a** Linear terminal complexes found in some cellulosic algae. For simplicity the number of subunits drawn is less than that actually observed. **b** Rosettes and globules found in some cellulosic algae and in lower and higher plants. Linear complexes as in (A) usually fracture with the *EF face* but sometimes fracture with the *PF face*; rosettes fracture with the *PF face* and globules, with the *EF* face. In double-replica fracturing, globules are sometimes found in complementary location to rosettes, leading to the speculation that rosettes and globules may be part of the same complex in vivo as shown in (**c**). *MF* = microfibril [14].

The terminal complex of the outer surface of the plasmalemma is connected with a tip of a newly formed microfibril, the rosette of the inner surface of membrane has a cavity (Fig. 1.43) which is presumably the site of binding of cellulose precursors and withdrawal of glucose "carriers".

The necessity of an influx of glucose carriers towards the inner surface of plasmalemma may account for the poor utilization of exogenous UDPG for the biosynthesis of cellulose in nondamaged cells. The slowly penetrating NDPS can be actively utilized only after violation of cell integrity when these compounds get into the cell and have access to rosettes.

At the same time, isolation of plasmalemma cellulose–synthetase particles from cells results in considerable loss of their activity not probably due to drop

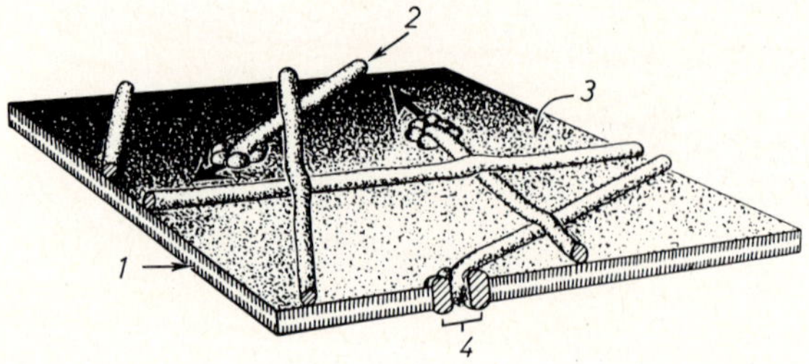

Fig. 1.43. Deposition of cellulose microfibrils in *Micrasterias* during secondary cell wall formation.

1 Lipid bilayer, *2* Microfibril, *3* Outer surface of plasmalemma, *4* "Rosette"

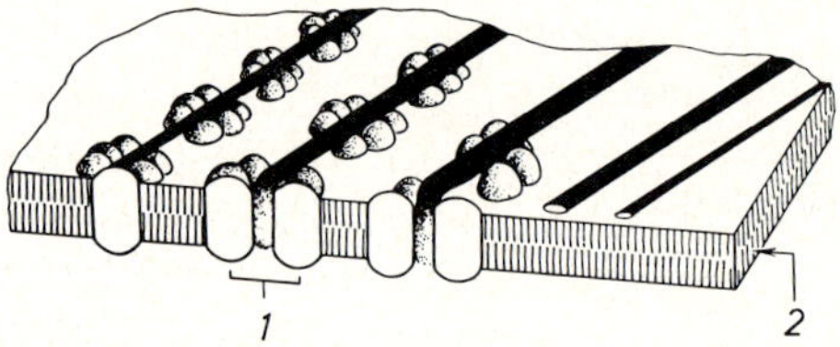

Fig. 1.44. Formation of cellulose microfibrils during secondary cell wall formation in *Micrasterias.*

1 "Rosette" (the apparatus for synthesis of microfibrils of cellulose) *2* Lipid bilayer

in transmembrane electric potential but more likely due to the disintegration of a complex rosette–terminal block.

Single rosettes together with respective terminal complexes are supposedly responsible for the synthesis of random microfibrils of the primary cell wall (Fig. 1.43) while an assembly of hexagonally oriented microfibrils (up to 175 in some regions of the plasmalemma) determines the formation of parallel sets of microfibrils of secondary-cell-wall cellulose (Fig. 1.44) [139]. Each series of rosettes is involved in the formation of a single fibril. The distance between rows of rosettes corresponds to that between microfibrils in the cell wall. It was supposed that the thickest microfibrils which are located in the centre of a layer of microfibrils are synthesized by the longest rows of rosettes located in the centre of a hexagonal complex and shorter rows of rosettes are responsible for formation of relatively thinner microfibrils (Fig. 1.44). Each rosette is believed to form a 5-nm microfibril (about 50 polyglucan chains constitute thin microfibrils) and they are combined to produce larger fibrils of the secondary cell wall, the largest diameter being 28.5 nm [139]. This is a product of 16 rosettes, and with hexagonal packing of 16 elementary, 5-nm microfibrils; the width of such aggregates should be 30 nm which is close to experimental data.

Two possible modes of participation of enzyme complexes in the biosynthesis of cellulose microfibrils have been considered [142]. They are: a) simultaneous synthesis by a single protein particle, of several chains of the cellulose fibril, including antiparallel glucan chains, which may be due to the presence of sufficiently large amount of the enzyme molecules in the protein particle; b) synthesis of one fibril by a relatively large number of protein particles, whereby each produces cellulose chains which are later combined, due to hydrogen bonds to form a fibril. The scheme of participation of the terminal complexes in the synthesis of cellulose fibrils is depicted in Fig. 1.42. It involves both these mechanisms as it provides for simultaneous synthesis of about 50 glucan chains in the elementary microfibril as well as subsequent lateral aggregation of several microfibrils into fibrillar structure of a higher order.

The role of protein particles of the plasmalemma in synthesis of cellulose can be judged by their "behaviour" in protoplasts isolated from higher plants and regenerating cell wall [107]. Following enzymatic digestion of cell walls and isolation of protoplasts accumulation of single rosettes begins at the inner surface of the plasmalemma. The next day, their packing into dense rows with the formation of hexagonal structures in certain regions of the plasmalemma occurs. At the corresponding areas of the outer surface of the plasmalemma the rows of terminal complexes protrude over the membrane. By this time the parallel lines of particles become distinctly visible on the inner surface of the plasmalemma besides hexagonal aggregates of rosettes. They number 12–14 in each line a day after isolation of protoplasts and 24–28 after two or three days. Distance between lines is larger than that in hexagonal aggregation of rosettes. Microfibrils of cellulose on the surface of protoplasts have the same direction as the parallel lines of particles, however the latter cannot serve as matrices for synthesis of cellulose because their size is too small compared to the length of a cellulose molecule which can be 7000-nm long [143].

Schnepf [144] studied the kinetics of cellulose synthesis and the turnover of the rosettes in moss protonemata. He concluded that rosettes occur only when cellulose is synthesized, that they have a short lifetime and that cellulose synthesis and pectin secretion are correlated. A rosette has a lifetime of about 15 min, during this time one elementary fibril per rosette is formed. A glucan chain elongates by about 2000 glycosidic linkages per minute. Experiments with plasmolysis show that the rosettes are extremely labile structures and disappear immediately when the cell is plasmolyzed, that plasmolysis equally interrupts cellulose synthesis, pectin secretion, and the production and migration of Golgi vesicles.

The question should be considered of the size of molecules of native cellulose, the so-called cellulose I (to distinguish it from cellulose II which is modified in the process of isolation and treatment). Primary wall cellulose is rather short, the amount of glucose residues varying from 2,000 to 6,000 [145]. This leads to the conclusion that molecular size is thermodynamically determined. Secondary wall cellulose possesses a higher degree of polymerization. What is especially noteworthy, is that this value (14,000) is constant (Fig. 1.45).

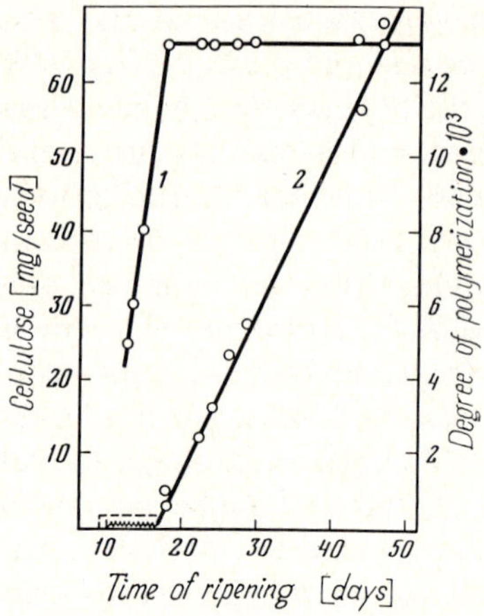

Fig. 1.45. Age-dependent in biosynthesis of cellulose in cotton plant pod [146]. *1* Degree of polymerization *2* Intensity of cellulose synthesis (Left ordinate, mg cellulose per seed. Right ordinate, degree of polymerization, $\times 10^{-3}$ Abscissa, time after anthesis, days)

This points to the fact that a template exists from which cellulose molecules are "printed". However, the search for such a die gave no positive results.

If one assumes that a microfibril is synthesized by a combination of cellulose synthetases which constitute terminal complexes and rosettes, the unanswered question arises what is the mechanism which "counts" 14,000 glucose residues to be attached to the growing cellulose terminus [147].

The role of the template which could determine the size of the cellulose molecule was ascribed to microtubules located in a cortical layer of cytoplasm directly beneath plasmalemma. However, this hypothesis was not justified and microtubules seem rather to serve for the orientation of newly forming cellulose microfibrils relative to the cell surface.

4.3 Inhibitors of Cellulose Biosynthesis

A specific role in studies on peculiarities of one or another physiologo-biochemical process is played by inhibitors. This applies totally to the biosynthesis of cellulose. The inhibitors can be subdivided into three groups which affect in a different manner the deposition of cellulose in cell walls [148, 149]. To the first group belong inhibitors which influence the regular change in microfibril orientation directed by microtubules. These inhibitors, e.g. colchicine, preclude aggregation of tubulin to form microtubules or bring about disintegration of the latter. Cellulose synthesis and microfibril formation are not considerably delayed but fibril orientation is violated.

The second group contains inhibitors of glucan chain synthesis such as coumarin. Even more effective inhibitors of cellulose synthesis were found [99]. They include 2,6-dichlorobenzonitrile, 2-chloro-6-fluorobenzonitrile, and chlorothiamide which inhibit biosynthesis of cellulose at concentrations by several orders of magnitude less than coumarin (Figs 1.46, 1.47). Mechanism of their action is unknown.

Calcium ionophores A-23187 and X-537A also inhibit cellulose synthesis [150]. Potent inhibition by cycloheximide [148] was unexpected. Treatment with cycloheximide for several hours results in the total disappearance of terminal blocks on the surface of the plasmalemma. It thus can be concluded that maintenance of the "population" of terminal complexes requires continuous synthesis of constituent proteins since they are in a status of permanent renewal (degradation and synthesis).

Belonging to the third group are inhibitors which do not slow down synthesis of cellulose chains but which prevent their interaction (crystallization). They are the dyes which bind specifically to polysaccharides, viz. Congo Red and Calcofluor White. In the presence of these dyes the formation of microfibrils stops and it is accompanied by a several-fold increase in the amount of terminal complexes (per area unit) as well as by thickening of each terminal complex, largely, due to the pairing of single terminal blocks.

It is known that Congo Red is not a promoter of protein synthesis and an increase in the number of terminal complexes can be accounted for by the prevention of their conventional breakdown either due to direct action of the

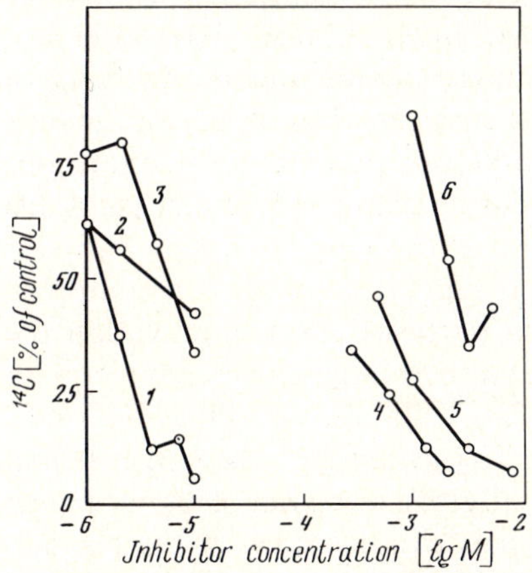

Fig. 1.46. Synthesis of cellulose from ^{14}C-glucose in cotton-seed hairs during in vitro culturing of seed bud as a function of inhibitor concentration.

1 2,6-Dichlorobenzonitrile (DCB), *2* Chlorothiamide, *3* 2-Chloro-6-fluorobenzonitrile, *4* EDTA, *5* Coumarin, *6* Bacitracin

(Abscissa, Logarithm of molar concentration of an inhibitor. Ordinate, Incorporation of ^{14}C, per cent of a control experiment)

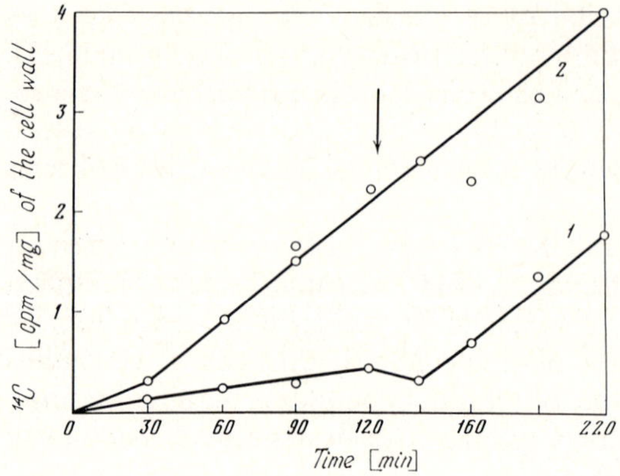

Fig. 1.47. Reversibility of inhibition of cellulose synthesis by DCB in cotton fiber

1 Experiment *2* Control (Abscissa, Time, min; Ordinate, ^{14}C, cpm per mg cell wall $\times 10^{-3}$) The inhibitor (10 μM) was added at 'zero' point, after 120 min the hairs not detached from seeds were transferred for 2.5 min into a ^{14}C-glucose- and inhibitor-free medium and then into ^{14}C-glucose-containing test medium. Control seeds were incubated, washed and reincubated in similar manner

dye, or due to the violation of biosynthesis of the cellulose microfibrils. Supposedly, Congo Red can regulate the ratio of rates of destruction and synthesis of microtubules by some feedback mechanism [148].

Microtubules belong to structures responsible for the synthesis of microfibrils. Also participating are protein particles on the inner surface of the plasmalemma and terminal complexes on its outer surface. The latter are directly responsible for the synthesis of cellulose chains and the forming of microfibrils. Elongation of microfibrils and their fixation in a cell-wall matrix exert growing pressure upon the terminal complex so that its movement begins within the outer lipid layer in a plane of the plasmalemma. The direction of this movement is governed by an orientation of rows of protein granules on the inner surface of the plasmalemma which, in turn, is regulated by the orientation of microtubules (Fig. 1.48). The rows of protein particles are located within the internal lipid layer of the plasmalemma and can therefore come in contact with microtubules. Change in orientation of extending cellulose microfibrils is conditioned by disaggregation of microtubules and by assembly of new ones which are disposed usually at right angles to the original direction. This results in alteration of orientation of rows of protein particles (rosettes) of the plasmalemma and in the start of deposition of a new layer of microtubules.

Violation of crystallization (loosening of cellulose chains) by the action of dyes results in the disappearance of the "pushing moment" which usually causes movement of the terminal complexes in the plasmalemma and they become immobile.

Cellulose ribbon in *A. xylinum* is formed in a single region of the cell surface with the aid of a linearly oriented assembly of cellulose-synthetase blocks.

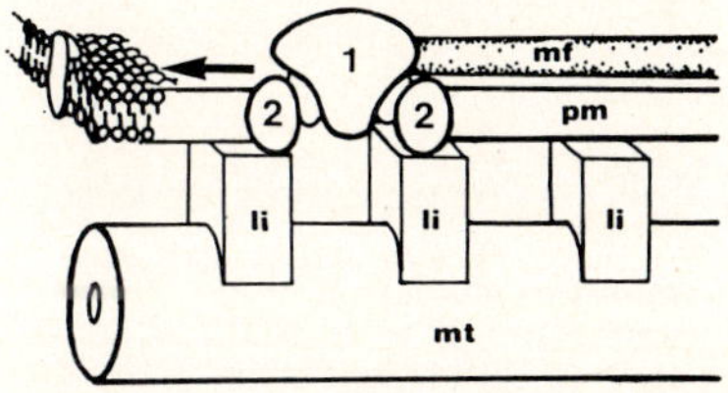

Fig. 1.48. Interrelation of orientation of a growing microfibril (MF) and microtubule (MT) disposed beneath plasmalemma (PM). Microtubule—membrane links (Li), microfibril—synthesizing complex (1 and 2). The arrow indicates the direction of movement of the particle complex [151]

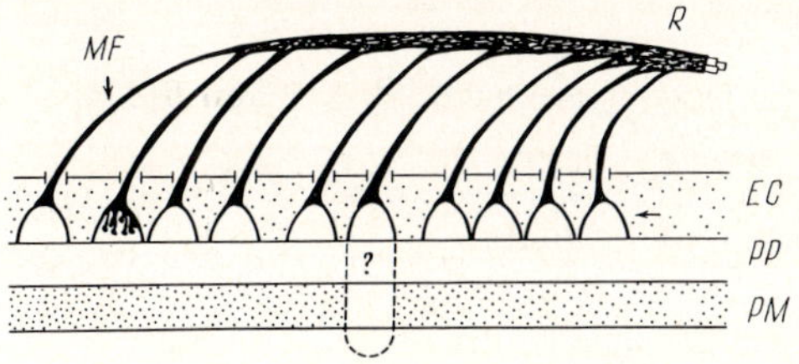

Fig. 1.49. Cellulose ribbon formation in *A. xylinum* [152]. R, ribbon; MF, cellulose microfibril; EC, cellulose-synthetase enzyme complexes; PM, plasma membrane; PS, periplasmic space

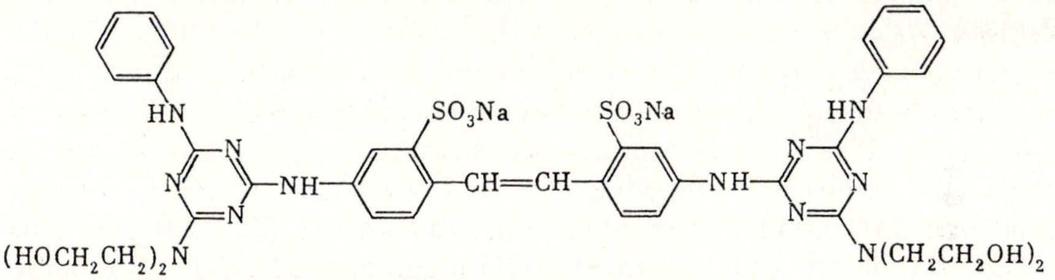

Fig. 1.50. Chemical structure of Calcofluor White dye

Cellulose microfibrils produced by each synthetase complex are later aggregated (Fig. 1.49) to form a flat and somewhat twisted ribbon.

The rate of formation of all microfibrils is the same and is determined, obviously, by the rate of the "slowest" enzyme complex. When bacterial cells are treated with a dye of the Calcofluor White type (Fig. 1.50) which binds to glucose residues of cellulose through hydrogen bonds, these being responsible for microfibril formation from the cellulose chains and ribbons, the rate of cellulose synthesis is increased by a factor of 4 [153]. The reason is that the regulatory pressure from higher structures is released and the majority of microfibrils or even cellulose chains can be synthesized at a higher rate. It is interesting to follow the change in the structure of a bacterial cellulose complex

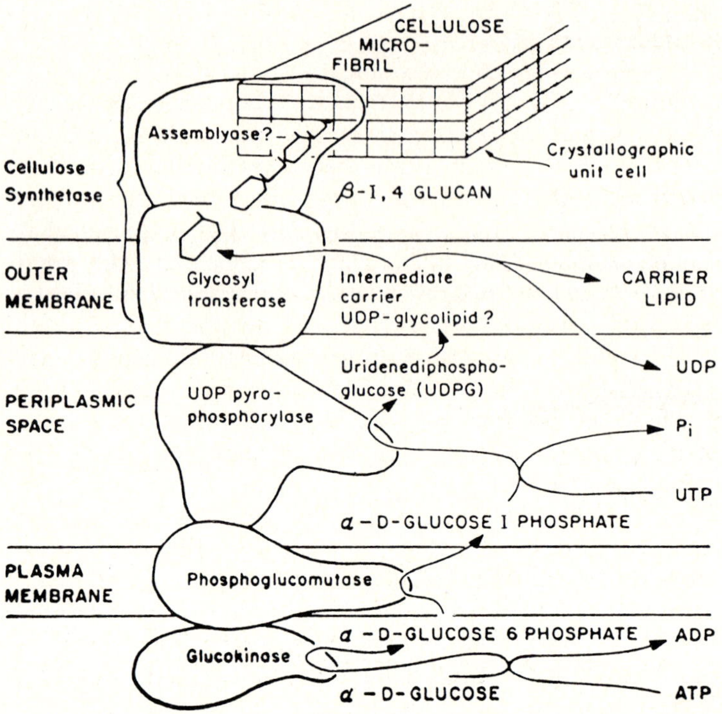

Fig. 1.51. A proposal for the location of enzymes involved in cellulose biosynthesis in *Acetobacter* [129]

after addition of the dye. The cellulose ribbon which is parallel to the longitudinal cell axis and exits from a linear cellulose synthetase complex at one side of bacterial cell is converted following the dye action into a "brush" of thin cellulose filaments oriented vertically to the cell surface [152]. It was thus concluded that the processes of polymerization and crystallization of cellulose are spatially close but separated and dyes of the calcofluor type can uncouple them. The former process is catalyzed by a glycosyl transferase and the latter by "assemblyase". Both are the components of the cellulose-synthetase complex and "assemblyase" is located at the very surface of a cell wall (Fig. 1.51).

4.4 Influence of Microtubules on the Orientation of Cellulose Microfibrils

The researchers who studied the ultrastructure of plant cells investigated a long time ago the parallel orientation of cellulose microfibrils and microtubules, the protein structures located in cortical layer of cytoplasm just beneath the plasmalemma [154–159].

Microtubules appear as fibers of about 25 nm in diameter and several-micrometers length, built of protein globules of 4 to 5 nm in diameter and a molecular mass of 55,000. The amount of globular protein tubulin in cell is rather high, it is composed of subunits of two types, and depending on the cell status, its aggregation can produce long microtubules whose disaggregation produces tubulin once more. One can say that microtubules and the constituent tubulin are in equilibrium (tubulin$\rightleftarrows$microtubules), the mechanism of preponderance of direct or reverse reaction is not sufficiently clear. Aggregation is promoted by the illumination of photosynthesizing cells, by plant hormones [160], NTP, trans-phosphorylases, NTPases, and a protein factor with a molecular mass larger than that of tubulin. Lowering the temperature of higher plant cells results in a sharp decrease in the amount and length of microtubules [161]. It is noteworthy that the recovery of microtubules in *Oocystis* is rapid (within 5 min) and almost quantitative upon heating to 25 °C following prior chilling to 1 °C whereupon many microtubules were disaggregated (Table 1.4) [148]. The aggregation rate of microtubules can be as high as 1–3 μm per min.

Microtubules can be disaggregated by inhibitors of two types [162]. Those of the first type can bind with tubulin thereby favouring the disappearance of microtubules from cells, among them are colchicin, colcemide, podophyllotoxin, griseofulvin, vinblastin [157]. Inhibitors of the second type disturb the intercellular level of calcium ions which participate in tubulin aggregation leading to formation of microtubules. They involve calcium ionophores [163] and herbicides aniprophosmethyl, trifluramine, oryzalin [164].

As the methods for observing microtubules become more perfect, attempts to explain their role and mechanism of participation in cell wall formation become more and more attractive. The use of immunofluorescent analysis made it possible to see a much more abundant population of microtubules in cells

Table 1.4. Recovery of amount of microtubules following chilling of the alga

Conditions	Cells with microtubules %	Microtubules: Average amount (per cell)	Microtubules: %
150 min, 25 °C	100	10.7	100
150 min, 1 °C	37	3.8	35
145 min, 1 °C + 5 min, 25 °C	73	6.5	61
135 min, 1 °C + 15 min, 25 °C	79	7.0	65
120 min, 1 °C + 30 min, 25 °C	25	7.3	68

(cultured carrot cells) [165] than before. Treatment with colchicin for 24 hours resulted in disaggregation of microtubules but the elongated shape of a cell was well preserved being conditioned by the outline of the cell wall. After some time the cell acquired a spherical shape. It points to the fact that microtubules constitute the part of internal cytoskeleton and participate together with the cell wall in cytomorphogenesis.

Disaggregation of microtubules induced by colchicin or other inhibitors does not stop cellulose synthesis but causes the appearance of randomly oriented cellulose microfibrils. According to other data, the parallel orientation of microfibrils inside the cell-wall layer remains but their ability for regular rotation ceased (e.g. by 90°) relative to the previously formed microfibrils of the neighbouring layer [166]. Washing the inhibitors off resulted in a normal, "sandwich-like" structure of the part of cell walls formed.

Despite the fact that coupling of orientation of cellulose microtubules and microfibrils was established long ago, the absence of such interrelation was also documented [167]. It was also concluded that microtubules disappear when synthesis of the secondary cell wall in the cell begins, thus they cannot play a role in the orientation of cellulose microfibrils [129, 131]. This caused scepticism with regard to the most interesting hypothesis put forward by Heath [168], according to which the synthesis of cellulose microfibrils is directed by a transmembrane conjugation of cellulose-synthetase complexes and microtubules. The devotees of this hypothesis have recently mentioned [162] that those who opposed it earlier have changed their point of view [169] and thus not only can it be regarded as refuted but it has been confirmed by more recent data.

In this respect, of especial interest are the facts of regulation not only of intensity of tubulin aggregation and microtubule disaggregation processes, but also of microtubule orientation. Indoleacetic acid, which activates the longitudinal growth of cells, caused reorientation of microtubules and they became oriented perpendicularly to the longitudinal axis of a cell, whereas kinetin, which stimulates cell thickening, brought about orientation of microtubules parallel to the long axis of a cell [170, 171, 172].

And how do microtubules determine the orientation of cellulose microfibrils? They can arrange the influx of Golgi-apparatus vesicles filled with pectic substances, hemicelluloses, and glycoproteins to those sites of the plasmalemma where synthesis of cellulose microfibrils takes place [173, 174]. According to the other hypothesis [168], microtubules, on account of transmembranous linkages, direct the movement of cellulose-synthetase complexes in a semiliquid plasmalemma and therefore the cellulose microfibrils "extruded" from the complexes are oriented along the microtubules situated beneath the plasmalemma. Inhibitor-induced disaggregation of microtubules during intensive cellulose formation will not affect essentially the orientation of microfibrils because the "preset" direction of the continuous growth of microfibrils is determined by strengthening bonds with various polysaccharides and protein components of the cell-wall layer which is being formed.

4.5 Chitin

For some time past much attention is being drawn to chitin as a possible, additional raw material for industry of high polymers. Chitin is a fibrillar, crystalline poly(*N*-acetylglucosamine). It is absent from higher plants and is found in cell walls of fungi, several diatomous algae and chrysoflagellates, in the cuticle of insects and arthropodes, in the chaeta of *Polychaeta*, and in bacterial cell walls [175, 176].

Several reviews on the mechanism of chitin microfibril formation have recently appeared [176, 177]. It was shown that UDP-*N*-acetylglucosamine is the substrate for chitin synthetase. This is activated by divalent cations, especially Mg^{+2}, Mn^{+2}, and Co^{+2} and by free *N*-acetylglucosamine (the substrate Michaelis constant is 4 to 12 mM, which is higher than its concentration in cytoplasm). It is assumed that *N*-acetylglucosamine is an allosteric activator of the enzyme though it only mimics the direct substrate for chitin synthesis, UDP-*N*-acetylglucosamine, the latter activating chitin synthetase at much lower concentrations. The most specific inhibitors of the enzyme are polyoxines, a family of antibiotics from streptomycetes.

Chitin synthetase is located in the cytoplasm in the form of microvesicles with diameter of 40 to 70 nm named chitosomes [178] but synthesis of chitin takes place not in cytoplasm but in the cell membrane. Chitosomes serve as means of transport and transfer chitin synthetase from its generation site (rough endoplasmic reticulum) to the site of its action.

There are data in favour of a transfer of *N*-acetylglucosamine from UDP-*N*-acetylglucosamine to a lipid acceptor, diglycosyl diglyceride and then to a polypeptide enriched in basic amino acids which is an actual initiator of chitin chain growth. The degree of polymerization of chitin is 1800 to 2200 [179]. The use of dolichol pyrophosphate *N*-acetylglucosamine labelled in the sugar moiety did not lead to incorporation of the label into chitin upon incubation with chitin synthetase in vitro. It was thus concluded [177] that a lipid carrier (if it participates in chitin synthesis at all) differs from dolichols.

Organization of the chitin-synthetase complexes depends on taxonomic position of the organism. Thus, for example, the alga *Cyclotella cryptica*

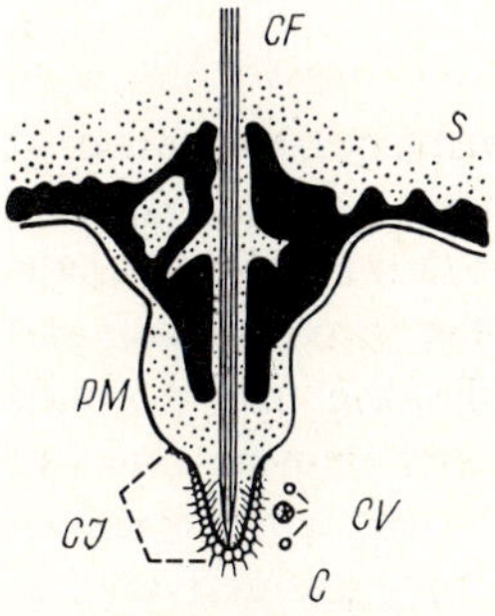

Fig. 1.52. Structure of a chitin-forming organ in a diatom [176] CF, chitin fibril; PM, plasmalemma; S, slime; CV, cytoplasmic vesicles; CI, conical invagination where the chitin-synthetase complex is located; C, cytoplasm

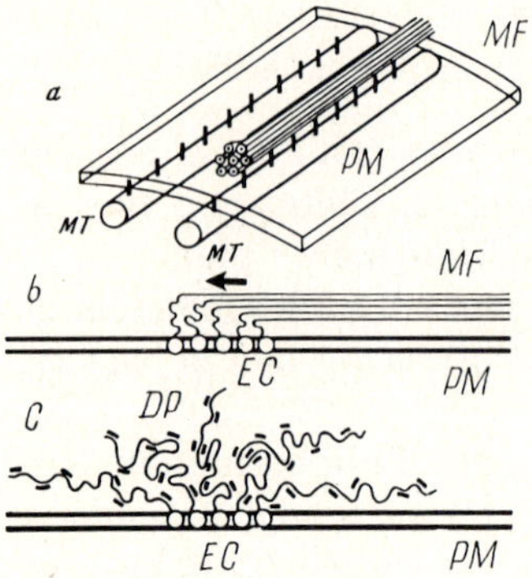

Fig. 1.53a–c. A scheme of movement of chitin-synthetase complexes in plasmalemma [176]. **a** Parallel arrangement of microtubules (MT) and of chitin microfibrils (MF) synthesized by the enzyme complexes (EC). **b** Directed movement of the chitin-synthetase complexes in plasmalemma (PM). **c** Formation of hydrogen bonds between dye particles (DP) and growing chitin molecules preventing their aggregation to microfibrils and regular movement of the enzyme complexes in plasmalemma.

possesses the specialized chitin-forming apparatus embedded in the plasmalemma (Fig. 1.52) in which synthesis of chitin ribbons built of several parallel subunits occurs. Quite another structural organization is responsible for chitin synthesis in the alga *Poterioochromonas stipetata* (Fig. 1.53). It resembles cellulose-synthetase blocks which "repel themselves" from the microfibrils formed and "hardened" within the cell wall and which move laterally in the plasmalemma. In both cases the orientation of the growing microfibrils is determined by the position of microtubules beneath the plasmalemma. Addition of Calcofluor White or Congo Red dyes which bind to chains of the newly formed polysaccharides (both cellulose and chitin) thereby preventing intermolecular hydrogen bonds' formation disturbs formation of microfibrils and results in irregular arrangement of chains (Fig. 1.53).

Fungal infection results in the accumulation of phenolic acids and in the lignification of the host tissues (cultured carrot tissues). The mycelium activates chitinase which liberates oligosaccharides of a different chain length. The latter, in turn, induce the accumulation of mostly caffeic and, to a lesser extent, ferulic acids which like lignin itself prevent the propagation of infection [180].

4.6 Changes in the Synthesis of Structural Polysaccharides Caused by Lead Nitrate and Pipolfen

Studies, in the authors' laboratory, of the effect of phytohormones, inhibitors of ATP synthesis, exogenous NTP's, retardants, and substrates at various concentrations, on the synthesis of structural polysaccharides revealed similar changes in intensity of label incorporation, from exogenous substrates, into cellulose and hemicelluloses (nitric acid–acetic acid–soluble fraction). Taking into account that cellulose is synthesized by protein particles of the plasmalemma one can assume that inhibition can occur under the action of substances which specifically modify plasmalemma proteins. As for hemicelluloses, they are delivered at the plasmalemma surface by Golgi-apparatus vesicles which fuse to the plasmalemma first of all due to the interaction of lipid components of membranes. Thus it is quite natural to expect that lipid-modifying substances inhibit synthesis of hemicelluloses.

It was anticipated that an alteration in the structure of plasmalemma proteins would affect, firstly, cellulose synthesis, while alteration in lipid structure would affect synthesis of hemicellulose. Experiments were carried out with lead nitrate which modifies proteins specifically and with pipolfen which preferentially modifies the lipid components of membranes [181]. These compounds do not enter the cell on relatively short exposure [182–185]. Therefore changes in the synthesis of structural polysaccharides caused by short-term incubation with these substances can be associated solely with changes in the plasmalemma structure.

Both these membranotropic compounds cause 50% inhibition of cellulose synthesis at concentrations of 10^{-3} M. Synthesis of hemicelluloses, unlike that of cellulose, was not necessarily inhibited by lead nitrate. In certain cases, it inhibited considerably incorporation of [^{14}C]-glucose in cellulose and promoted synthesis of hemicelluloses (Table 1.5).

As can be seen, pipolfen affected biosynthesis of both cellulose and hemicelluloses in a similar manner. When UDPG was used as a substrate the addition of lead ions did not decrease the intensity of cellulose formation but, indeed, somewhat stimulated this process.

The effect of membranotropic substances on the synthesis of polysaccharides can be indirect, through alteration of the supply of substrates across the

Table 1.5. The influence of $Pb(NO_3)_2$ (1 mM) and pipolfen (1 mM) on ^{14}C-incorporation (μg/h/g cell mass) into wheat-root polysaccharides

Substrate + additive	Cellulose		Hemicelluloses	
Glucose (control run, 1 h)	35.2	(100)	61.0	(100)
Glucose + $Pb(NO_3)_2$ (1 h)	15.5	(44)	99.4	(163)
Glucose + pipolfen (1 h)	15.8	(45)	29.3	(48)
Pipolfen (15 min), then glucose (1 h)	20.4	(58)	45.1	(74)
UDPG (control run, 1 h)	6.2	(100)	3.5	(100)
UDPG + $Pb(NO_3)_2$ (1 h)	7.1	(114)	4.6	(133)
UDPG + pipolfen (1 h)	3.8	(61)	2.9	(83)
Pipolfen (15 min), then UDPG (1 h)	5.4	(87)	2.6	(75)

Standard deviation $\leqslant 6\%$; figures in parentheses refer to percentages

Table 1.6. The influence of lead nitrate (1 mM) and pipolfen (1 mM) on uptake of different substrates (μm/h/g cell mass by wheat roots)

Substrate + additive	Substrate uptake
Glucose (control run, 1 h)	2.3 (100)
Glucose + $Pb(NO_3)_2$ (1 h)	1.8 (78.3)
Glucose + pipolfen (1 h)	1.3 (56.5)
Pipolfen (15 min), then glucose (1 h)	2.1 (91.3)
Sucrose (control run, 1 h)	1.2 (100)
Sucrose + $Pb(NO_3)_2$ (1 h)	0.6 (50.0)
Pipolfen (15 min), then sucrose (1 h)	1.7 (141.7)
UDPG (control run, 1 h)	0.6 (100)
UDPG + $Pb(NO_3)_2$ (1 h)	0.8 (133.3)
Pipolfen (15 min), then UDPG (1 h)	1.5 (250.0)

Standard deviation $\leqslant$ 10%, figures in parentheses refer to percentage

plasmalemma inside cells to the sites of their utilization. Special experiments were therefore carried out aimed at studying the consumption, by root tissues, of different substrates and their participation in the synthesis of structural polysaccharides. In the absence of the aforementioned additives the most intensity consumable was glucose, followed by sucrose, and finally UDPG (Table 1.6).

It can be seen that lead nitrate inhibited the uptake of glucose and, especially of sucrose and favoured UDPG uptake. Short-term incubation with pipolfen resulted in the minor inhibition of glucose transport and strongly activated uptake of sucrose and, to an even greater extent, of UDPG.

Comparison of changes in consumption of substrates in the presence of membranotropic substances with those in intensity of synthesis of polysaccharides allows one to conclude that there is no direct connection between the influence of membrane-modifying agents on transport of substrates inside cells and synthesis of polysaccharides therefrom (Table 1.7).

In the same experiment the proportion of ^{14}C incorporation (in percent of the total ^{14}C consumed by cells) from different substrates in polysaccharides could be determined.

It can be seen that pipolfen inhibited the incorporation of ^{14}C into polysaccharides to the largest extent, consumption of substrates being enhanced in the majority of cases (Table 1.8). This conclusion holds for both cellulose and hemicelluloses. Estimation of the incorporation of ^{14}C (in percent of the total substrate consumption) reveals that the lead salt inhibits cellulose synthesis from all the substrates tested with concomitant enhancement of hemicellulose synthesis.

Table 1.7. The influence of lead nitrate (1 mM) and pipolfen (1 mM) on incorporation of ^{14}C (nM/h/g cell mass) from substrates specified in cellulose and hemicelluloses under special conditions

Substrate + additive	Cellulose	Hemicellulose
Glucose (control run, 1 h)	256 (100)	689 (100)
Glucose + $Pb(NO_3)_2$ (1 h)	139 (54)	656 (95)
Glucose + pipolfen (1 h)	72 (28)	200 (29)
Pipolfen (15 min), then sucrose (1 h)	94 (37)	333 (48)
Sucrose (control run, 1 h)	129 (100)	389 (100)
Sucrose + $Pb(NO_3)_2$ (1 h)	67 (52)	316 (81)
Pipolfen (15 min), then sucrose (1 h)	61 (47)	216 (56)
UDPG (control run, 1 h)	25 (100)	267 (100)
UDPG + $Pb(NO_3)_2$ (1 h)	29 (116)	521 (195)
Pipolfen (15 min), then UDPG (1 h)	20 (80)	124 (46)

Standard deviation $\leqslant$ 6%, figures in parentheses refer to percentage

Table 1.8. The influence of lead nitrate (1 mM) and pipolfen (1 mM) on incorporation of ^{14}C (in percent of the total substrate consumption by tissues) from substrates specified in cellulose and hemicelluloses under special conditions

Substrate + additive	Cellulose	Hemicelluloses
Glucose (control run, 1 h)	11.2	30.2
Glucose + $Pb(NO_3)_2$ (1 h)	7.6	35.8
Glucose + pipolfen (1 h)	5.6	15.7
Pipolfen (15 min), then glucose (1 h)	4.5	16.2
Sucrose (control run, 1 h)	11.2	33.7
Sucrose + $Pb(NO_3)_2$ (1 h)	10.4	49.4
Pipolfen (15 min), then sucrose (1 h)	3.5	12.5
UDPG (control run, 1 h)	4.3	45.9
UDPG + $Pb(NO_3)_2$ (1 h)	3.8	65.8
Pipolfen (15 min), then UDPG (1 h)	1.3	8.2

Standard deviation $\leqslant$ 10%

The fact that lead nitrate affects processes of synthesis of polysaccharides from different substrates in a similar fashion (inhibition of cellulose synthesis and enhancement of hemicellulose production) can point to the final and not the initial steps of a metabolic chain to be responsible for these effects. Otherwise the strongest deconjugation effect for cellulose and hemicellulose synthesis would have been observed with glucose and not at all with UDPG. It is known that the final steps of cellulose synthesis are membrane-linked-transglycosylase mediated, when glucose from UDPG is transferred to a lipid carrier (supposedly dolichyl phosphate) and then to a protein which participates in the formation of cellulose molecules [97]. Inhibition of cellulose biosynthesis caused by lead nitrate can be explained by selective binding of lead ions by protein particles of the plasmalemma, which represent transglycosylases or β-glucan synthetases, and conformational changes. According to the generally accepted point of view, hemicelluloses are synthesized in a microsomal fraction inside the cell and lead nitrate, which cannot penetrate plasmalemma, does not inhibit their synthesis. Activation of hemicellulose synthesis observed in the above experiments under the action of lead nitrate can probably be connected with incorporation of those substrate molecules (ATP and UDPG) that remained nonconsumed due to the inhibition of cellulose synthesis.

The marked inhibition of the synthesis of cellulose and hemicelluloses caused by brief (15-min) incubation with pipolfen is possibly due to the modification of a lipid fraction of the plasmalemma as a result of the 'liquefication' of phospholipids and a lowering in their melting point [185–187]. This can hamper the transglycosylation reaction involved in a metabolic chain of cellulose synthesis and associated with the participation of lipids, as well as the fusion of vesicles (microsomal fraction) with the plasmalemma. Another explanation for pipolfen action is also possible. Inhibition of cellulose synthesis can be accounted for by the fact that pipolfen-induced modification of lipids disturbs the structure of plasmalemma proteins, transglycosylases and β-glucan synthetase, which contact with lipids.

Enhancement of the inhibitory effect of pipolfen on polysaccharide synthesis with an increase in incubation time can be explained by both its more profound action and its penetration into the cell where it acts as a specific inhibitor of flavine enzymes [188]. It can also interact nonspecifically with lipids of membranes of the Golgi apparatus thereby disturbing their structure and hampering synthesis of hemicelluloses.

Pipolfen favoured consumption of both sucrose and UDPG, the plasmalemma being a barrier for them. Probably, pipolfen-induced modification of lipid phase of the plasmalemma deprives the latter of selective permeability thus making possible enhanced penetration of sucrose and, especially, of UDPG into a cell along the concentration gradient.

It is quite possible that the concentration of glucose established in the above experiments (10 mM) is too low with respect to that of endogenous glucose and therefore its consumption was not intensified upon addition of pipolfen. On the contrary, even exit of endogenous glucose from cells might be expected.

The influence of lead nitrate on a classical cellulose-synthetase test-system, cotton-seed hairs is illustrated in Fig. 1.54. Synthesis of cellulose, in the case of wheat, as well, is inhibited while synthesis of hemicelluloses is either unaffected or even enhanced by lead nitrate.

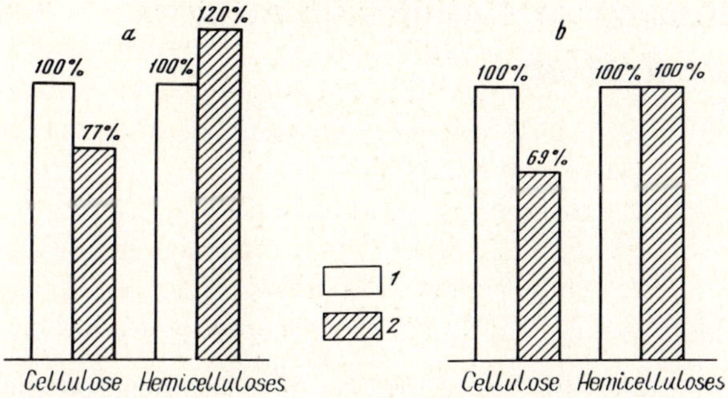

Fig. 1.54a, b. Influence of lead nitrate (1 mM) upon intensity of formation of polysaccharides in primary (**a**) and secondary (**b**) cell wall of cotton fiber.
1 No lead nitrate added, *2* in the presence of lead nitrate
Subscripts: Cellulose, Hemicelluloses, Cellulose Hemicelluloses

Thus it can be concluded that in addition to the earlier approach to desynchronize synthesis of cellulose and hemicelluloses by changing concentrations of Mg^{+2}-UDPG [106] were able to offer, for the first time, a way for differently-directed change in the intensity of synthesis of cellulose and hemicelluloses. This is achieved by the use of lead nitrate which serves as a modifier of the plasmalemma protein components. Our results demonstrate that the mechanism of synthesis of cellulose differs from that of hemicelluloses and confirm the concept of different localization of synthetic sites for cellulose and matrix polysaccharides in plant cells.

5 Dependence of Intensity of Cellulose Synthesis on Substrate Concentration

The study of the relationship between the concentration of substrates and the intensity of synthesis of cellulose is not only of theoretical interest, it is of practical interest as well, since it allows one to evaluate possibilities and to outline the ways to increase the harvest of raw cotton. It was anticipated that the synthesis of cellulose should reach a maximum at concentrations of substrates above those in the metabolism of cells [189].

5.1 Experiments with Cotton Fibers

Studies of carbohydrate metabolism in cells of cotton-seed hairs [80] revealed that in the initial period of secondary cell wall formation the concentration of glucose in a metabolic pool is 50 mM, of glucose-6-phosphate 4 mM, of glucose-1-phosphate 1 mM, and of UDPG 7 mM.

An increase in the concentration of UDPG from nanomoles to micromoles results in a considerable enhancement of its utilization in synthesis of cellulose and callose in cotton-seed hairs [190, 191]. Analogous enhancement of polysaccharide synthesis in the cell walls of cotton fiber was observed when substrate concentration was raised from 5 μM to 1 mM (Table 1.9) [72]. The highest increase (4850-fold!) in glucan synthesis during primary cell wall formation was connected with the rise in UDPG concentration while with other substrates the activation was increased only 62 to 173-fold. Synthesis of secondary-wall glucans requires sucrose and glucose to a much higher extent (more than hundred-fold and twenty-fold) than that of primary-wall polysaccharides. It should be emphasized that low substrate concentration used (it is by one to two orders of magnitude less than that coming to cells via conducting ways) casts doubt on the authors' conclusion on the absence of competition between UDPG and sucrose for their transfer into cells and utilization in synthesis of glucans.

One of the important consequences of changes in substrate concentration can be the change in ratio of rates of synthesis of different glucans. It was shown [188] that enzyme preparations from oat coleoptiles produce cellulose from UDPG at low concentration (1 μM) while at a higher concentration (of about 1 mM) the main product was callose-like β-1-3-glucan. Analogous results were obtained with enzyme preparations from wheat seedlings [66], peas epicotyl

Table 1.9. Utilization of glucose from different substrates in synthesis of cellulose of primary and secondary cell walls of cotton-seed hairs

Substrate	Concentration, μM	Incorporation of glucose in cellulose (nM per cell)		Intensity of glucose incorporation in secondary vs primary cell wall
		Primary cell wall	Secondary cell wall	
UDPG	5	0.8	16.7	20.9
	1000	3666	5764	1.6
GDPG	5	1.5	2.6	1.7
	1000	92.5	161.7	1.8
Glucose	5	0.3	8.3	27.6
	1000	52.0	1197	23.0
Sucrose	5	0.6	81.2	135
	1000	80.0	11061	138

[69, 192], cotton fiber [72, 193]. The last-mentioned authors have shown that cellulose constituted 25 to 30% of UDPG-derived cell-wall glucans at a concentration of 5 μM and only 4 to 7% at a 200-fold higher concentration.

Magnesium ions were shown to direct β-glucan synthesis from UDPG [192]. At micromolar concentrations of the latter cellulose synthesis was greatly enhanced by $MgCl_2$ while β-1-3-glucan synthetase was suppressed. Synthesis of cellulose from UDPG was accelerated upon addition of cellobiose [190]. Incubation of 17 to 21-day-old, seed-detached hairs with cellobiose (12.5 mM), UTP (1.25 mM), and UDPG (5 μM) resulted in a more than 20-fold increase in incorporation of [^{14}C]-glucose from UDPG into cellulose. Cellobiose and UTP, added separately, had very little effect on the formation of cellulose from UDPG, as did combinations of cellobiose with ATP or GTP. Addition of sucrose, maltose, or glucose together with UTP also favoured utilization of UDPG but to a lesser extent (5 to 10-fold). Higher concentrations of UDPG (1 mM) cellobiose + UTP enhanced cellulose formation by a factor of 2.8; cellobiose alone—by a factor of 4.5; while UTP alone inhibited the process. The above pattern of the influence of cellobiose and UTP, added together or separately, into incubation mixtures containing 1 mM UDPG was also observed with membrane-bound enzyme preparations isolated from cotton fiber [191].

Kinetic parameters of the synthesis of cellulose and callose are evidence in favour of the involvement of different enzyme systems in these processes. In higher plants they are localized in the plasmalemma, while in *Fungus saprolegnia* the β-1,4- and β-1,3-glucan synthetases differ in localization within cells as followed from their separation upon ultracentrifugation in the sucrose gradient

[106]. The highest β-1,3-glucan synthetase activity was associated with membranes of the Golgi apparatus which were able to form polysaccharide microfibrils in vitro. The above data suggest activation of β-1,3-glucan synthetase by increasing concentrations of UDPG. It was concluded that interaction of UDPG with the enzyme alters its structure. Addition of magnesium ions prevents enzyme activation at high concentrations of UDPG.

Solution of the problem of the relationship between substrate concentration and the intensity of synthesis of polysaccharides, including cellulose, is faced with many difficulties. Usually, exogenous substrates (glucose, sucrose, NDPS) labelled with ^{14}C or ^{3}H are used for the determination of the rate of polysaccharide synthesis. It should be taken into account that exogenous, labelled sugars entering the cell are diluted with nonlabelled endogenous substrates whose concentration can vary with plant ontogenetic and growth conditions. This 'dilution effect' requires certain corrections to be made to conclusions about the stimulation factors for the synthesis of cellulose and other structural polysaccharides by increasing the concentrations of substrates.

According to a great deal of data summarized in a monograph by A. L. Kursanov "Transport of Assimilates in Plants" [82], the exogenous glucose and sucrose rapidly penetrate into the cell proportional to their concentration. Then their consumption follows different pathways including participation in the synthesis of structural polysaccharides. The carbon 'flows' from exogenous and endogenous sources are mixed and the incorporation of exogenous, radioactive glucose or sucrose into polysaccharides is inversely proportional to the concentration of endogenous substrates. With equal concentrations of endogenous and exogenous substrates the specific radioactivity of their mixtures is half that of the starting substrate and radioactivity of cellulose synthesized therefrom is also half as much provided no enhancement in intensity of biosynthesis occurs as a result of increased substrate concentration.

In our studies on the influence of a substrate concentration on intensity of biosynthesis of cellulose and other polysaccharides in cell walls of cotton-seed hairs and in wheat leaf and root tissues the range of concentrations used covered approx. two orders of magnitude (by an order of magnitude above and below that of endogenous substrates). Incorporation of [^{14}C]-glucose into cellulose of the primary cell wall of cotton-seed fibers revealed the following intensity vs concentration dependence: for 1, 5, 10, and 50 mM concentrations the reciprocal intensities were 4, 19, 33, and 176 μg glucose/g dry fiber per hour; in another experiment the 10, 50, and 500 mM concentrations gave correspondingly 30, 105, and 750 μg glucose/g dry fiber per hour [194]. Dilution of the starting [^{14}C]-glucose with a nonlabelled one was employed to adjust the concentration of the exogenous substrate to the required value, in calculations of the rates of cellulose synthesis the isotope effect was neglected. The data obtained allow us to conclude that the biosynthesis of cellulose can be intensified by increasing the substrate concentration.

It should be noted that overestimation of the intensification of biosynthesis due to the 'dilution effect' is only possible for glucose concentrations below those specific for the endogenous metabolic level. Considerable increases in rates of cellulose formation at much higher concentrations of exogenous glucose can be accounted for only by a 'stimulation effect' due to either substrate activation of enzymes or high 'throughput' of a channel of cellulose biosynthesis (undersaturation of reactions with substrates), both reasons being likely.

Due to the 'stimulation effect' the increase in intensity of the cellulose biosynthesis is the more pronounced as the concentration of the exogenous glucose is increased until the process is substrate-saturated. A subsequent increase in concentration of the exogenous substrate results in the formation of hypertonic solutions and nonspecific inhibition of the cellulose synthesis. The lower the exogenous substrate concentration the more the intensity of cellulose labelling is affected by the 'dilution effect' of the high level of endogenous substrate on low concentrations of [^{14}C]-glucose.

On the basis of indirect data it was concluded [82] that sugars of a metabolic pool (outside vacuoles) amount to approx. 10% of their total content in a cell, about 20% being localized in a free space of the cells. The remaining sugars are located within the vacuole and escape from it very slowly. The latter is accounted for by the fact that the main function of vacuoles is to maintain turgor.

In direct experiments it was shown that vacuoles of isolated protoplasts contain more than 90% sugars, some of which could not be detected outside the vacuole [195]. It should be taken into account, however, that these objects are devoid of the main receptacle of free-space sugars, viz. cell walls, and besides, the out-of-vacuole sugars could be washed off into a culture medium that results in lowering their concentration. It can be concluded that the concentration of sugars in the metabolic pool (outside vacuoles) is of the order of tens of millimole.

Analogous experiments with cotton fibers at the stage of secondary cell wall formation have led to the conclusion that this period was also characterized by a considerable undersaturation of a cellulose-synthetase system with exogenous substrates. Increase in exogenous glucose concentration from 10 to 250 mM (with concomitant decrease in specific radioactivity) has led to an increase in cellulose radioactivity by 50% rather than a decrease. Studies on the influence of substrate concentration on cellulose biosynthesis are usually conducted with [^{14}C]- or [$^{-3}$H]-labelled compounds. There are data on the increase of the mass of cellulose upon increase of the concentration of nonlabelled substrates. Thus, the change in UDPG concentration from 0.6 to 4.8 mM resulted in an increase in mass of cellulose produced in cell culture of soybean from 1.4 to 4.0 mg per g dry mass [81]. Analogous effect was observed in our studies with cotton fibers at the stage of secondary cell wall formation with sucrose as a substrate. The influence of exogenous substrates on cellulose biosynthesis was shown by the following fact. Incubation of seed hairs in a buffer solution for 20 h resulted in cellulose fiber gain by 6.7% in the absence of exogenous sucrose and by 21.9% in the presence of 1 M sucrose. It should be mentioned that quantitation of the 'stimulation effect' by this method gives only approximate values, nevertheless the information is of considerable importance.

The material presented above allows one to deduce that studies of the peculiarities of biosynthesis of cell wall polysaccharides from exogenous [^{14}C]-labelled substrates requires the complicating role of endogenous substrates to be taken into consideration since it can vary with their concentration in metabolic resources. In connection with this, additional studies on concentration dependence of characteristic features of cellulose biosynthesis from exogenous and endogenous substrates were undertaken.

5.2 Experiments with Wheat Leaves

The photosynthesizing subject, wheat leaves, was employed to facilitate regulation of the level of endogenous substrates in metabolic resources. Photosynthesis could be 'switched on' or 'switched off' thereby replenishing or exhausting endogenous metabolic pools and this approach was used in an attempt to estimate competition between endogenous and exogenous substrates for their participation in cellulose biosynthesis.

Increase in concentration of exogenous glucose resulted, as in the case of cotton fiber, in increase in cellulose radioactivity (Table 1.10). The unusually high inhibition of cellulose biosynthesis (by a factor of 2.5) upon two-fold increase in exogenous glucose concentration (from 250 to 500 mM) is probably accounted for by the osmotic action of glucose since on average the osmotic pressure of leaf tissue sap determined by the Shardakov procedure [196] was equivalent to 380 mM sucrose concentration.

Table 1.10. Effect of concentration of exogenous [^{14}C]-glucose on label incorporation into cellulose in leaves

Concentration of exogenous glucose, mM	Radioactivity of cellulose, cpm/min per g wet mass per hour	Intensity of cellulose synthesis from exogenous glucose, nmol/g wet mass per hour
10	26.3×10^3	14.4
50	29.3×10^3	80.5
250	16.5×10^3	227.7
500	6.0×10^3	165.5

Leaf fragments were exposed in a solution of [^{14}C]-glucose for 2 h, intensity of illumination was 20 w/m^2. Initial concentration of [^{14}C]-glucose was 0.2 μM. The intensity of cellulose biosynthesis presented in this and subsequent tables was calculated from incorporation of exogenous glucose without correction for its dilution with [^{12}C]-substrates from endogenous metabolic pools. The starting specific radioactivity of [^{14}C]-glucose samples could be different in different runs

It should be kept in mind that sugars are located mainly in vacuoles and play a role of an osmotic agent which enables maintenance of turgor. Carbohydrates may leave this reservoir and pass into the cytoplasm. Sugars in metabolic pools of leaves constitute, as has already been mentioned, about one tenth of their total content.

In subsequent experiments the concentration of exogenous sugars did not exceed 300 mM which was much higher than that in metabolic pools but osmotic stress was not very pronounced. It was of interest to establish the influence, on utilization of exogenous [^{14}C]-glucose, of exhaustion or enrichment of endogenous metabolic pools which can be achieved by a prior prolonged exposure of plants to darkness or to light.

Data presented in Table 1.11 allow one to conclude that prolonged exposure in darkness results in a more intensive utilization of exogenous glucose than does prior illumination, especially at low concentrations. This dependence holds in the dark-synthesis of cellulose as well.

Increase in concentration of CO_2 in the incubation atmosphere to a level much above the threshold of carbon-dioxide saturation leads, like prior illumination, to inhibition of cellulose formation from exogenous glucose (Table 1.12). Difference in utilization of glucose depend on light conditions levels, to a certain extent, at high glucose concentrations while with changes in conditions of CO_2 supply, this difference remains independent on the level of exogenous glucose. This distinction leads to the suggestion that it is the nature of substrates formed directly in photosynthesis that makes them competitive with exogenous sugars (not only glucose but sucrose as well as shown by studies on this substrate) and with those formed during previous prolonged illumination. To check this suggestion the intensity of [^{14}C] incorporation from HCO_3^- into cellulose was studied with different glucose concentrations and different light conditions. It was shown (Table 1.13) that the concentration of exogenous glucose has no influence on the direct utilization of photosynthetic products for cellulose production. Prior illumination does not diminish (as was the case with exogenous [^{14}C]-glucose) incorporation of ^{14}C into cellulose but rather favours it.

Intensity of photosynthetic fixation of $^{14}CO_2$ could affect the utilization of the photosynthesized substrates and this factor was also taken into account (Table 1.14). Regularities observed in the previous experiments hold true in this one.

The suggestion on different competitiveness of exogenous glucose with respect to directly formed photosynthetic products ('young' products according to nomenclature of O. V. Zalensky and coworkers) and preformed ones was confirmed in special experiments. Taken into account was the utilization of preformed photosynthetic products on cellulose formation in subsequent dark or light phases (Fig. 1.55). In these experiments the variables were the concentrations of exogenous glucose or of preformed endogenous substrates, the latter being mainly free sugars of cytoplasm and the vacuole as well as starch.

Table 1.11. Synthesis of cellulose from exogenous glucose as a function of changes in endogenous substrate concentration caused by prior exposure in the dark or in the light

Entry	Conditions	Concentration of exogenous glucose, mM	Radioactivity of cellulose, cpm/g wet mass per hour	Intensity of cellulose biosynthesis, nM glucose/g wet mass per hour
1	Exposure in dark, 24 h	5	53.8×10^3	7
		100	23.5×10^3	64
	Illumination, 24 h	5	28.6×10^3	4
		100	20.2×10^3	55
2	Exposure in dark, 48 h	5	27.7×10^3	7
		200	9.7×10^3	90
	Illumination, 48 h	5	20.2×10^3	5
		200	8.3×10^3	80

Leaf fragments were incubated in a solution of [^{14}C]-glucose for 1 h, intensity of illumination was 20 w/m^2.

Table 1.12. Effect of CO_2 concentration on biosynthesis of cellulose in photosynthesizing wheat leaves from exogenous glucose

Concentration of			
CO_2, %	exogenous glucose, mM	Radioactivity of cellulose, cpm/g wet mass per hour	Intensity of synthesis of cellulose, nM glucose/g wet mass per hour
0.04	5	40.3×10^3	4
	100	26.1×10^3	47
1.00	5	25.2×10^3	2
	100	20.2×10^3	31

Leaves were exposed in a solution of [^{14}C]-glucose for 30 min, intensity of illumination is above 120 W/m^2

Table 1.13. Effect of concentration of exogenous glucose on cellulose formation from $^{14}CO_2$ assimilated during photosynthesis depending on prior light conditions

Light conditions	Concentration of exogenous glucose, mM	Radioactivity of cellulose, cpm/g wet mass per hour	Intensity of cellulose biosynthesis, nM CO_2/g wet mass per hour
Exposure in the dark, 24 h	5	99.1×10^3	81.9
	100	100.8×10^3	81.1
Illumination, 24 h	5	117.6×10^3	97.2
	100	121.0×10^3	100.0

Incubation for 1 h; illumination, 20 W/m^2

Concentration of the directly produced (phosphorylated) photosynthetic products could be changed by alteration of the CO_2 content.

This approach made it possible to conclude, in precursory form that a considerable increase in CO_2 concentration makes the 'photosynthetic' production of cellulose 'sensitive' to the pressure of excessive concentrations of exogenous glucose.

To prove this suggestion it was necessary to take into consideration the influence of elevated concentrations of exogenous glucose on the intensity of $^{14}CO_2$ fixation. With corrections for the decrease in intensity of $^{14}CO_2$ fixation, it was shown that the utilization of intermediates of photosynthesis for cellulose production was not inhibited even at concentrations of exogenous glucose as high as 300 mM (Table 1.15).

Table 1.14. Effect of concentration of exogenous glucose on cellulose production from $^{14}CO_2$ assimilated in photosynthesis depending on light conditions

Conditions	Concentration of exogenous glucose, nM	Intensity of photosynthesis, $\mu M CO_2$/g wet mass per hour	Radioactivity of cellulose, cpm/g wet mass per hour	Intensity of cellulose biosynthesis	
				nM CO_2/g wet mass per hour	in per cents to radio-activity of assimilates
Exposure in the dark, 48 h	5	38.4	58.8×10^3	48.7	0.13
	200	33.3	52.1×10^3	43.0	0.13
Illumination, 48 h	5	45.1	83.2×10^3	68.7	0.15
	200	38.4	73.3×10^3	63.8	0.17

Exposure, 1 h; Illumination, 20 W/m^2

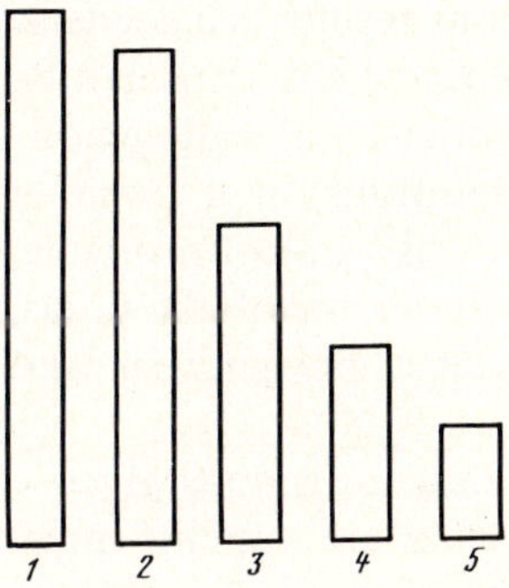

Fig. 1.55. Intensity of cellulose formation from $^{14}CO_2$ (arbitrary units)

1 Photosynthesizing leaves, 1 h, *2* after second hour, under illumination, 5 mM glucose solution, *3* as above, 200 mM glucose solution, *4* after second hour in the dark, 5 mM glucose solution, *5* as above, 200 mM glucose solution

Table 1.15. Effect of exogenous glucose concentration on cellulose formation from [^{14}C]-labelled photosynthesis products at high concentration of carbon dioxide (1%)

Concentration of exogenous glucose mM (exposure 2 h)	Intensity of photosynthesis, μM CO_2/g wet mass per hour	Radioactivity of cellulose cpm $\times 10^{-3}$/g wet mass per hour	Intensity of cellulose biosynthesis, nM CO_2/g wet mass per hour	Cellulose ^{14}C-labelling % of total label assimilated during photosynthesis
—	25.4	490	280	1.1
5	25.4	510	291	1.1
300	10.1	180	100	1.0

Photosynthesis was performed for 30 min in an atmosphere containing 1% CO_2; intensity of illumination, 80 W/m^2

It has already been mentioned that enhancement of ^{14}C incorporation into cellulose stimulated by elevated concentrations of exogenous [^{14}C]-glucose is accounted for by the 'dilution effect', i.e. by endogenous, nonlabelled substrates and, on the other hand, by the 'stimulation effect' on cellulose biosynthesis by high concentrations of exogenous substrate. Experiments with leaves previously exposed to illumination or kept in the dark and with different CO_2 levels show that the 'stimulation effect' rather than the 'dilution effect' to be primarily responsible for an increase in cellulose labelling. This effect was studied in detail with cotton fibers. It was especially high in cotton-seed hairs where formation of the secondary cell wall occurred. According to some data the proportion of glucose in the metabolic pool during this period amounts to 1/40–1/10 of the total glucose content [80].

At the stage of primary cell wall formation the 'dilution effect' is of importance due to the higher content of sugars in cells including metabolic pools.

Data presented in Tables 1.11 and 1.12 indicate that enrichment of metabolic pools with glucose caused either by prolonged, continuous illumination of leaves or by photosynthesis at elevated CO_2 concentrations results in a decrease in the contribution of exogenous glucose to cellulose biosynthesis. This can be explained by a larger contribution, to cellulose biosynthesis, of endogenous sugars (which are accumulated in cells under these conditions) as a result of either inhibition of exogenous glucose utilization or, and this is most probable, higher dilution of exogenous, labelled glucose in endogenous metabolic pools, decrease in specific radioactivity of substrates for cellulose biosynthesis and consequently of intensity of cellulose labelling.

Literature data available [82] allow us to suggest that exogenous glucose and glucose of endogenous metabolic pools of leaf cells are rapidly interchangeable. It is caused, on one hand, by easy penetration of exogenous glucose into the cell (where it is combined with metabolic pools) and, on the other hand, by its rapid exit for the most usual destination of the photosynthesizing leaf cells, delivery of assimilates to nonphotosynthesizing organs. The accumulated glucose can escape the cell, on a large scale, in the dark. Exogenous glucose is rapidly transformed into sucrose in illuminated photosynthesizing tissues and it is sucrose that is transported from cells (Fig. 1.56). It is by these reasons that can be explained the detected by us rapid drop in growth of cellulose radioactivity with time independent on exogenous glucose concentration (Fig. 1.57).

Products formed during photosynthesis (probably, phosphorylated ones) are preferable to endogenous glucose for cellulose biosynthesis. This was shown by the fact that under conditions of increasing concentrations of endogenous metabolic pools of carbohydrates (cf. Tables 1.13, 1.14) the intensity of $^{14}CO_2$ utilization for cellulose biosynthesis is higher than under conditions of exhaustion of endogenous pools.

Enhanced formation of structural polysaccharides with overcrowding of cells with assimilates is expedient since it is the means for removal of excessive, osmotically active substances (sugars). Massive production of a reserve polysaccharide, starch, is also directed to this end, though its deposition can damage structure and functioning of chloroplasts.

It is with the intensification of the synthesis of structural polysaccharides in plants grown in open places and before drought, when increase in sugar concentration in cell sap of leaves is observed, that is connected with the appearance of xeromorphous (or heliomorphous) structure. One of its main features is the thickening of cell walls. Preferential utilization of 'early' phosphorylated products of photosynthesis for cellulose synthesis is manifested not only in their enhanced ^{14}C-labelling from $^{14}CO_2$ rather than from exogenous [^{14}C]-glucose, but also in the independence of the metabolic flow of $^{14}CO_2$ to cellulose on increase in the exogenous glucose concentration. Of these products, phosphorylated sugars appear to play a primary role in the biosynthesis of cellulose. They are converted into the final products of photosynthesis, free sugars, and the latter become almost interchangeable competitors with exogenous sugars for the use in the synthesis of structural polysaccharides. Advantages of assimilates are finally lost providing sugars formed during photosynthesis enter the vacuoles (cf. Fig. 1.56).

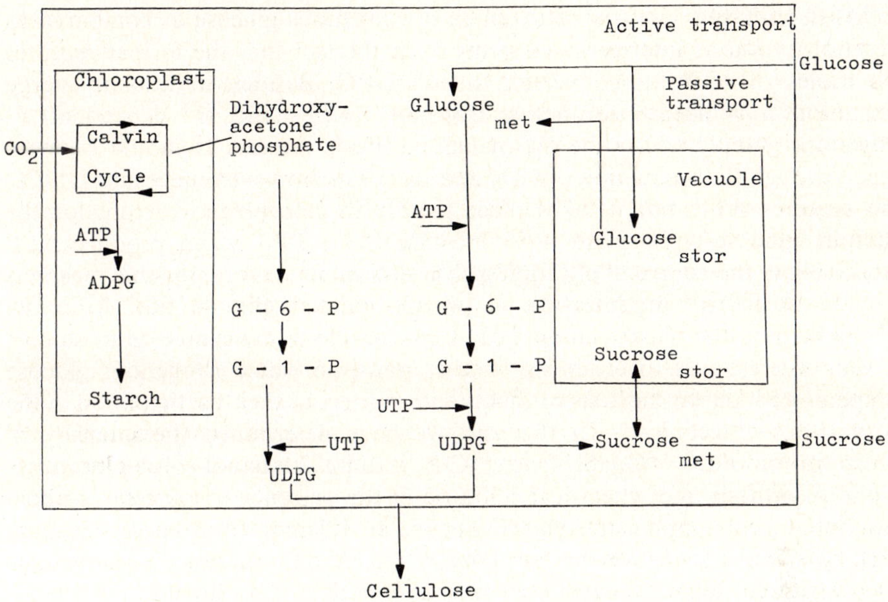

$\text{Glucose}_{\text{stor}}$, $\text{Glucose}_{\text{met}}$ refer to storage pool and metabolic pool of glucose

$\text{Sucrose}_{\text{stor}}$, $\text{Sucrose}_{\text{met}}$, refer to storage pool and metabolic pool of sucrose.

Fig. 1.56. Synthesis of cellulose from carbon dioxide and exogenous glucose [80]

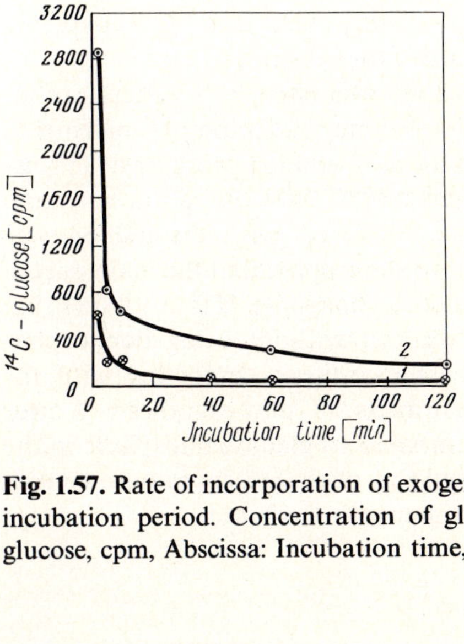

Fig. 1.57. Rate of incorporation of exogenous [^{14}C]-glucose into wheat-leaf cellulose vs. incubation period. Concentration of glucose: 1, 0.6 mM; 2, 6 mM. Ordinate: [^{14}C]-glucose, cpm, Abscissa: Incubation time, min

Most probably, less competitiveness of exogenous glucose as compared to the photosynthesis intermediates stems from the fact that the former requires two moles of ATP for conversion into UDPG. Besides, additional energy consumption is needed for active transport (ATP- and H^+-dependent) of exogenous glucose across the plasmalemma inside the cell. It should be mentioned though that reduction of CO_2 and its conversion eventually into UDPG also requires ATP, but it takes place mainly in chloroplasts (except for the reaction: glucose-1-phosphate + UTP → UDPG + PP_i) which produce ATP intensively in the course of photophosphorylation. One more possible reason is that the structural organization of phosphorylated glucose flow from the chloroplasts to the plasmalemma [82] is preferable to exogenous glucose.

Our interest was to establish whether phosphorylated exogenous glucose competes with photosynthesis products with respect to their participation in the biosynthesis of cellulose. To this end we have determined the intensity of incorporation of the ^{14}C-label from $^{14}CO_2$ into 80%-ethanol-soluble products of photosynthesis and wheat-leaf cellulose in the presence of exogenous phosphorylated and nonphosphorylated glucose at different concentrations. Data listed in Table 1.16 prove the fact that a considerable increase in exogenous glucose (up to 200 mM) results in retardation of $^{14}CO_2$ fixation by wheat-seedlings' leaves (obviously, due to osmosis), intensity of utilization of photosynthesis intermediates for cellulose production remaining at the same level. In contrast, an increase in the concentration of exogenous glucose phosphate inhibits the transfer of the label from $^{14}CO_2$ to cellulose and its inhibitory effect is the more pronounced the higher its concentration increase in length of incubation. Radioactivity of cellulose synthesized in the presence of 200 mM exogenous glucose phosphate (incubation time 2.5 h) is almost one third of that with glucose.

Results obtained allow us to suggest a preference for exogenous sugar phosphates over nonphosphorylated sugars as substrates for cellulose synthesis. It can easily be seen if cellulose labelling from [^{14}C]-glucose and [^{14}C]-glucose phosphate is compared, the latter being much more potent.

It was necessary to learn whether or not glucose phosphate is dephosphorylated during its passage through the plasmalemma (or on the plasmalemma). Analysis of the aqueous-ethanol-soluble fraction of isolated protoplasts following their incubation in a medium containing [^{14}C]-glucose phosphate and filtration through a layer of silicon oil has revealed, with the aid of two-dimensional paper chromatography and autoradiography, that the main part of the label in protoplasts is connected with glucose phosphate. It thus follows that exogenous glucose phosphate penetrates the cell without being dephosphorylated and is actively utilized in cellulose biosynthesis competing with the endogenous substrates, phosphorylated products of photosynthesis. Lower competitiveness of exogenous glucose as compared to glucose phosphate in the process of cellulose biosynthesis is accounted for, probably, by requirement in ATP consumption for conversion of the former into the latter.

Table 1.16. Intensity of ^{14}C-incorporation from $^{14}CO_2$ into photosynthesis products and cellulose

Conditions	Substrate concentration, mM	Photosynthesis, mg CO_2/g wet mass per hour	Cellulose, μg CO_2/g wet mass per hour	Incorporation of ^{14}C into cellulose relative to that assimilated in photosynthesis, %
Glucose or glucose phosphate + 1% $^{14}CO_2$, 30 min	Glucose			
	10	1.56	8.84	0.57
	200	1.32	6.74	0.51
	Glucose phosphate			
	10	1.69	9.77	0.60
	200	1.40	4.93	0.35
Glucose or glucose phosphate, 2 h, then 1% $^{14}CO_2$, 30 min	Glucose			
	10	1.45	8.83	0.59
	200	1.40	6.84	0.50
	Glucose phosphate			
	10	1.48	7.39	0.49
	200	1.12	2.09	0.18

One can conclude from the above data that there is considerable undersaturation of a system of cellulose synthesis with substrates. Channels for synthesis of both structural polysaccharides and other products of photosynthetic carbohydrate metabolism possess a much higher capacity than is usually observed. It follows that there is a possibility of increasing several-fold the efficiency of CO_2 fixation, compared to the normal one, with carbon-dioxide saturation of photosynthesis and the more so when the maximum potential efficiency of photosynthesis is achieved. It was shown that cell status, in particular, degree of saturation of endogenous metabolic carbohydrate pools can regulate consumption of exogenous sugars and their utilization in the synthesis of structural polysaccharides. This is in keeping with the idea that '. . . introduction of exogenous metabolites into autotropic tissues . . . requires care in evaluation of the results obtained, since this permits cell potentialities with respect to the substance introduced to be displayed rather than real rates and the direction of its metabolization upon its endogenous production in the cell' (Kursanov, [82] p. 77).

Structural polysaccharides can be regarded as direct products of photosynthetic assimilation of CO_2 only in illuminated chlorophyll-containing cells. When plants are placed in darkness or illumination is switched off the preformed, phosphorylated products of photosynthesis undergo rapid dephosphorylation to produce sucrose and hexoses which replenish the endogenous (intracell) pools of these sugars.

Dark synthesis of structural polysaccharides in chlorophyll-bearing cells proceeds with the utilization, as starting substrates, of a reserve product of photosynthesis, starch, localized in plastids and sugars localized in cytoplasm, in metabolic pools capable of replenishment with sugars present in vacuoles (Fig. 1.56). The role of substrates for polysaccharides in nonphotosynthetic cells, tissues and organs is played by transport products of CO_2 assimilation, principally sucrose. Some features of a 'remote' transport of sucrose from photosynthesizing organs to nonphotosynthesizing ones is treated in detail by Kursanov [82].

5.3 Approaches to Increase the Content of Substrates of Cellulose Synthesis

Of essential importance in the synthesis of structural polysaccharides in non-assimilating cells is that the flow of incoming substrates does not cease, and this can happen in some cases [197–199]. Deficiency in substrates for synthesis of cell wall polysaccharides can be overcome either by an increase in intensity of photosynthesis that can result in intensification of delivery of sucrose or other transport forms of carbohydrates to their consumption sites or by increase in the mass of photosynthesizing organs with respect to consuming ones.

Potentialities of the first route can be exploited rather completely by making mineral nutrition, water supply, and, in some cases, temperature conditions optimal for photosynthesis. The main natural factor that limits the intensity of photosynthesis is the low CO_2 content in the atmosphere and this prevents increase independent of other factors. Also taken into account should be the fact that solubility of CO_2 in water (or in protoplasmic sap) drops at higher temperatures which results in an even higher limiting role of low CO_2 content. These circumstances are beneficial for plants capable of concentrating CO_2 in some tissues, though at the expense of additional energy consumption (ATP). These plants belong to the C_4-group, which is characterized by initial introduction of the label from $^{14}CO_2$ not into C_3-compounds (3-phosphoglycerate, 3-phosphoglyceraldehyde, hydroxyacetonephosphate) typical for the majority of algae and higher plants of moderate climate, but into oxalacetate, malate, or aspartate, the C_4-acids. The mechanism of photosynthesis in C_4-plants is treated in detail in several manuals [73, 74]. C_4-plants have advantages over C_3-plants as regards intensity of photosynthesis and yields only in dry and hot climates, under conditions of moderate climate, C_3-plants gain the lead [8].

Attempts to increase essentially the intensity of photosynthesis in plants by selective breeding has not yet given positive results due to the complexity of the process which is under the control of numerous genes. Alteration of the nature of one or even several of them does not result in an increase in the intensity of photosynthesis since the overall process is limited by many other reactions the rates of which were not affected during mutations.

There are other possibilities to change the exterior of plants, the proportions of different organs including those producing and consuming assimilates, to improve conditions of transportation of photosynthesis products at the expense of either increase in cross-section of conducting bundles (thereby increasing their carry capacity) or diminishing distances between organs which produce and consume assimilates (thereby decreasing sugar consumption for respiration and transport along the conducting channels).

Peculiarities of photosynthesis and consumption of assimilates in the system of the whole plant can be best exemplified with the cotton plant, one of the most important cellulose producers. These plants are multicarpous, the age of fruits (in days after anthesis) varies over a broad time range, and this determines the character of relationships between leaves and bolls. One of the most detailed studies on photosynthesis and transport of assimilates in cotton plant was carried out by [200]. Photosynthesis in *Gossipium hirsutum* (108-F) can be expressed in the following figures: 33% of the total CO_2 consumption falls within main stem leaves, 55% with leaves of branches, 4% with bracts, 2.1% with fruit organs, 4.4% with stalks, and 1.5% with bark. Prior to and including anthesis the major part of the assimilates is localized in the leaves and stem, during fruit bearing and ripening they are directed mainly towards fruit elements. The flow of assimilates into bolls depends on their age. This dependence is expressed as a one-peak curve with maximum at 20–27 days after anthesis. Supply of assimilates into bolls lasts until their opening. What is most interesting is that the attracting ability of bolls whose seed hairs form intensely secondary cell wall (20–30 days after anthesis) is so great that they 'draw off' assimilates not only from the nearest leaves but from remote ones as well thereby hampering the carbon nutrition of young bolls situated close to these leaves. As a result of such competition among bolls for assimilates, up to 80–85% of younger fruit organs fall off (Fig. 1.58).

According to tracer studies [200] up to 25–30% of the assimilates are consumed nonproductively by the cotton plant during fruit bearing. This is used in formation of the fruit elements which fall off later and for the late apical growth. This can result in a decrease of the raw-cotton harvest from 6 to 20%.

An increase in the concentration of substrates in cellulose-producing, non-photosynthesizing organs may be achieved by various means. The main one is the intensification of the influx of photosynthetic products (sucrose) from the leaves and other photosynthesizing organs to bolls. This can be caused either by intensification of photosynthesis or by an increase in the mass of photosynthesizing tissues with respect to consuming ones (the increase in source-to-sink, i.e. producer-to-consumer, ratio) providing alteration in intensity of photosynthesis is insignificant.

In practice selective breeding did not make it possible to intensify photosynthesis and so increase productivity under optimum growth conditions. Under conditions when intensity of photosynthesis and, hence, the yield is limited by an insufficient concentration of nutrients or lack of water in the soil, this can be improved by fertilization or irrigation. The ratio of source-to-sink

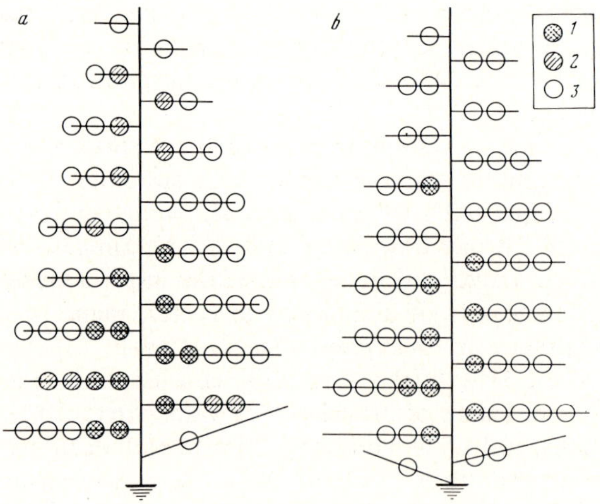

Fig. 1.58a, b. Position of remained and fallen off fruit elements on model cotton bushes. **a** August **b** end of vegetation; *1* bolls remained; *2* buds and ovaries remained; *3* fallen off fruit elements

can be readily altered using breeding methods. This ratio is increased in highly productive lines due to a higher mass of the assimilating organs compared to consuming ones. In the case of multicarpous plants such as cotton this can be achieved, in contrast, by a decrease in the number of fruit elements. This can be performed either using breeding methods or by removal of the less mature bolls which attract a part of the assimilates but finally die off because they cannot compete with the ones already formed. Removal of the 'excessive' bolls can be done by mechanical or chemical chopping, e.g. with dimethylmorpholinium chloride or its analogues [201].

6 The Effect of Energy Conditions of Cells on the Biosynthesis of Cellulose

It has already been mentioned (Chapter 4) that a number of reactions involved in cellulose biosynthesis require ATP consumption. With CO_2 as a substrate, ATP is consumed in a cycle of regeneration of the CO_2 acceptor, in reducing 3-phosphoglyceric acid to 3-phosphoglyceraldehyde, in formation of NDPG from glucose-1-phosphate, and phosphorylation of dolichol or dolichyl phosphate. With exogenous glucose or sucrose as substrates, ATP is used for the active transport of sugars from free space inside the cell and for phosphorylation of glucose, subsequent reactions being similar to the aforementioned ones. Large amounts of ATP are doubtless consumed in the processes of formation and translocation inside the cell by elements of the Golgi apparatus which participate in the formation of matrix polysaccharides and their transport towards the cell wall and by the endoplasmic reticulum which delivers enzymes and proline-rich proteins to the plasmalemma.

Several mechanisms exist for the production of ATP and they differ in substrates and localization in cells. In photosynthesizing cells the three main processes are as follows: 1) photosynthetic phosphorylation in thilacoids of chloroplasts; 2) oxidative phosphorylation in mitochondria; 3) substrate phosphorylation during glycolysis in cytoplasm and their intensity decreases in this series as can be seen from Table 1.17 [202].

Functioning of all these mechanisms in cells leads to the establishment of an ATP concentration sufficient for the supply of different energy-consuming processes including synthesis of structural polysaccharides. The highest concentration of ATP is established in photosynthesizing cells at a high level of illumination due to realization of noncyclic photophosphorylation. The ATP level drops dramatically in photosynthesizing cells when they are placed in the dark [203, 204] which can result in the inhibition of the synthesis of cellulose and other polysaccharides.

6.1 Experiments with Wheat Leaves

These experiments were carried out in the laboratory [205] for photosynthesis in the Kazan Institute of Biology, the Academy of Sciences, U.S.S.R. It was shown that the intensity of cellulose biosynthesis in the dark from the exogenous

Table 1.17. Types of phosphorylation in plant cells, substrates, and products

Type of phosphorylation	Substrates	Products	Intensity of phosphorylation, μM/mg chlorophyll per hour
Noncyclic photo-phosphorylation	Light, water, ADP, P_i, $NADP^+$	ATP, NADPH, O_2	2000
Cyclic photo-phosphorylation	Light, ADP, P_i	ATP	200
Oxidative photo-phosphorylation	NADH, ADP, P_i, O_2	ATP, H_2O, NAD^+	200
Glycolysis	1,3-Diphospho-glyceric acid, ADP Phosphoenol-pyruvate, ADP	ATP, 3-phospho-glyceric acid, ATP, pyruvic acid	50

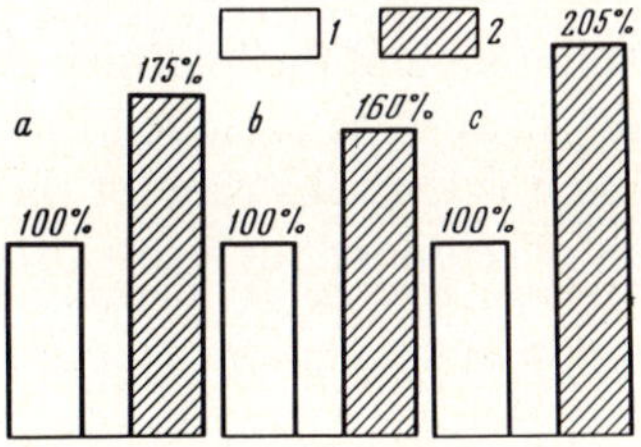

Fig. 1.59a–c. Effect of illumination on intensity of cellulose biosynthesis.
a intact leaves, intensity of illumination 36 W/m^2; **b, c,** cut leaves, 36 W/m^2 (**b**) and 60 W/m^2 (**c**) *1* in the dark (control) *2* illumination

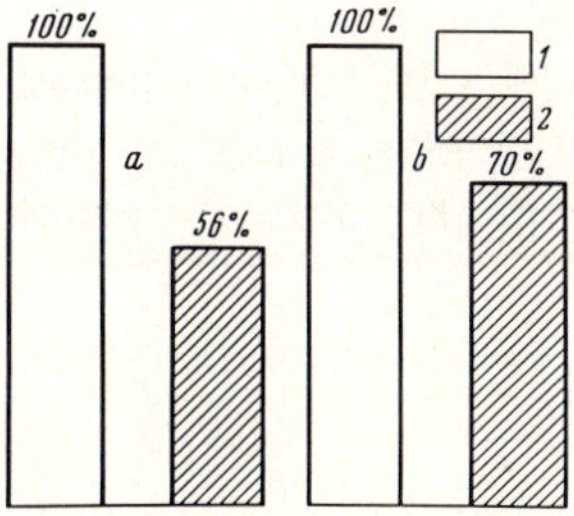

Fig. 1.60a, b. Intensity of cellulose formation in wheat seedlings' leaves under the action of diuron (**a**) and antimycin A (**b**) (10^{-5} M, intensity of illumination 20 W/m^2). *1* Control; *2* experiment

[^{14}C]-glucose is much lower than with illumination, even at a level below that necessary for saturation of photosynthesis (Fig. 1.59). Action of diuron, the inhibitor of noncyclic transport of electrons in the chloroplasts and of the associated photophosphorylation, results in the inhibition of cellulose synthesis (Fig. 1.60) and is similar to switching the light off. Under the action of antimycin

A, an inhibitor of the cyclic transport of electrons and of the associated phosphorylation, the intensity of cellulose synthesis also decreased but to a lesser extent (cf. Fig. 1.60).

It should be noted that quercetin, which is the inhibitor of energy transfer, hampered cellulose formation in wheat leaves when exposed to light but not when kept in the dark (Fig. 1.61) that shows its chloroplast-specificity.

Exogenous ATP favours the incorporation of exogenous glucose in cellulose both in the dark and with illumination of the leaves (Fig. 1.62). Of note is that the activating effect of exogenous ATP depends inversely on the intensity of illumination of the leaves (cf. Fig. 1.62). The reason is that the increase in intensity of illumination intensifies ATP formation in the course of photophosphorylation (especially, noncyclic) and, hence, increases ATP level in cells which results in a diminished uptake of exogenous ATP.

The relatively low stimulating effect of exogenous ATP on cellulose biosynthesis in the dark is probably due to the photosensitivity of several enzymes of glucose-to-cellulose metabolism. At a low illumination level which is favourable for photoactivation of the enzymes and which is characterized by moderate ATP productivity, the effect of exogenous ATP is the largest. At high illumination intensity together with high activity of the enzymes and a sufficient supply of endogenous ATP, the exogenous ATP only exerts a small effect on [^{14}C]-glucose consumption.

The mechanism of the relationship between ATP content and the intensity of cellulose biosynthesis remains unclear. The possibility cannot be excluded that ATP activates cellulose biosynthesis not only because it is the substrate for the formation of metabolites participating in the synthesis of cellulose but it can also

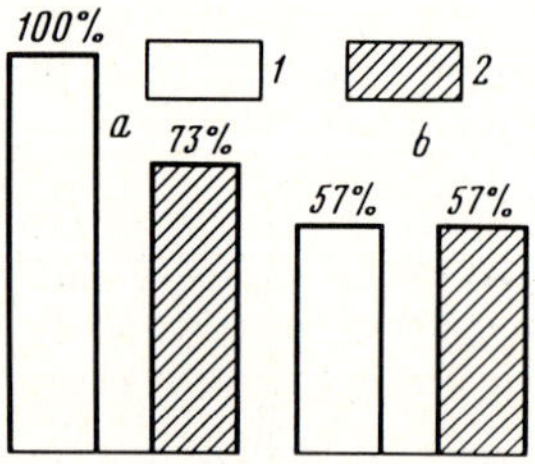

Fig. 1.61a, b. Effect of quercetin (5×10^{-6} M) on intensity of cellulose formation in wheat seedlings' leaves

a in the light (36 W/m^2); **b** in the dark, *1* Control; *2* experiment

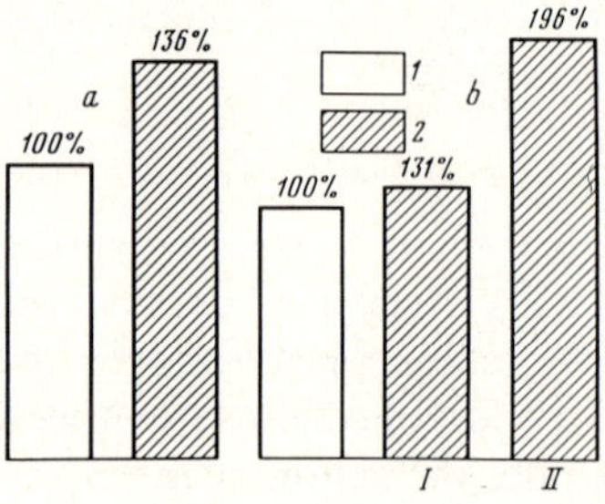

Fig. 1.62. Effect of exogenous ATP (1 mM) on intensity of cellulose formation in wheat seedlings' leaves in the dark (a) and in the light (b).

I, intensity of illumination 60 W/m^2, II, 36 W/m^2
1 Control, *2* experiment

activate enzymes which catalyze this process and the transport of exogenous glucose inside the cell and of metabolites formed to the site of cellulose synthesis, i.e. to the plasmalemma surface.

In fact, ATP affects formation of phosphorylated products (glucose-6-phosphate from glucose, UDPG and GDPG from glucose-1-phosphate) rather than glucose transport. This can be judged from the following data. Leaves were treated with exogenous, labelled glucose for 0.5 h and then incubated (3.5 h) in a glucose-containing medium either in the presence or the absence of ATP. The results showed that exogenous ATP has a pronounced stimulating effect on cellulose biosynthesis. Labelling of cellulose in the presence of ATP was 1.5 times more than that in a control experiment. It is possible that a portion of exogenous [^{14}C]-glucose still remained in then free cell volume when leaf fragments were transferred from glucose solution to ATP solution but it can certainly be concluded that ATP affected mainly glucose consumed by cells, i.e. incorporated in the endogenous pools.

6.2 Experiments with Cotton Fibers

Of considerable interest was to see if we could establish the relationship between intensities of ATP formation and cellulose production, from exogenous sugars, in cotton-seed hairs in which photosynthetic phosphorylation does not take place and where ATP is synthesized in the course of glycolysis and oxidative phosphorylation. Several aspects of this problem have been studied [72, 195, 206]. In our paper special emphasis was made on how the synthesis of cotton cellulose is provided with energy-rich phosphates [206] and it was shown that variations in the synthesis of cellulose depend on the type of energy-rich phosphates and the 'age' of the cotton fibers. Thus, for example, the GTP activation effect was the highest at the stage of primary cell wall formation and declined thereafter.

The results of additional experiments (presented here for the first time) demonstrate that GTP does stimulate cellulose synthesis at the stage of secondary cell wall formation but its activity is approx. one third of that of UTP. Addition of GTP (1 mM) to a medium with [^{14}C]-glucose (10 mM) enhanced cellulose labelling by 96% while with UTP (1 mM) this value was 270%. The obvious rationale for this is that GDPG is the main substrate for primary-cell-wall cellulose and UDPG is the one for secondary-cell-wall cellulose [79, 207]. In most of our experiments the activation effect of ATP was manifested to a greater extent in the biosynthesis of cotton-seed secondary-cell-wall cellulose (Table 1.18).

Inhibitors of oxidative phosphorylation (rotenone and antimycin A) and glycolysis (sodium fluoride, iodoacetate) strongly inhibited cellulose synthesis, rotenone and iodoacetate being suppressors of primary-wall cellulose synthesis rather than secondary (Table 1.19). Intensity of cellulose synthesis, especially at

Table 1.18. Effect of ATP on incorporation of exogenous [^{14}C]-glucose in cellulose of primary and secondary cell walls of cotton-seed hairs (expressed as % relative to the control)

Glucose concentration, mM	Primary cell wall		Secondary cell wall	
	1 mM ATP	5 mM ATP	1 mM ATP	5 mM ATP
		Run 1		
10	126	169	241	246
250	148	154	222	193
		Run 2		
10	156	—	319	—

Table 1.19. Effect of phosphorylation inhibitors on the synthesis of cellulose from [^{14}C]-glucose (10 mM) in cotton-seed hairs (expressed as % relative to the control)

Inhibitor	Concentration	Primary cell wall		Secondary cell wall	
		without ATP	with ATP (1.25 mM)	without ATP	with ATP (1.25 mM)
Rotenone	5×10^{-5}	49	69	66	140
Iodoacetate	1×10^{-3}	50	66(58)*	64	97
Antimycin	1×10^{-5}	47	—	—	—
Fluoride	1×10^{-2}	55	—	—	—

* Effect of GTP (1.25 mM) instead of ATP, in parentheses

the stage of secondary cell wall formation, could be restored (in the case of iodoacetate) or even surpassed (rotenone) by the introduction of exogenous ATP together with the inhibitor.

The absence of inhibition of the synthesis of secondary-cell-wall cellulose from UDPG under action of rotenone on cotton-seed hairs is surprising. On the contrary, a 12%-enhancement of incorporation of [^{14}C]-glucose from UDPG into cellulose was observed with added rotenone (10^{-5} M). Incorporation, into cellulose, of [^{14}C]-glucose from UDPG was inhibited by iodoacetate (10^{-3} M) but to a lower extent (14%) compared to that from free [^{14}C]-glucose ($>$ 50%). Likewise, severe soil drought causing stable wilting of cotton leaves affected cellulose formation from exogenous UDPG and glucose (10 mM each) to varying extent: its intensity was to 82 and 55% respectively (compared to the control).

The idea that dolichyl phosphate participates in biosynthesis of cellulose should imply that the regeneration of dolichyl pyrophosphate, which is a glucose acceptor from UDPG, is ATP-dependent. The fact that inhibition of the UDPG-to-cellulose conversion by iodoacetate or under drought conditions is much less pronounced than that of glucose utilization could be accounted for by the stronger inhibition of glucose phosphorylation or glucose-1-phosphate-to-UDPG conversion with a decrease in ATP concentration than that of dolichol (or dolichyl phosphate) phosphorylation. A suggestion of unequal sensitivity of kinase reactions to ATP level has been forwarded by the author [51, 208] on the basis of indirect data on the effect of drought on label distribution from $^{14}CO_2$ among different products of photosynthesis. This was confirmed in direct experiments with enzyme systems [209] which have shown, for example, that phosphoribulose kinase is less ATP-sensitive than 3-phosphoglycerate kinase.

Data presented in Chapter 4 on the necessity of maintaining definite potential ψ on the plasmalemma imply a consumption of energy-rich phosphate for its creation and maintenance. And, indeed, the potential difference is caused by the difference in concentrations of cations and anions between the outer and inner surfaces of the plasmalemma. Ionic gradients across the plasmalemma are determined either directly by the activities of membrane-bound cation-dependent ATPases or indirectly by the creation of a transmembrane proton gradient with the help of a plasmalemma-embedded proton pump. The proton gradient then forms ionic gradients on plasmalemma and in either case it requires energy consumption.

The material of this chapter gives convincing evidence that an insufficient level of energy-rich phosphates in cells can limit the intensity of cellulose biosynthesis. It means that one of the possible routes of regulating (intensifying) cellulose production is the enhancement of energy supply to cells by the stimulation of ATP-forming processes.

7 The Effect of Hormones on the Biosynthesis of Structural Polysaccharides

7.1 The Mechanism of Hormone Action

Phytohormones represent powerful regulators of growth processes in plants. Mechanism of their action on plant cells presumably involves activation of reading off information from DNA (transcription), production thereby of additional matrix RNA molecules and then, in turn, of enzymes synthesized with RNA intermediacy thus activating cell metabolism as a whole. Phytohormone-induced alterations of cell-membrane permeability is of considerable importance as is also enhanced intensity of ATP formation during respiration and photosynthesis. Phytohormones, being natural metabolites, regulate the level of growth and development processes [210–214]. Credit should be given to the Soviet scientists [215, 216] for their contribution to the development of our knowledge of phytohormones.

Experimentally, studies on the mechanism of their action are usually performed in the following way. Plant segments (stem or root sections, hypocotyl or epicotyl) or isolated cells are treated with exogenous phytohormone and the time-dependence of changes in physiologobiochemical processes is then followed to establish the sequence of cell-responses to a certain hormone.

Analysis of the effect of phytohormones on cell-production of structural polysaccharides and cell growth, i.e. on irreversible increase in size and mass of cells (Fig. 1.63), allows one to distinguish at least three steps in cell response. They are 1) rapid acidification of cell walls due to proton release which usually begins after 5–7 min (Fig. 1.64); 2) 'softening' (loosening) of cell walls after 14–15 min, probably, due to pH drop; 3) activation of enzymes involved in the synthesis of structural polysaccharides and growth promotion (18–20 min after hormone treatment). These steps are exemplified by cell responses to auxin action.

7.2 Acidification of Cell Walls

Acidification of cell walls (Fig. 1.64) as the earliest cell response to auxin action has been demonstrated by many authors [211, 219–224].

To prove that it is cell-wall protonation that is the reason for subsequent transformations it was necessary to demonstrate that acidification of intercell

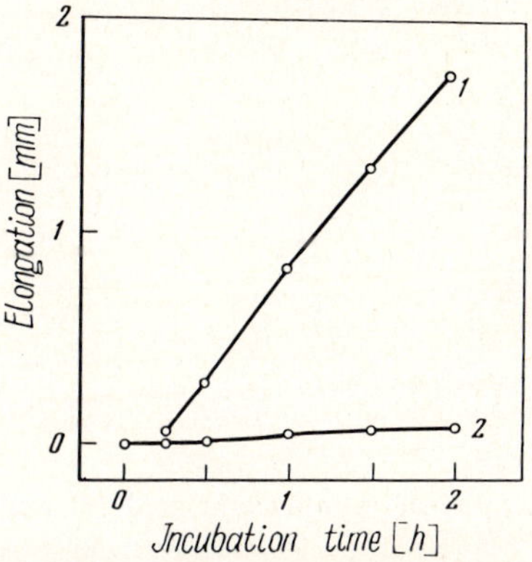

Fig. 1.63. Effect of indoleacetic acid on elongation of segments of bean epicotyl as a function of incubation time [217]

Twenty segments of epicotyl were incubated in the light in 10 mM K-phosphate buffer (pH 6.0) in the presence of indoleacetic acid (10^{-4} M) (*1*) or in its absence (*2*) ordinate: Elongation, mm, abscissa: Incubation time, hours

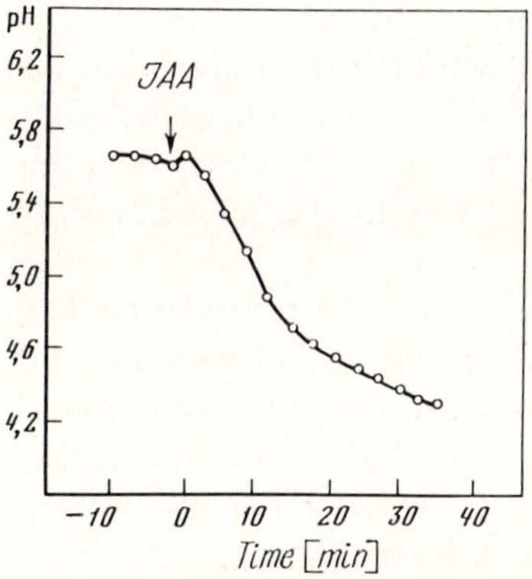

Fig. 1.64. pH-Lowering in free space in indoleacetic acid-treated maize coleoptiles [218]. Abscissa: Time, min. Arrow: Indoleacetic acid treatment

space precedes or, at least, does not lag behind other alterations. It should be noted that the detection of the beginning of acidification depends on macro- or microelectrodes used and on the structure of the object. In cuticle-layer-free *Avena* coleoptiles proton excretion induced by fusicoccin could be detected within the first minute (Fig. 1.65) when an electrode was attached directly to the segment of coleoptile under study [225]. The optimum pH value for cell wall growth, i.e. 5, was established in 5 to 8 min. When indoleacetic acid was used the lag-phase of initial pH-change was extended to 14 min and the pH-5-value was established in 20 min. This is much longer than found previously [223] due to a

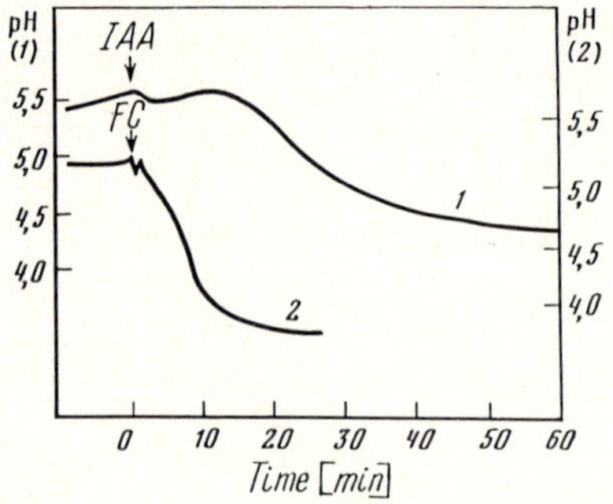

Fig. 1.65. Rate of proton release following treatment with indoleacetic acid (*1*) and fusicoccin (*2*). Abscissa: Time, min.

lower temperature. The most precise pH-values can be obtained with electrodes implanted in a tissue although this approach is faced with certain difficulties.

Artificial acidification of tissues can mimic the action of auxins on cell growth whereas alkalification inhibits growth [211]. At high concentrations of indoleacetic and 2-naphthaleneacetic acids which act as inhibitors the pH-value is increased and the interval between auxin addition and cell-wall alkalification is less than that at the beginning of growth inhibition [224].

Cyanide acts as an inhibitor as well and auxin-induced growth was inhibited to a larger extent than that stimulated by artificial acidification of cell walls [222]. It points to dependence of a proton-pump action on respiratory metabolism (ATP supply) rather than on acid-mediated cell-wall loosening followed by increase in its mass [222].

These data support those obtained earlier [225] which showed that fusicoccin-induced cell-wall acidification and extension can be suppressed with cyanide after 1 and 3–5 min. Auxin-induced proton excretion was inhibited in 30 S and a complete arrest of growth occurred in 5–7 min. All these data indicate that auxin activate a membrane-bound proton pump possessing ATPase properties, i.e. consuming ATP for transmembrane proton transport.

Cycloheximide, the known inhibitor of protein synthesis on 80 S ribosomes, acts unexpectedly as an inhibitor of auxin-induced cell-wall acidification as well (Fig. 1.66) with a lag-phase of 5–8 min [226]. Inhibition of growth began in 8 to 10 min. This could be accounted for by auxin-induced synthesis of specific proteins, normally present at low concentration, named growth-limiting proteins. However, in special experiments no alteration, under the action of phytohormones, in the spectrum of protein synthesis was observed [220]. It was suggested therefore that auxin converts inactive growth-limiting proteins in to active ones or alters their compartmentalization. These proteins are also supposed to be a part of a mechanism responsible for proton-pump action. This is in accord with a vesicular-secretion model [219]. Auxins do not affect the intensity of endoplasmic-reticulum-vesicles formation but activate the related proton pump so that the content of vesicles is acidified and following its

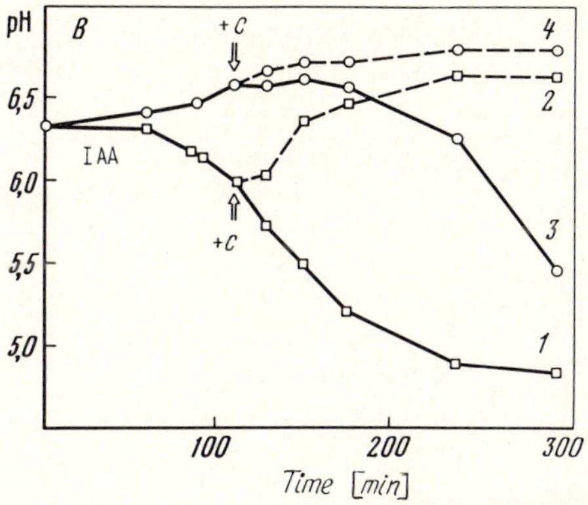

Fig. 1.66. Effect of cycloheximide (C) on indoleacetic acid-induced proton excretion. *1* Indoleacetic acid, *2* Indoleacetic acid + cycloheximide, *3* Control, *4* Control + cycloheximide. Abscissa: Time, min

transfer to the outer surface of the plasmalemma the pH-drop of cell walls takes place. This concept is supported by the detection of auxin-binding sites in the endoplasmic network. The growth-limiting proteins are probably necessary to enable functioning of a proton-secreting mechanism.

7.3 Loosening of Cell Walls

Some data indicate that activity of structural polysaccharides hydrolases is enhanced following auxin treatment, probably due to cell wall acidification (Fig. 1.67). Auxin was shown to activate one of the two forms of cellulases [227–229]. The other form of the enzyme was activated with ethylene [227, 230]. Cellulases are known to hydrolyze not only cellulose but xyloglucans as well (the latter possess a β-1,4-glucan skeleton and xyloglucan side chains) and one can expect hormone-induced degradation of these polymers.

NMR data indicate that it is matrix polysaccharides rather than cellulose which underwent changes [231]. Auxin action results in liberation of neutral xyloglucan from the cell wall of segments of pea epicotyl thereby loosening it [232]. Liberation of this main component of hemicelluloses begins 15 min later after the hormone treatment and this is considered as a test for analysis of auxin action on the cell wall.

The principal role in cell-wall extension is played by xyloglucan chain cleavage, splitting off arabinogalactans followed by its scission, and finally, a loosening of the noncovalent bonds between xyloglucans and cellulose [33].

Treatment with indoleacetic acid brings about, in as little as 30 min, depolymerization (decrease in molecular mass) of xyloglucans of *Vigna augularis* epicotyls and a lowering of the arabinogalactan level together with an increase

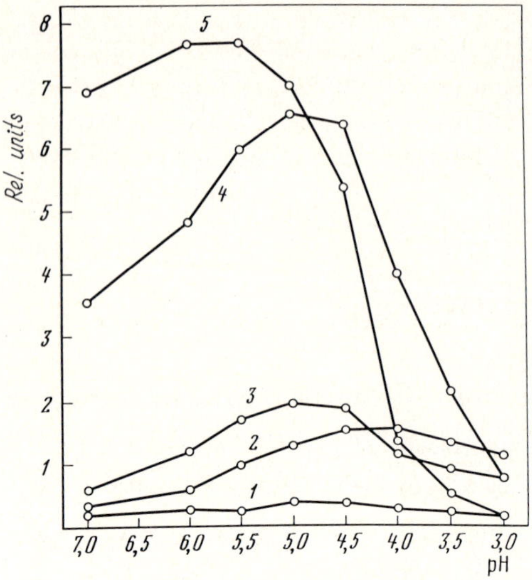

Fig. 1.67. pH-Dependence of activity of glycosidases associated with coleoptile cell-walls [226]. *1* Xylosidase *2* mannosidase *3* galactosidase *4* glucosidase *5* fucosidase. Ordinate–arbitrary units.

in the molecular mass of pectin, this is observed even under conditions of growth inhibition with mannitol as osmoticum [217]. It was suggested that polymerization of pectin, built of rhamnogalacturonans and arabinogalactans, is a result of transglycosylation of pectic polysaccharides originally present in the cell wall. The former concept of auxin action [33] was supplemented with the idea of cell-wall loosening due to pectic polysaccharide elongation or increased branching, the latter lowering interpolysaccharide interaction [233, 234].

It thus follows that loosening of intermolecular bonds between hemicelluloses, pectic substances and cellulose leads to turgor-induced longitudinal slipping of cellulose microfibrils or their lateral detachment due to the decreased cementing effect of matrix polysaccharides.

What should also be taken into consideration is the possibility of a direct effect of acidification on loosening of links between cell-wall polysaccharides by means of cleavage of xyloglucan-cellulose microfibril hydrogen bonds, rather than activation of hydrolytic enzymes.

Structural modifications of hemicelluloses which result in a drop of molecular mass can occur prior to the decrease of their content [235].

Glycosidases which depolymerize structural polysaccharides thereby loosening cell walls are inhibited by abscisic acid and cycloheximide [226], fluoride and arsenite [232], and increased calcium content which being bound to pectic substances renders them more 'rigid', thus preventing the liberation of a xyloglucan [217, 232]. Inhibitors of DNA and RNA synthesis prevent action of gibberellic acid and auxins, impart rigidity to cell walls, and encourage elongation of lettuce cells [236]. It should be noted that auxin-induced cell elongation is caused by decrease in cell wall resistance to the turgor pressure rather than to change in the water permeability of the plasmalemma. This was conclusively proved by collaborative studies of three laboratories [237] which put an end to discussion on this matter.

7.4 Enhancement of Cellulose Biosynthesis

The third stage of the effect of auxins on cell growth is the enhancement of cellulose biosynthesis. This results finally in increased size of cells and of plant organs built from them as well as the mass of cell walls and of constituent structural polysaccharides.

A great deal of information on the effect of phytohormones on plant growth is summarized in numerous reviews [210–214, 238–243].

The specific activating effect of phytohormones on polysaccharide biosynthesis is dealt with in numerous studies with coleoptiles of cereals and dicotyledonous epicotyls, hypocotyls, and stem segments. Extensive studies in this field have been carried out by Ray and coworkers [244–246] and it was concluded that rapid induction, by auxins, of polysaccharide synthesis is caused by the activation of cellular polysaccharide synthetases. Listed below is the effect of auxin analogues on glucan synthetase activity expressed as cpm for [^{14}C]-glucose incorporated in cellulose (Table 1.20).

Examination of the effect of auxins on the synthesis of polysaccharides by pea stem segments has revealed some increase in activity of hexokinase and a

Table 1.20. Effects of auxin analogues (30 μM) on glucan synthetase activity

Test compound	Glucan synthetase assay
	dpm
None	3150
Indoleacetic acid	7730
Indole-3-isobutyric acid	5100
L-Tryptophan	4020
Naphthalene-1-acetic acid	6720
2,4-Dichlorophenoxyacetic acid	5440
2,6-Dichlorophenoxyacetic acid	3720
4-Chlorophenoxy-isobutyric acid	2410
Phenylacetic acid	3080
Triiodobenzoic acid	3300

considerable increase in the activity of β-glucan synthetase with UDPG as a substrate (Fig. 1.68) [244]. Activation of UDPG-dependent β-glucan synthetase under the action of indoleacetic acid becomes noticeable after 10–15 min, this being the most pronounced in tissues with a lowered synthetase activity (Fig. 1.69) and in the presence of sucrose added together with the auxin [245]. Some activation was also observed for UDP-Gal- and UDP-Xyl-dependent synthetases and, to a lesser extent, of a GDPG-dependent β-glucan synthetase.

Actinomycin D and cycloheximide did not suppress rapid, auxin-induced activation of glucan synthetases [246]. Gibberellic acid induced two-fold promotion of both growth and synthesis of polysaccharides in cell walls of oat internodes [247] the lag-phase lasting for approx. 1 h. At the same time, a considerable decrease in the cell content of glucose, glucose-6-phosphate, and fructose-6-phosphate was observed. Incubation of internode segments in 100 mM glucose prevented gibberellic acid-induced changes in concentrations of these metabolites as well as in the energetic adenylate charge, the growth rate being greatly enhanced. No changes in activity of polysaccharide synthetases which utilize UDPGlc, UDPGal, UDPXyl, and UDPAra as substrates was observed under the action of gibberellic acid while that of GDPG-dependent β-glucan synthetase was increased 2 to 4-fold.

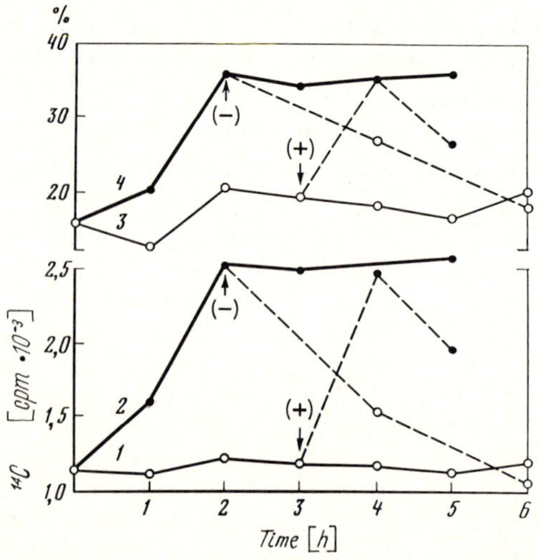

Fig. 1.68. Effect of indoleacetic acid on synthesis of polysaccharides in pea stem cell walls.

Stem segments were incubated in water (*1* and *3*) or in a solution of indoleacetic acid (*2* and *4*) and then, for 1 h, in a medium with UDP-[^{14}C]-glucose. Arrows indicate the time at which the stem segments were transferred to (+) or from (−) the auxin solution.

Abscissa: Time, h, Ordinate: cpm × 10^{-3} (for lower part) and labelling of cell-wall polysaccharides expressed as per cents of total label incorporation (for upper part)

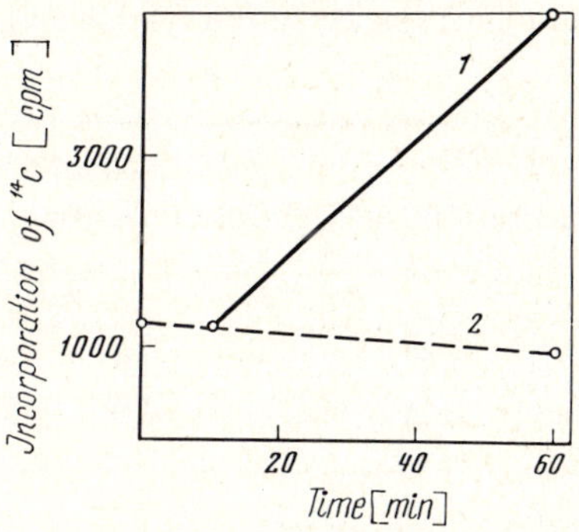

Fig. 1.69. Indoleacetic acid-induced changes in glucan synthetase activity in pea tissues. Pea tissues were incubated in 30 mM sucrose at 35 °C in the presence (*1*) or in the absence (*2*) of the auxin (17 μM).

Abscissa: Time, min. Ordinate: Incorporation of ^{14}C (cpm) from labelled sucrose into cellulose.

It may thus be concluded that growth-promoting phytohormones activate at least a part of cell-wall polysaccharide synthetases, this action being mediated by other processes. The latter follows from the fact that cell-free enzyme preparations of β-glucan synthetases were not activated by the auxin [246].

Strong inhibitory effect of cycloheximide on the induction, by auxin of β-glucan synthetase activity is noteworthy [248, 249]. In these experiments the activating effect of auxins on RNA-mediated synthesis of enzymes has been, most probably, exhibited. This can be deduced from some other data [211] e.g., from the increase in mRNA level [250]. Actinomycin D and cycloheximide did not prevent rapid activation of β-glucan synthetases, although in other plant systems and under different conditions auxins can supposedly [246] promote synthesis of new molecules of β-glucan synthetases.

7.5 The Effect of Phytohormones on the Synthesis of Cotton-Fiber Cellulose

Data on the effect of phytohormones on the biosynthesis of cellulose in cotton fiber are unfortunately scarce. They characterize primarily the level of endogenous hormones in cells of cotton-seed hairs cultured in vitro [251–255]. Experiments in the author's laboratory have shown that indoleacetic and gibberellic acids and kinetin favour the incorporation of [^{14}C]-glucose into cellulose which is usually the most intensive in seed hairs where formation of secondary cell wall is proceeding. Phytohormone-induced promotion of cellulose biosynthesis increases with time as can be seen from the data presented below.

Table 1.21. Effect of phytohormones on utilization of [^{14}C]-glucose in the synthesis of cellulose in cotton-seed hairs

Phytohormones	10 mM glucose	50 mM glucose	100 mM glucose	500 mM glucose
IAA, 5×10^{-5} M	114	122	125	126
GA, 5×10^{-5} M	152	161	187	130
Kinetin, 5×10^{-5} M	152	166	125	119

Seed hairs were incubated in water for 2 and 4 h and then in 10 mM [^{14}C]-glucose for 1 h and glucose incorporation served as a control (100%). When indoleacetic acid solution replaced water in the first incubation, the incorporation rates were 138 and 168%, with gibberellic acid the corresponding values amounted to 148 and 207%. When glucose concentration was increased to a certain level, the stimulating effect was first raised and then reached a plateau or even diminished. The following doses of glucose as a substrate were found to be optimum for displaying the activation effect by the hormones: 50 mM for indoleacetic acid, 100 mM for gibberellic acid, and 50 mM for kinetin (Table 1.21).

The cumulative effect of phytohormones and GTP has been shown to take place at the stage of secondary wall formation. It can be seen from the data given below where intensity of labelling of cotton-seed hair secondary cell-wall cellulose from ^{14}C-glucose (10 mM) is expressed as a percentage of a control:

Indoleacetic acid	137
Gibberellic acid	203
GTP	140
Indoleacetic acid + GTP	246
Gibberellic acid + GTP	248

Both auxins exerted a stimulating effect on the formation of primary and secondary cell-wall cellulose while gibberellic acid inhibited synthesis of cellulose at an intermediate stage (10–11 days after anthesis) and even addition of GTP did not restore the original synthetase activity. It can be seen from label incorporation rates (expressed as per cent of the control, 100%): gibberellic acid, 66%; GTP, 110%, gibberellic acid + GTP, 86%. This effect was reproducible for two years, its reasons remain unclear.

Phytohormones exhibited a much higher stimulating effect on the biosynthesis of cellulose in the case of UDP-[^{14}C]-glucose as a substrate rather than glucose. This can be accounted for by the more distant route taken by free glucose to 'pass' into cellulose (and by a larger diversion for various metabolic reactions) compared to that of glucose from UDPG.

Treatment with phytohormones activates the synthesis not only of cellulose but also of other cell-wall constituents [256]. Thus, ethylene promotes the formation of hydroxyproline-rich proteins and gibberellic acid that of polysaccharides from UDPAra. Formation of hydroxyproline-rich proteins, hemicelluloses, and pectic substances takes place in the Golgi apparatus and endoplasmic reticulum. Of note are data in favour of the possible proliferation of rough endoplasmic reticulum in cells of aleurone tissue under the action of ethylene as well as appearance of a large number of Golgi bodies in lettuce hypocotyl cells following treatment with gibberellic acid [256].

8 Effects of Cultivation Conditions on Cell Wall Regeneration by Isolated Protoplasts

It has already been mentioned in Chapter 1 that isolated protoplasts which regenerate cell walls represent a unique model for studying the mechanism of the formation of structural polysaccharides (for a review see [257]). To date, in the studies of the mechanism of cell wall regeneration by protoplasts isolated, in a majority of cases, from the leaf mesophyll, the cultivation is usually performed under heterotrophous conditions, viz. with exogenous carbohydrate substrates and in the dark, as a rule (or in some cases under light of very low intensity).

Moreover, it is believed that a substrate quantity of light exerts negative effect on development of isolated protoplasts [59, 258]. However, the original leaf mesophyll cells are photosynthesizing ones and according to the author's data obtained with wheat-seedling leaves, the products of photosynthetic fixation of CO_2 are involved in the synthesis of cell-wall polysaccharides more intensely than are exogenous sugars. It was demonstrated that illumination resulted in approx. 2-fold activation of cellulose synthesis from exogenous [^{14}C]-glucose at the expense of ATP produced upon photophosphorylation [205].

Thus, the essential role of illumination in the synthesis of cellulose and other structural polysaccharides in leaf tissues is doubtless. This leads to a suggestion that cultivation of isolated protoplasts regenerating cell walls in the light is preferable to that in the dark which is generally accepted.

The photosynthesizing protoplasts can regenerate cell walls in light with CO_2 as a sole carbon source [259]. At the same time the dependence of cell wall regeneration on photosynthetic activity of protoplasts has not yet been investigated. The following problems had to be solved: 1) the effect of illumination on the intensity of the biosynthesis of cell-wall polysaccharides and on the photosynthetic activity of protoplasts (isolated from bean and tobacco leaf mesophyll) in connection with their role in polysaccharide formation, and 2) the dynamics of ultrastructural changes of protoplasts and the intensity of biosynthesis of structural polysaccharides of different groups in the course of cell wall regeneration under different cultivation conditions.

8.1 Culturing of Protoplasts in the Light and its Influence on the Synthesis of Structural Polysaccharides

All the standard media for culturing protoplasts contain high concentrations of sucrose or glucose (1–3%) as carbohydrate substrates, sometimes together with other monosaccharides. We have studied the effect of sugar concentration in the incubation medium on photosynthetic activity of isolated protoplasts. This was considerably decreased, more than by 30%, upon even brief (30 min) incubation in 3% sucrose. High concentration of a carbohydrate substrate seems to be necessary only for prolonged culturing of protoplasts and is not required for a 1 to 2 day period.

The effect of a sharp decrease in carbohydrate content was studied when protoplasts were cultured in the dark. Tobacco protoplasts were incubated for 3 days in 1% and 0.1% sucrose solutions. No visual changes could be detected by optical microscopy: no increased destruction of protoplasts was observed at low sucrose concentration, conversion of protoplasts into ellipse-shaped ones was synchronous, in both cases division of protoplasts occurred on the third day. This allowed us to culture protoplasts in our subsequent experiments at a lowered glucose level (10–50 mM) which does not affect the intensity of photosynthesis.

Culturing of protoplasts has been carried out at illumination intensity of 20 W/m^2. An important point to be checked was whether this illumination caused destruction of protoplasts bearing in mind data [258, 260] on the decrease of colony-forming protoplasts at an illumination intensity of 12 W/m^2 and of the percentage of live protoplasts even at 5.2 W/m^2. Determination of live protoplasts performed in our work by staining them with phenosaphranin led to the conclusion that there was no negative effect on their viability on culturing for 24 h. The chlorophyll level remained practically the same in the suspension culture of protoplasts after 1 day in the dark and in the light. Intensity of the photosynthetic fixation of $[^{14}C]$-$NaHCO_3$ by protoplasts cultured in the light for 1 day did not differ from that of freshly-isolated ones.

It is thus shown that photosynthesis is actively functioning during cell wall regeneration during culturing of protoplasts in the light. Electron-microscopy studies of ultrathin slices showed that culturing of bean and tobacco protoplasts in the light at standard or low glucose concentration did not disturb the course of cell wall regeneration.

The main ultrastructural differences between light- and dark-protoplasts concern the chloroplasts and mitochondria. Electron microscopy disclosed a normal structure of chloroplasts of bean protoplasts at the early period of culturing, the grains were well-developed, thylacoids of grains and stroma were clearly manifested. The plastid stroma contained starch granules. After 12-h-culturing in the dark, the peripheric reticulum became apparent in the stroma of chloroplasts, its network became well-developed on the third day. Starch granules are rarely detected in the dark variants and disappear completely later on. Chloroplasts of light-cultured protoplasts are characterized by a developed lamellar system with stroma thylacoids predominating, starch granules being detectable even after 7 days of culturing.

Starch accumulation at the early stages of dark-culturing of protoplasts was documented earlier [261, 262]. Formation of starch, which can serve as an important substrate for the synthesis of structural polysaccharides, during the first days of dark-culturing of protoplasts can be conditioned by a high glucose level in the incubation medium and its easy penetration across the plasmalemma. In later periods the starch resource thus formed can be consumed in the course of the synthesis of structural polysaccharides whose rate increases with time.

It is necessary to emphasize that dark-cultured protoplasts, unlike light-cultured ones, manifest alterations in configuration of mitochondria in as little as 18 h and elongated dumbbell-like organelles appear. On subsequent culturing these changes are retained and become more pronounced. The number of mitochondria close to cell organelles participating in the synthesis of structural polysaccharides is increased, their division begins. These changes in mitochondrial structure seem to be connected to the energy-consuming character of a process of formation of structural polysaccharides and to the fundamental role of mitochondria (oxidative phosphorylation) in its supply with energy-rich phosphates when photophosphorylation does not operate.

Thus our ultrastructural studies have revealed several essential drawbacks characteristic of the dark culturing of isolated protoplasts and the unfavourable effect of heterotrophous conditions on ultrastructure of cell organelles, up to their pathological changes.

We have also studied the effect of illumination on the intensity of utilization of exogenous [^{14}C]-glucose in the synthesis of structural polysaccharides. We subjected to analysis the acid-soluble fraction of cell walls (mainly, hemicelluloses), cellulose, and extracellular polysaccharides composed of pectic substances, some hemicelluloses and arabinogalactan glycoprotein and secreted by protoplasts into the culture medium [62, 263]. The main bulk of structural polysaccharides is represented by cell-wall matrix polysaccharides and extracellular polysaccharides (ECP). Radioactivity of cellulose amounts to about 1–2% of total radioactivity of polysaccharides.

Analogous results were obtained with tobacco protoplasts [264] and soybean protoplasts [62]. A much higher incorporation of exogenous sugars into ECP was observed for carrot protoplasts [87] and the reasons for this difference remain unclear. The proportion of cellulose synthesized in the whole tissues and cell cultures is much higher and amounts to 20–40% of the total structural polysaccharides [13, 145, 265] (Laschek et al., 1981, 1982). It was established that culturing of protoplasts in the light results in a considerable enhancement of exogenous glucose incorporation into cell-wall polysaccharides, the size of the effect depending on the glucose concentration (Table 1.22). It is significant that illumination changed the proportion of different polysaccharide fractions because only the formation of cell-wall hemicelluloses and cellulose was stimulated and labelling of extracellular polysaccharides remained at the same level or even decreased (Table 1.22). These changes in the ratio of different fractions can be accounted for either by light-induced alteration in the 'quality' of polysaccharides or, more probably, their fixation in the cell wall under conditions of an increased level of synthesis of the cellulose microfibrils.

Light-stimulation of the utilization of exogenous [^{14}C]-glucose for the synthesis of structural polysaccharides seem to be linked to a better supply of energy-rich phosphates to the cell due to photosynthetic phosphorylation.

Table 1.22. Effect of light on incorporation of exogenous glucose in different fractions of structural polysaccharides of bean leaf protoplasts as a function of glucose level (culturing for 14 h in the dark or in the light)

Culturing conditions	Hemicelluloses		Cellulose		Extracellular polysaccharides	
	A	B	A	B	A	B
Glucose, 15 mM, in the dark	118	57.4	5.5	2.6	82	40.0
Glucose, 15 mM in the light	857	89.3	15	1.6	87	9.1
Glucose, 200 mM, in the dark	382	49.6	8.7	1.2	379	49.2
Glucose, 200 mM, in the light	1002	72.0	16	1.2	373	26.8

A, nM/mg of chlorophyll per h
B, % of total radioactivity of polysaccharides

Our results show that the stimulating effect of light is more pronounced at a low glucose level (15 mM) (Table 1.22), this value being below or close to that in the metabolic pool of cytoplasm [81, 82]. Penetration of glucose into the cell is brought about mainly by active transport which requires energy consumption. Light-induced activation of a process of the synthesis of structural polysaccharides cultured in a medium with a high glucose level (200 mM) is probably conditioned to a large extent by the substrate role of ATP, formed upon photophosphorylation, in polysaccharide synthesis.

Inhibition of ATP production with diuron and antimycin A (Fig. 1.70) results in a sharp decrease in radioactivity of the cell-wall polysaccharides, thus proving an important role of photosynthetic phosphorylation is providing the processes of synthesis of structural polysaccharides with energy-rich compounds. Addition of ATP (10^{-5} to 10^{-4} M) to an incubation medium, increased label incorporation on dark culturing and had practically no effect on these processes in the light (cf. Fig. 1.70).

It should be noted that the isolated protoplasts cultured in the light do not, in practice, suffer from undersaturation of carbohydrate substrates for the synthesis of structural polysaccharides at the early stages of cell wall regeneration. An increase in glucose concentration positively influences synthesis of structural polysaccharides (matrix ones to a large extent) in the dark. The enhancement observed due to higher concentrations of substrates is however much inferior to that with an other photosynthesizing subject—leaves of wheat seedlings. This may be possibly accounted for by low rates of the synthesis of structural polysaccharides at the early stages of cell wall regeneration by

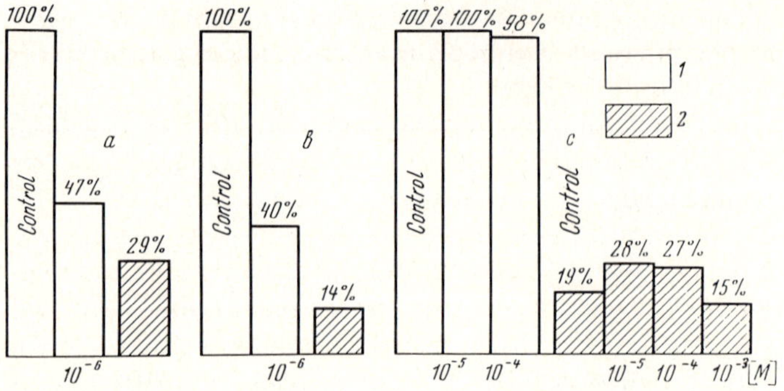

Fig. 1.70a–c. Incorporation of exogenous [^{14}C]-glucose (15 mM) by bean mesophyll protoplasts into cell-wall polysaccharides at various levels of energy-rich phosphates. **a** Effect of diuron (exposure for 24 h, illumination for 12 h); **b** effect of antimycin A (exposure for 14 h, illumination for 14 h); **c** effect of ATP (exposure for 24 h, illumination for 14 h).

1 in the light; *2* in the dark.

isolated protoplasts. The existence of such a lag-phase for synthesis of cellulose, for example, has been documented.

Thus in our experiments the culturing of protoplasts of bean leaf mesophyll was carried out until cell walls were regenerated under conditions of relatively high illumination. The results obtained, allow us to conclude that photosynthesis can participate in cell wall regeneration at early stages of culturing of protoplasts in the light by providing ATP for the synthesis of structural polysaccharides. This makes timely the problem of elaboration of optimum conditions for light culturing of protoplasts and search of optimum hetero-autotrophy ratio in isolated protoplasts which regenerate cell walls.

Considerable amounts of matrix polysaccharides and glycoproteins produced upon regeneration of the cell walls are not retained by a surface of the protoplast and the newly formed cell wall but enter the 'environment' and are to be found in a fraction which is not absolutely correctly called 'extracellular polysaccharides'.

The ratio of fractions formed can vary considerably with culturing conditions. In our experiments [266] the proportion of extracellular polysaccharides dropped when protoplasts of bean mesophyll tissue were cultured in the light. Of note is the low percentage of cellulose in regenerating cell wall (1–2% vs 20–30% in the original mesophyll tissues) and this has been observed by others.

In the course of cell wall formation the rate of cellulose production grew steadily during a 7-day-experiment unlike that for hemicelluloses which reached its maximum value in the first two days of culturing and remained almost constant thereafter [266].

The difference in rates of synthesis of structural polysaccharides can be rationalized on the basis of the suggestion advanced by Willison and Brown

[267] and Roland and Vian [268] that the cell wall is required to enable formation of the ordered structure of cellulose microfibrils and their high rate of synthesis.

The effect of the cell wall on cellulose synthesis was studied in a set of special experiments [266]. Isolated bean protoplasts were cultured for 24 h (then they become oval-shaped and the rate of cellulose production grew considerably in comparison with that at the beginning of culturing) and divided into two groups. Protoplasts of the first group were treated with cellulase (1/5 its usual concentration) for 1 h whereupon the protoplasts reacquired their original spherical shape. The cellulase was washed off and the cells recultured. Protoplasts of the second group were treated analogously (centrifugated, resuspended) but with the cellulase treatment omitted, total treatment time amounting to 2 h, Then the protoplasts of the both groups were cultured for 4 h in the presence of [^{14}C]-glucose, the material was fixed, and the radioactivity of the cellulose produced was measured. Protoplasts of both groups were shown to synthesize cellulose at higher and almost equal intensity.

These data are in accord with those of Burgess et al. [269]. These authors detected, with the help of electron scanning microscopy, the appearance of fibrils after 10-h-culturing of isolated tobacco protoplasts. Following removal of the fibrils by cellulase treatment and washing the enzyme off, the protoplasts were recultured whereupon fibrils appeared without any lag-period.

It thus follows that the differences observed in the dynamics of the synthesis of matrix polysaccharides by isolated protoplasts do not relate to the absence of the cell wall. More probably, it relates to the extreme complexity of organization of cellulose synthetase complexes whose formation apparently takes a considerable length of time. It is noteworthy that the intensity of cellulose production during the first-day culturing could not be increased by an increase in glucose level, nor by treatment with phytohormones, nor by the addition of ATP, i.e. by the factors which usually increase the synthesis of polysaccharides.

It should be noted that the increase in the rate of cellulose formation in isolated protoplasts with time could only be observed in the presence of phytohormones. The enhanced synthesis of cellulose seems to require an assembly of new enzyme complexes whose formation is stimulated by phytohormones.

From the above data it is possible to conclude that the isolated protoplasts possess some operative mechanisms which ensure, at early stages of culturing, intensification of cellulose synthesis comparable to activation of synthesis of matrix polysaccharides. It seems probable that the de novo formation of additional synthetase complexes constitutes the step which limits cell potential for the rapid activation of cellulose synthesis. It should be kept in mind that removal of the cell wall by enzymatic treatment disturbs the established interrelationship with the cell contents, feedback between cell-wall components and intracellular compartments where its synthesis is localized. The mechanism of this feedback is completely obscure. It could be suggested that specific signal molecules exist which are formed during cell-wall assembly and are directed inside the cell to correct the rates of synthesis of cellulose, hemicelluloses, pectic substances, and proteins. In this connection a new type of regulatory substances has attracted attention, the so-called 'oligosaccharines' [36] these are the fragments of cell wall hemicelluloses and pectic substances.

8.2 The Dynamics of Ultrastructural Changes and Intensity of Structural Polysaccharide Formation in the Course of Cell Wall Regeneration

It has already been mentioned in Chap. 4 that the synthesis of cellulose and matrix components is differently localized in the cell. According to the accepted concepts, the formation of matrix polysaccharides is controlled by the functional activity of the Golgi apparatus and the endoplasmic reticulum while synthesis of cellulose is related to enzymic complexes located in the plasmalemma [151, 270–273]. In this connection it was of interest to compare the dynamics of ultrastructural changes and the intensity of synthesis of various groups of structural polysaccharides during culturing of protoplasts in the light and in the dark.

Ultrastructural reorganizations became visible in our experiments with bean protoplasts after 10 to 12 hours of culturing. The processes which were apparent were the formation of the endoplasmic reticulum from nuclear envelope, movement of microtubules of the endoplasmic reticulum from the nucleus to the plasmalemma, development of elements of the Golgi apparatus, simultaneous localization of the endoplasmic reticulum and Golgi apparatus near the plasmalemma, production of Golgi vesicles, and secretion of their content on the surface of the plasmalemma by way of exocytosis. The vesicles separating from the Golgi bodies are regularly observed close to the plasmalemma in the first two days of culturing of bean protoplasts in the light and in the dark, all these changes being observed earlier in the former case. Intensive production of matrix polysaccharides is evident from our cytochemical analysis with a contrasting stain 'Ruthenium Red'. An intensively stained layer of matrix polysaccharides was thus revealed on the surface of the protoplasts. These data possibly indicate the principal role of the Golgi bodies at this stage of the matrix polysaccharide synthesis. Elements of the endoplasmic reticulum are permanently located close to the Golgi apparatus and the plasmalemma and the endoplasmic reticulum itself seems to participate in the processes of formation of membranes of the Golgi apparatus and biosynthesis of enzymes which are involved in cell wall genesis.

Our observations fit well the endomembrane concept *sensu strictu* [274–277] according to which the formation of membranes of the Golgi apparatus is necessarily accompanied with a localization of endoplasmic reticulum microtubules at the site where elements of the Golgi apparatus are preformed and is governed by the endoplasmic reticulum. This concept originally forwarded for animal cells proved to be valid for higher plants [278].

Biochemical analyses of polysaccharides of growing cell walls of bean protoplasts at the early stage of culturing support the electron microscopy and cytochemical data and give evidence of a sharp increase in the incorporation of [^{14}C]-glucose in matrix polysaccharides parallel to the development of the endoplasmic reticulum and Golgi apparatus. In these experiments the label ([^{14}C]-glucose) was given to protoplasts as a 'gulp' for 4 h after different intervals. It turned out that the intensity of its incorporation into the acid-soluble fraction after 14 h of culturing is 10 times that after 4 h. The rate of [^{14}C]-glucose utilization for cellulose synthesis increases much more slowly during this period. In subsequent culturing, the rate of labelling of matrix polysaccharides remains at the same level whereas that of cellulose increases considerably.

Microphotographs allowed us to detect, in our experiments, the first cellulose microfibrils after 18 h of culturing of bean protoplasts. Duration of a lag-phase varies, according to different estimates, from several minute to three days [269, 279, 280]. It may possibly depend on isolation conditions and type of tissue employed, as well as on the functional state of the system of production of microfibrils by the cell wall at the moment of isolation of protoplasts [257, 269, 271, 281, 282].

Sensitivity of methods used should also be taken into consideration. With the use of tracer analysis it was established by us and other authors [62, 283] that synthesis of cellulose begins without a lag-phase.

After 48 h of culturing of bean protoplasts the microphotographs reveal the appearance of a random layer of microfibrils on a surface of protoplasts. It is probable that the microfibrils' termini are bound to the plasmalemma or to structures incorporated which correspond to the terminal complexes (cellulose synthetases) revealed by a freeze-etching technique [134]. It should be noted that extensive development of the endoplasmic reticulum takes place in that period (3–7 days) and microtubules are oriented in parallel and close to the plasmalemma. These rearrangements may be caused by the increasing role of the endoplasmic reticulum in cell wall regeneration during intensive cellulose formation. Massive development of the endoplasmic reticulum may be related to a formation of cellulose-synthetase complexes on the plasmalemma [284] or to enhanced synthesis of proteins (structural ones and enzymes) necessary for cell wall formation [285].

Exocytosis of the Golgi vesicles is seen in the microphotographs as before. It is significant that a Golgi apparatus is disclosed at this period producing vesicles which are in contact with the endoplasmic reticulum microtubules rather than being directed to the plasmalemma. The possibility of such a functional interaction of the Golgi apparatus with the endoplasmic reticulum in cells of higher plants has recently been demonstrated [120, 273, 286], these facts are in contrast to the established, endomembrane concept on unidirectional transport of elements of the endoplasmic reticulum to the Golgi apparatus.

The dynamics of the development and localization of the endoplasmic reticulum and Golgi apparatus allow one to conclude that it is the latter which in bean protoplasts plays the primary role in the first stages of cell wall formation during the most intensive synthesis of matrix polysaccharides. The activity of the former becomes more and more important during subsequent stages—in the course of intensive cellulose formation.

In our works as well as in some others [138, 272, 287] it was shown that orientation of microfibrils formed upon prolonged culturing of protoplasts (for 72 to 168 h) becomes more and more ordered and stratified. It was in this period that we detected the microtubules oriented in parallel which are supposedly involved in the orientation of microfibrils on the protoplast surface [165, 168, 288].

When protoplasts were cultured for a long period (from 3 to 7 days) the studies of intensity of formation of different fractions of structural polysaccharides were performed using two approaches: [^{14}C]-glucose was introduced into a medium either for the total culturing period or at its particular stage for a limited time (the so-called 'gulp'). In the former case the two-fold increase in culturing duration resulted in more than a three-fold increase in cellulose radioactivity which points to a rate enhancement of cellulose synthesis with time. With a [^{14}C]-glucose level of 200 mM during the whole culturing period the radioactivity of cellulose, hemicelluloses, and extracellular polysaccharides (expressed as uM per mg chlorophyll) changed from 0.2 to 0.7, from 28.1 to 31.0, and from 9.3 to 13.5 respectively from the third to the seventh day. Analogous results with a several-fold increase in cellulose productivity in the course of prolonged culturing were obtained with suspension cultures of carrot protoplasts [283] and tobacco protoplasts [145]. The increase in radioactivity of hemicelluloses (days 3–7) was very low and amounted to about 10%, and that of extracellular polysaccharides was 1.6-fold that demonstrates retardation of the labelling of extracellular polysaccharides with time.

'Gulp'-culturing has confirmed the data on the increase in intensity of processes of cellulose formation and on the decrease in labelling of extracellular polysaccharides with culturing time (Table 1.23), the rate of hemicellulose production being relatively permanent.

Comparison of electron microscopy and biochemical data leads to the conclusion that cellulose synthesis rate enhancement parallels formation of an ordered cell wall structure. This agrees with data from other authors [267, 268] which show that the ordered structure of a newly formed cell wall and the high rate of its synthesis requires the preformed cell wall.

Matrix polysaccharides are synthesized, according to modern concepts, inside the cell and the absence of a wall does not prevent its formation.

The observation we made concerning the decrease in radioactivity of extracellular polysaccharides during culturing of bean protoplasts show that it may be conditioned by the following reasons. The high radioactivity of extracellular polysaccharides at the early stages of cell wall regeneration is accounted for by the fact that the nutrition medium may contain the soluble pectic

Table 1.23. Incorporation of [^{14}C]-glucose (200 mM) into structural polysaccharides expressed as nM glucose per mg of chlorophyll per hour upon culturing in the light of bean protoplasts (glucose 'gulp' for 24 hours)

Feeding at the day	Extracellular polysaccharides	Hemicelluloses	Cellulose
First	167 (100)	450 (100)	3 (100)
Fourth	142 (86)	280 (62)	4 (131)
Seventh	114 (68)	320 (71)	8 (238)

Percentage is given in parentheses

substances and, in addition, vesicles separated from the plasmalemma containing matrix polysaccharides since they are not retained in the cell wall due to the absence of a microfibril network. At the later stages of cell wall regeneration when cellulose production is accelerated the observed decrease in the intensity of extracellular polysaccharide production seems to be connected to the retention of matrix polysaccharides (probably together with vesicles which contain them) by the cell wall formed. The construction of cell walls of complete composition by *Vinca rosea* protoplasts during prolonged culturing was studied by Takeuchi and Komamine [289, 290]. Analysis of the intensity of accumulation of [^{14}C]-glucose in acid-soluble polysaccharides with 'gulp' culturing and with permanent glucose feeding allowed us to understand the possible metabolization of hemicelluloses during cell wall regeneration. The hemicellulose fraction of protoplasts was shown to be composed mainly of arabinogalactans, xyloglucan, and β-1,3-glucan [62, 87, 263, 291].

The presence of β-1,3-glucans in a fraction of structural polysaccharides of cell wall protoplasts unlike the original tissues is well documented [62, 263, 264]. The β-1,3-glucan was supposed to serve as a cellulose precursor acting as a glucosyl donor [102]. Our results can be explained only to a small extent by its participation in cellulose synthesis because the increment in cellulose radioactivity constitutes a small proportion of that of hemicelluloses subject to degradation. A slight increase in the radioactivity of hemicelluloses with time of culturing may be conditioned by their possible conversion into soluble degradation products which on being washed out into a medium are incorporated into extracellular polysaccharides. Thus, for example, the 3,6-arabinogalactan in *Vinca rosea* cell culture is intensively synthesized and incorporated into the cell wall while, on the other hand, it is rapidly degraded and excreted to the medium [291, 292, 293]. These data as well as those accumulated in our works indicate the important role of hemicelluloses in cell wall metabolism.

Analogous biochemical and ultrastructural analyses have been performed by us with protoplasts isolated from tobacco leaf mesophyll, the classical object for studies on cell wall de novo regeneration. Tobacco protoplasts, unlike bean ones, undergo intensive division on days 2–3 of culturing and produce cell colonies on more prolonged culturing.

Results of biochemical studies confirmed the main regularities of the process of structural polysaccharides formation established in studies of cell wall regeneration in bean protoplasts. Here again the initial stages of culturing were characterized by a low rate of cellulose synthesis from exogenous [^{14}C]-glucose which sharply increased upon subsequent culturing (Table 1.24). The rate "jump" with tobacco protoplasts was even higher, the intensity of cellulose formation on day 7 was 40 times that on the first day of protoplasts' culturing. Presumably this is due to the intensive division of the newly formed tobacco cells. And this may probably account for the more than 8-fold increase in the rate of [^{14}C]-glucose incorporation into matrix polysaccharides from day 7 of culturing.

Comparison of glucose accumulation in matrix polysaccharides with 'gulp' and permanent feeding has confirmed the conclusion on the possible metabolism of matrix polysaccharides. With tobacco protoplasts, the activation of the endoplasmic reticulum, its localization close to the plasmalemma, and vesicle formation on its surface on day 3 is analogous to bean protoplasts. The vesicles formed come in contact with the plasmalemma. Unlike bean protoplasts the development of the Golgi apparatus was not observed in the case of tobacco protoplasts. After 72 h of culturing a layer of uniformly oriented and tightly packed cellulose microfibrils is observable on a surface of tobacco protoplasts. The electron-microscopic pattern displayed similarity for tobacco protoplasts cultured in the light and in the dark.

No activation of the Golgi elements during cell envelope regeneration was observed in tobacco protoplasts cultured under standard conditions [272, 284, 294]. Differences in ultrastructure of bean and tobacco protoplasts during cell

Table 1.24. Incorporation of exogenous [^{14}C]-glucose into cell wall polysaccharides of tobacco leaf protoplasts (60 mM glucose, in the light)

	Hemicelluloses		Cellulose	
Feeding at the day	μM glucose per mg chlorophyll	Relative units	nM glucose per mg chlorophyll	Relative units
	'Gulp' for 24 h			
1st	4	1	40	1
3rd	6	2	133	3
7th	34	8	1683	42
	Permanent feeding			
3rd	19	4	173	4
7th	52	13	3513	87

Table 1.25. Synthesis rate and percentage of cellulose fractions of different DP during cell wall formation in tobacco isolated protoplasts and cell culture

Feeding conditions	Cellulose content in cell wall		Percentage of cellulose fractions		
	%	cpm	DP 500	DP 500–2000	DP 2000–3000
	Protoplasts				
[^{14}C]-glucose, 16 h	4.2	13,000	66.9	—	33.1
[^{14}C]-glucose, 16 h + [^{12}C]-glucose, 24 h	7.1	49,280	34.2	13.1	52.7
[^{14}C]-glucose, 40 h	6	118,500	69.9	5.8	24.3
[^{14}C]-glucose, 16 h + [^{12}C]-glucose, 48 h	6.6	52,300	33.3	8.2	58.5
[^{14}C]-glucose, 64 h	7.6	200,700	59.7	8.5	31.8
	Cell culture				
[^{12}C]-glucose, 4 d + [^{14}C]-glucose, 6 h	33.9	17,100	29.8	39.1	31.1
[^{12}C]-glucose, 4 d + [^{14}C]-glucose, 6 h + [^{12}C]-glucose, 6 h	34.7	33,120	24.2	38.9	36.9
[^{12}C]-glucose, 4 d + [^{14}C]-glucose, 6 h + [^{12}C]-glucose, 12 h	39.9	39,590	21.8	38.1	40.1
[^{12}C]-glucose, 4 d + [^{14}C]-glucose, 6 h + [^{12}C]-glucose, 24 h	37.3	31,840	17.7	34.6	47.7

envelope regeneration may be explained by individual features of plants and protoplasts isolated therefrom.

It has been shown recently that the ratio of cellulose species of different degrees of polymerization (DP) depends on the duration of cell wall regeneration by tobacco protoplasts [145]. [^{14}C]-Glucose was incorporated into two main cellulose species, those of DP below 500 and DP 2000–3000. In cell culture the main part of labelled glucose was found in cellulose of intermediate DP values (Table 1.25). The mesophyll cell wall contained less of this fraction and a lot of DP-4000-cellulose was present.

9 The Effect of Climate and Soil Conditions on Cellulose Biosynthesis

Cellulose is synthesized by numerous organisms living in the oceans and on every continent of our planet, i.e. under a broad diversity of temperature, illumination, mineral nutrition, and irrigation conditions, etc. Of course, all these factors as well as species themselves strongly affect the intensity of synthesis of cellulose and other cell wall polysaccharides. It is quite natural to suggest that the highest cellulose productivity is realized for each species under optimum in situ conditions. However it is not always the case. The intensity of cellulose synthesis can considerably be enhanced under experimental conditions by varying certain factors, e.g. atmosphere composition. Maximum productivity of the cellulose synthesis per area unit increases from polar regions to subtropics and then slightly drops in the equatorial region.

The highest productivity of structural polysaccharides per mass of the producent cell is specific for microorganisms, e.g. for *A. xylinum* and, of higher plants, for cereals, e.g. sugar cane, maize, etc. It should be emphasized that relatively few studies have been specially devoted to the peculiarities of the biosynthesis of structural polysaccharides related to the effect of changes in the environment. Discussed below are the effects of temperature, water supply, illumination, and some other factors on cellulose biosynthesis.

9.1 The Effect of Temperature

The temperature dependence of intensity of cellulose biosynthesis is similar to that of other physiologo-biochemical processes. Those who have specially studied this problem came to the conclusion on a normal distribution curve dependence for cells of cotton seed hairs with the optimum temperature of about 30 °C. In our experiments on analysis of the daily dynamics of cellulose synthesis in the cotton fiber secondary cell wall the maximum rate was observed at 1 p.m. and at 40 °C (Fig. 1.71). Most probably, the temperature interval of 30 to 40 °C is optimum for the cellulose synthesis in cotton pods and variations are caused mainly by strain differences. The temperature optimum of cellulose synthesis seems to lie at higher values than that of photosynthesis. And therefore elevated temperatures which are most favourable for cellulose synthesis can inhibit cellulose formation through indirect action. It was mentioned in Chap. 5 that

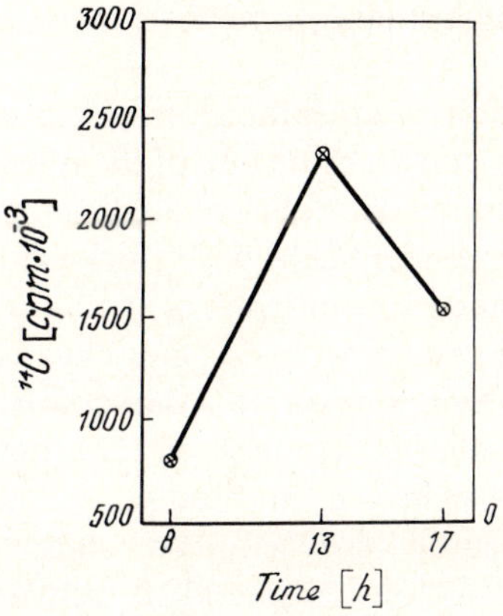

Fig. 1.71. The intensity of the cellulose synthesis from [^{14}C]-glucose during secondary wall formation in a cotton fiber as a function of day time. Abscissa: Day time, hour, Ordinates: [^{14}C], cpm

synthesis of cellulose depends greatly on the starting substrate level and therefore the decrease in intensity of photosynthesis due to superoptimal (for this process) temperatures can slow down the synthesis of structural polysaccharides due to lowering of the level of such products of photosynthesis as phosphorylated hexoses. If supply to polysaccharide synthesis of nucleoside triphosphates is considered, the role of the principal energetic 'engine' for the cell is probably transferred to the mitochondria under the above temperature conditions.

It is known that the temperature optimum for respiration greatly exceeds that of photosynthesis. It is more important for nonphotosynthesizing cells where respiration is the only source of energy-rich phosphates consumed in the process of the synthesis of structural polysaccharides.

9.2 The Effect of Dehydration

A large number of plants live in regions of unbalanced aqueous soil conditions and in arid lands. This necessitates that we pay especial attention to the study of the effect of drought on the synthesis of structural polysaccharides. It was demonstrated long ago [208] that severe soil drought can result in an increase of cellulose content in wheat leaves compared to other products of photosynthesis. This is especially pronounced under conditions when the absolute rate of photosynthetic consumption of carbon dioxide drops sharply. One should say therefore that the possibility exists of an indirect effect of the soil drought on cellulose synthesis caused by the inhibition of the formation of the photosynthesis products which serve as substrates for cellulose synthesis. In other words, synthesis of cellulose is less affected by the soil drought than production of some other products of photosynthesis.

One more indirect route for soil drought action is an inhibition of plant growth processes. The growth through extension is known to be one of the most vulnerable processes of a plant which is violated as soil drought becomes more severe and results in the formation of relatively small-sized cells with thickened cell walls and in morphological changes. The so-called xeromorphous structure appears and this is characterized by smaller dimensions of various plant organs and the plant as a whole and by higher 'rigidity'. This can result in a decrease in the absolute rate of cellulose synthesis whereas relative content of cellulose (expressed as percentage to dry mass) is much enlarged in plants with a xeromorphous structure. Besides cell-wall thickening and cellulose content increase, also typical are more extensive wax deposition compared to that in plants under conditions of optimum water supply, as well as a more pronounced development of the cuticle as a whole. However, despite a higher cellulose level in tissues of plants with a xeromorphous structure, the total productivity of cellulose formation is diminished per planting area unit or that of natural phytocenosis.

It has already been mentioned in Chap. 6 that soil drought lowers the intensity of cellulose synthesis from exogenous glucose in cotton fibers. This can be restored to a considerable extent with the aid of energy-rich phosphates. The effect of elevated temperatures, atmospheric drought, and soil drought is similar in that the role of respiration is increased in comparison to that of photosynthesis as a means of supplying energy-rich phosphates for the cell's needs. The unfavourable effect of soil drought on cellulose synthesis in cotton plants results in considerable losses in the raw-cotton crop thus necessitating large-scale irrigated cotton-growing.

9.3 The Effect of Illumination

An increase in the intensity of illumination favours considerably the formation of structural polysaccharides in wheat-leaf tissues (cf. Chap. 6 and 8 where energetic aspects of cellulose synthesis are discussed). This is accounted for by an illumination-induced enhancement in photosynthetic production of phosphorylated hexoses and NDPS, as well as of energy-rich phosphates formed by photophosphorylation. When an illumination effect is considered with respect to the synthesis of structural polysaccharides outside the photosynthesizing organs, the possible indirect effect of illumination has to be taken into account. This can depend primarily on changes in intensity of photosynthesis and, therefore, in the rate of transport of assimilates from chlorophyll-bearing tissues to the nonphotosynthesizing organs. Analysis of an age-dependence of intensity of the cellulose synthesis in cotton fibers has revealed that the increase in

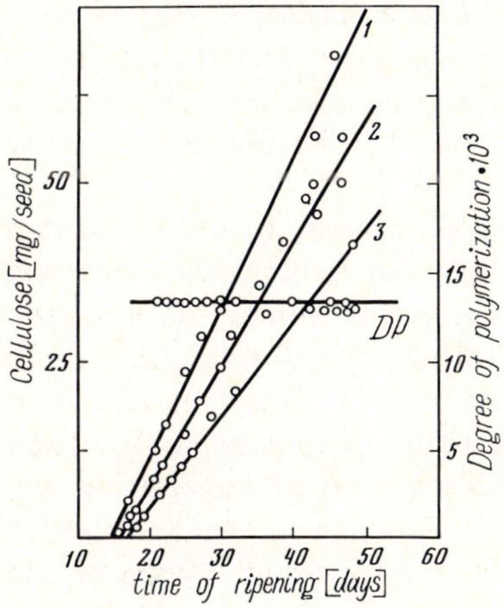

Fig. 1.72. Effect of illumination on degree of polymerization of cotton-fiber cellulose and its accumulation.

Illumination intensity: *1* 60 klx; *2* 15 klx; *3* 4 klx. Degree of polymerization is the same. Abscissa: Maturing time, days. Ordinate (left): Cellulose, mg per seed; (right): D.P. $\times 10^{-3}$

illumination intensity does not affect the degree of polymerization of cellulose but, on the other hand, strongly promotes its biosynthesis [146] (Fig. 1.72).

Since cotton seed hairs are chlorophyll-free it may be concluded that the effect of illumination on cellulose synthesis is mediated by an alteration in the level of transportable photosynthesis products which flow to hairs from leaves and photosynthesizing tissues of other plant organs. As has already been mentioned, the insufficient concentration of substrates for cellulose can be the limiting factor in the raw-cotton crop. This is even more clearly manifested in plants grown under insufficient illumination conditions.

The growth processes related to the increase in cell-wall dimensions and mass were greatly intensified in years with favourable temperature, water supply, and illumination conditions. This is evident, for example, from year-ring enlargement in tree stems.

9.4 The Effect of Mineral Nutrition

The intensity of synthesis of cell-wall polysaccharides is greatly affected by mineral nutrients. The effect of mineral nutrition on cotton fiber production can be studied better than on other plants employed as industrial sources for cellulose. A high-level of soil fertility increases the grade and strength of cotton fibers [295].

Experimental data show that the development and quality of fibers depend both on dose and the form of the fertilizers. Technological properties of cotton fibers were best with ammonium nitrate as the nitrogen nutrient compared to nitrate and amide [296]. Elevated concentrations of phosphorus positively affect fiber quality [297].

It should be noted that there is an optimum for mineral nutrition and that excessive doses of fertilizers ($N_{400}P_{250}K_{75}$) do not change the raw-cotton crop beneficially and fiber quality (strength, maturity, metric number) deteriorates [298]. Optimization of mineral nutrition of cotton requires a complex application of macro- and microelements. Shortage of the latters lowers fiber productivity and quality. Efficiency of the use of wastes of mining and its related industries, which contain microelements necessary for cotton fertilization, has been studied [297], and some of them have been shown to improve the length and denier number of fibers with the retention of their normal strength and tearing length [297]. In general, the microelements required for cotton plant can be arranged in the series: B > Zn > Cu > Mn > Mo > Co [299].

One of the reasons for a decrease in cotton crop productivity and quality is the ionic imbalance caused by soil salinization [300]. A sufficient calcium level considerably reduces the toxic effect of salinization on cotton plant growth and productivity. This is accounted for by a relative restoration of ion equilibrium in tissues due to lower ingress of chloride, phosphate, sodium, and magnesium ions [300].

Fertilizers can vary the proportions of structural polysaccharides in the vegetative organs of plants. Excessive nitrogen nutrition lowers the hemicellulose content most remarkably, the cellulose level being affected to a lesser extent if at all. Experiments with plants grown in nutrient solutions have shown that production of cellulose dropped with deficiencies of nitrogen, phosphorus, magnesium, calcium, and especially potassium. Of microelements boron influences cellulose synthesis to the greatest extent [71].

Also taken into account should be the fact that it is the yield of cellulose or other polysaccharides, per area unit, rather than their level that can essentially be changed with alterations in agricultural technology or climate conditions as it depends on the plant biomass.

10 Conclusion

The material presented in the section 'Biosynthesis of Cellulose' demonstrates the progress made in understanding the mechanisms of the synthesis of structural polysaccharides, viz. establishing intermediate stages in the glucose-to-polysaccharides conversion, cell localization of synthetases responsible for the synthesis of cellulose, hemicelluloses, pectic substances, and cell-wall proteins. Studies began with molecular mechanisms of the synthesis of cellulose microfibrils, on the role of plasmalemma components in this process. Importance of the protein microtubules in orientation of the newly-formed cellulose microfibrils in the cell wall is demonstrated.

Presented in this section are also the results obtained in the author's laboratory from studies of the effects of changes in substrate concentration; of membranotropic substances; nucleoside triphosphates; hormones; and other physiologically active substances; and climatic and agrotechnical conditions on the synthesis of structural polysaccharides. These data together allow us to build up a general impression of the peculiarities of the synthesis of cellulose primarily, and of other structural polysaccharides to some extent, in order to bring nearer the solution to the problem of mechanisms of endogenous regulation of these processes which are the most important for plant activity and, of course, for agriculture and cellulose-utilizing branches of industry.

In writing this section the author's subjectivism has been manifested in that preference was given to those studies whose authors are distinguished by creative imagination and putting forward daring hypotheses. Less attention has been paid to unresolved problems, to contradictory facts, insufficiently grounded ideas related to some essential aspects of the mechanism of the cellulose synthesis. Thus, for example, it seems doubtful that the 'rosettes' of the internal part of the plasmalemma and terminal cellulose-synthetase globules of its outer surface are necessarily complementary, that 'rosettes' are directly involved in the synthesis of cellulose microfibrils. This makes necessary the analysis of the role of 'rosettes' as organelles which participate in establishing the transmembrane connection between microtubules and terminal cellulose-synthetase complexes or in ensuring functional conditioning of plasmalemma [301]. Furthermore, it was suggested that the formation of native microfibrils in bacteria [302] and isolated tobacco protoplasts [303] did not require direct involvement of the terminal synthetase globules of the plasmalemma, thus their role remains obscure.

Several other important questions are not sufficiently or even partially solved. There is nonconclusive agreement on the existence of lipid- or protein-linked intermediates of cellulose synthesis. The mechanism of the synchronization of differently compartmentalized syntheses of cell wall components (cellulose, hemicelluloses, pectic substances, proteins) remains unclear as well as changes in their proportions during ontogenesis.

The reasons for the absence of a statically sound 'dispersion' of the degree of polymerization of the secondary cell-wall cellulose are puzzling. If there is a template which 'counts' a definite number of glucose residues for each cellulose molecule or 'measures' its length, what is then its nature and mechanism?

It is not clear what the relation between cellulose-synthetase activity and the potential difference in higher plant plasmalemma or the cell membrane of *Acetobacter xylinum*, whether the transmembrane electric potential creates the most reactive conformation of membrane-embedded glucan synthetases or just facilitates delivery of substrates.

It is very important to determine the level of substrate saturation for the process of cellulose synthesis and limiting steps in the most important cellulose producers such as cotton plants, flax, and arboraceous cultures. The solution to this problem will allow us to evaluate the prospects for the biotechnological production of cellulose using *A. xylinum*. An important problem posed for those who study cellulose synthesis is the development, primarily for the aforementioned plants, of a complex system for governing cellulose production with the use of not only fertilization and irrigation but also of hormones, retardants, and other physiologically active compounds.

References

1. Metzler D (1980) Biochemistry: Chemical reactions in living cell. Mir, Moscow, vol 2
2. Loktev S (1980) Zhurn DI Mendeleev Vsesoyusn. Khim. Obshch. (Sov. Journal of All-Union Chemical Society) 25: 536
3. Svirezhev YM, Tarko A (1979) Vestnik Akad. Nauk SSSR 12: 95
4. Nichiporovich AA (1982) Foreword. In: Physiology of Photosynthesis. Nauka, Moscow, p 3
5. Bazilevich NI, Rodin LE, Rozov NN (1971) Priroda (English transl.: Sov Nature) 1: 46
6. Gruber E (1976) Papier /BRD/ 30: 533
7. Nichiporovich AA (1973) The basis for photosynthesis productivity of plants. In: Current problems of photosynthesis. MGU Publishers, Moscow, p 17
8. Bidwell RGS (1979) Plant physiology. Second Edition. Macmillan, New York
9. Gollerbach MM (1977) The main types of the morphological structure of Algae body. In: Life of Plants. "Prosveshcheniye", Moscow, p 32
10. Sedova TV (1977) Algae Cell. In: Life of Plants, Prosveshcheniye, Moscow, vol 3, p 16
11. Takhtadjyan AL (1980) Microsporogenesis In: Life of plants. Prosveshcheniye, Moscow, p 43
12. Delmer D (1983) Adv Carbon Chem and Biochem 41: 105
13. Darvill A, McNeil M, Albersheim PA, Delmer DP (1980) The primary cell walls of flowering plants. In: The Biochemistry of Plants. Academic Press, New York, p 91
14. Delmer DP (1987) Ann Rev Plant Physiol 38: 259
15. Hayashi T, Read SM, Bussel J et al. (1987) Plant Physiol 83: 1054
16. Fry CS (1986) Ann Rev Plant Physiol 37: 165
17. Pochinok KhN (1976) Methods for Biochemical Analysis of Plants. Naukova Dumka, Kiev
18. Selvendran RR (1985) J Cell Biol Suppl 2: 51
19. Lamport DTA, Northcote DH (1960) Nature 188: 665
20. Lamport DTA, Hsung JC (1974) The shape extensin. In: Plant research. MSU, Michigan, USA, p 86
21. Smith MA (1981) Plant Physiol 68: 956
22. Smith MA (1981) Plant Physiol 68: 964
23. Kristen U (1985) Progress in Botany 47: 1
24. Van Holst GJ, Varner JE (1984) Plant Physiol 74: 247
25. Pope D, Lamport DTA (1974) A hydroxyproline–polysaccharide linkage. In: Plant Research MSU, Michigan, US, p 90
26. Brysk MM, Chrispeels MJ (1972) Biochim Biophys Acta 257: 421
27. Hori H, Fujii T (1980) Plant and Cell Physiol 21: 293

28. Mort A, Lamport DTA (1974) Chemical extraction of large glycopeptides from the cell walls of tomato suspension cultures. In: Plant research MSU, Michigan, US, p 102
29. Pope D, Lamport DTA (1974) Precursors of extensin in cell suspension cultures of *Acer pseudoplatanus.* In: Plant Research. MSU, Michigan, US, p 94
30. Klis FM, Lamport DTA (1974) Possible precursors of extensin in stem tissue of *Pisum sativum.* In: Plant Research. MSU, Michigan, US, p 92
31. Nimz K (1974) Angew Chem 13: 313
32. Kretovich VL (1971) Foundations of Plant Biochemistry. Vysshaya Shkola, Moscow
33. Keegstra K, Talmadge KW, Bauer WD, Albersheim P (1973) Plant Physiol 51: 188
34. Davies HM, Delmer DP (1982) Protein glycosylation in higher plants: Recent developments. In: Cellulose and Other Natural Polymer Synthesis. Plenum Press, New York, London, p 363
35. Polevoy VV (1986) Auxin role in regulation systems of plants.—Timiryazevskie Chtenia. Nauka, Leningrad
36. Albersheim P, Darvill AG (1985) V Mire Nauki 11: 16
37. Metlitsky LV, Ozeretskovskaya OL, Vasyukova NL, Sabirdina MS, Chalenko GI (1984) Doklady Akad Nauk SSSR 274: 1020
38. Vasyukova NI, Medvedeva TE, Chalenko GI, Ozeretskovskaya OL (1985) Doklady Akad Nauk SSSR 283: 502
39. Vasyukova NI, Ozeretskovskaya OL, Leont'eva GV, Chalenko GI, Elyakova LA et al. (1987) Bioorganic Chemistry (USSR) 13: 825
40. Thornber JP, Northcote DH (1961) Biochem J 81: 455
41. Meinert MC, Delmer DP (1977) Plant Physiol 59: 1088
42. Huwyler HR, Franz G, Meier H (1978) Plant Sci Lett 12: 55
43. Huwyler HR, Franz G, Meier H (1979) Planta 146: 635
44. Thornber JP, Northcote DH (1962) Biochem J 82: 1075
45. Shubert VD (1968) Biochemistry of lignin. Lesnaya Promyshlennost, Moscow
46. Bailey RW (1973) Chem and Biochem 1: 157
47. Kulaeva ON (1973) Cytokinins, their Structure and Function. Nauka, Moscow
48. Kulaeva ON (1982) Hormonal Regulation of Physiological Processes in Plants at a Level of Synthesis of RNA and Proteins. Nauka, Moscow
49. Delmer DP, Cooper G, Alexander D, Cooper J, Hayashi T, Nitsche C, Thelen M (1985) J Cell Sci Suppl 2: 33
50. Keerberg OF, Viil YuA (1982) Regulatory systems and energetics of the reductive pentose phosphate cycle. In: Physiology of Photosynthesis. Nauka, Moscow, p 104
51. Tarchevsky IA (1982) Mechanism of drought effect on photosynthetic assimilation of CO_2. In: Physiology of photosynthesis, Nauka, Moscow, p 118
52. Voskresenskaya NP (1982) Regulatory role of blue light in photosynthesis. In: Physiology of photosynthesis. Nauka, Moscow, p 203
53. Delmer DP, Beasley CA, Ordin L (1974) Plant Physiol 53: 149
54. Bajaj GPS (1977) Protoplast isolation, culture and somatic hybridization. In: Applied and fundamental aspects of plant cells, tissue and organ culture. Springer, Berlin, p 467
55. Cocking EC (1960) Nature 187: 962
56. Pojnar E, Willison JHM, Cocking EC (1967) Protoplasma, 1967, 4: 460
57. Nagata T, Takebe T (1970) Planta 92: 301
58. Ericsson T (1977) Technical advances in protoplast isolation and cultivation. In:

Plant tissue culture and its biotechnological application. Proc. 1st Intern congr on medicinal plant research, Sect. B, Munich, 1976
59. Butenko RG (1978) Physiologiya Rastenii (English transl.: Sov Plant Physiol) 25: 1009
60. Cocking EC (1972) Ann Rev Plant Physiol 23: 29
61. Williamson FA, Fowke LC, Weber C (1977) Protoplasma 91: 213
62. Klein AS, Montezinos D, Delmer DP (1981) Planta 152: 105
63. Roberts H, Butt VS (1969) Planta 84: 250
64. Updegraff DM (1969) Anal Biochem 32: 420
65. Talmadge KW, Keegstra K, Bauer WD, Albersheim P (1973) Plant Physiol 51: 158
66. Peaud-Lenoel C, Axelos M (1970) FEBS Lett 8: 224
67. Herth W, Franke WW, Bittinger H (1974) Cytobiologie, 9: 344
68. Heiniger U, Delmer DP (1977) Plant Physiol 59: p 719
69. Raymond J, Fincher GB, MacLachlan JA (1978) Plant Physiol 61: 938
70. Maltby D, Caprita NC, Montezinos D, Kulow C, Delmer DP (1979) Plant Physiol 63: 1158
71. Dugger WM, Palmer RL (1980) Plant Physiol 65: 266
72. Pillonel C, Buchala AJ, Meier A (1980) Planta 149: 306
73. Rubin BA, Gavrilenko VF (1977) Biochemistry and physiology of photosynthesis. MGU Publishers, Moscow
74. Tarchevsky IA (1977) Fundamentals of photosynthesis. Vysshaya shkola, Moscow
75. Barber GA, Elbein AD, Hassid WZ (1964) J Biol 239: 4056
76. Franz G, Meier H (1969) Phytochemistry 8: 579
77. Stepanenko BN, Morozova AV (1969) Doklady Akad Nauk SSSR 187: 1425
78. Stepanenko BN, Morozova AV (1970) Physiologiya Rastenii (English transl.: Sov Plant Physiol) 17: 302
79. Delmer DP (1977) Recent Adv Phytochem 11: 45
80. Carpita NC, Delmer DP (1981) J Biol Chem 256: 308
81. Brett CT (1978) Plant Physiol 62: 377
82. Kursanov AL (1976) Transport of Assimilates in Plants. Nauka, Moscow
83. Dalessandro G, Northcote DH (1981) Planta 151: 53
84. Loewus F (1965) Fed Proc 24: 855
85. Roberts H, Loewus F (1966) Plant Physiol 41: 1489
86. Asamizu T, Nishi A (1979) Planta 146: 49
87. Asamizu T, Nishi A (1980) Physiol Plant 48: 207
88. Verma DC, Dougall DK (1979) Planta 146: 55
89. Harran S, Dickinson DB (1978) Planta 141: p 77
90. Hopp HE, Daleo GK, Romera PA, Lezica RP (1978) Plant Physiol 61: 248
91. Hopp HE, Romero PA, Daleo GR, Lezica RP (1978) Europ J Biochem 84: 561
92. Hopp HE, Romero PA, Lezica RP (1978) FEBS Lett 86: 259
93. Forsee WT, Elbein AD (1975) J Biol Chem 250: 9283
94. Delmer DP, Kulow C, Ericson MC (1978) Plant Physiol 61: 25
95. Elbein AD (1979) Ann Rev Plant Physiol 30: 239
96. Martin HG, Thorne KJS (1974) Biochem J 138: 277
97. Lezica RP, Romero PA, Hopp HE (1978) Planta 140: 177
98. Helsper GPFG (1979) Planta 144: 443
99. Montezinos D, Delmer DP (1980) Planta 148: 305
100. Haas D, Hackspacker G, Franz G (1985) Plant Sci Lett 41: 1
101. Currier HB (1957) Ann J Bot 44: 478

102. Meier H, Buchs L, Buchala AJ, Homewood T (1981) Nature 289: 821
103. Waterkeyn L (1981) Protoplasma 106: 49
104. Pillonel CH, Meier H (1985) Planta 165: 76
105. Kauss H (1987) Ann Rev Plant Physiol 38: 47
106. Fevre M, Rongier M (1981) Planta 151: 232
107. Robenek H, Peveling E (1977) Planta 136: 135
108. Olson AC (1964) Plant Physiol 39: 543
109. Chrispeals MJ (1969) Plant physiol 44: 1187
110. Lamport DTA (1970) Ann Rev Plant Physiol 21: 235
111. Dashek WV (1970) Plant Physiol 46: 831
112. Roberts H, Northcote DH (1972) Planta 107: 135
113. Gardiner M, Chrispeels MJ (1975) Plant Physiol 55: 536
114. Kawasaki S, Sato S (1979) Bot Mag, Tokyo 92: 305
115. Kawasaki S (1981) Plant and Cell Physiol 22: 431
116. Frey-Wyssing A, Muhlethaler K (1968) Ultrastructure of plant cell, Mir, Moscow
117. O'Kelley JC (1953) Plant Physiol 28: 281
118. Setterfield G, Bayley ST (1958) Exp Cell Res 14: 622
119. Brown RM jun., Franke WW, Kleing H et al. (1969) Science 166: 894
120. Robinson DG (1980) The role of endomembrane organels in plant cell wall synthesis. In: Cell compartmentation and metabolic chanelling, VEB Gustav Fischer, Jena, p 485
121. Shore G, Maclachlan A (1975) J Cell Biol 63: 557
122. Ray PM, Eisinger WR, Robinson DG (1976) Ber Dt Bot Ges 89: 121
123. Harris PJ, Northcote DH (1971) Biochim Biophys Acta 237: 56
124. Wright K, Northcote DH (1976) Protoplasma 88: 225
125. Bacic A, Delmer DP (1981) Planta 152: 346
126. Delmer DP, Benziman M, Padan E (1982) Proc Nat Acad Sci USA 79: 5282
127. Aloni J, Delmer DP, Benziman M (1982) Proc Nat Acad Sci USA 79: 6448
128. Ross P, Weinhouse H, Aloni J, Michaeli D, Weinberger-Ohara et al (1987) Nature, 325: 279
129. Brown RM jun., Willison JHM (1976–1977) Golgi apparatus and plasma membrane involvement in secretion and cell surface deposition, with special emphasis on cellulose biogenesis. In: International cell biology. Rockfeller Univ. Press, New York, p 267
130. Brown RM jun., Montezinos J, Montezinos D (1976) Proc Nat Sci USA 73: 143
131. Montezinos D, Brown RM jun. (1976) J Supramol Struct 5: 277
132. Mueller SC (1982) Cellulose-microfibril assembly and orientation in higher plant cells with particular reference to seedlings of *Zea maize*. In: Cellulose and other polymer systems. Plenum press, New York, London, p 87
133. Willison JHM, Brown RM (1977) Protoplasma 92: 21
134. Willison JHM, Grout BWW (1978) Planta 140: 53
135. Mueller SC, Brown RM jun., Scott TK (1976) Science 194: 949
136. Mueller SC, Brown RM jun., (1980) J Cell Biol 84: 315
137. Burgess J, Linstead PJ (1976) Planta 139: 85
138. Burgess J, Linstead PJ (1977) Planta 133: 267
139. Giddings TH jun., Brower DL, Staehelin LA (1980) J Cell Biol 84: 327
140. Wada M, Staehelin LA (1981) Planta 151: 462
141. Kiermayer O, Sleitr UB (1979) Protoplasma 101: 133
142. Frey-Wyssling A (1969) Fortschr Chem Org Naturst 27: 1
143. Marx-Figini M, Schulz GV (1966) Naturwissenschatten, 1966, 53: 466

144. Schnepf C (1988) Cell wall formation in moss protonemata. In: Abstracts of 6th FESSP Congress. Split, Yugoslavia
145. Blashek W, Koehler H, Semler U, Franz G (1982) Planta 156: 550
146. Marx-Figini M (1966) Nature 210: 754
147. Marx-Figini M (1982) The control of molecular weight distribution in the biogenesis of cellulose. In: Cellulose and other natural polymer systems. Plenum press, New York, London, p 243
148. Robinson DG, Quader H (1980) Europ J Cell Biol 84: 229
149. Quader H (1981) Naturwissenschaften 67: 428
150. Quader H, Robinson DG (1979) Eur J Cell Biol 20: 51
151. Willison JHM (1981) Secretion of cell wall material in higher plants. In: Encyclopedia of plant physiology. Plant carbohydrates, vol. 13b. Springer, Berlin, p 513
152. Haigler H, Benziman M (1982) Biogenesis of cellulose. 1. Microfibrils occurs by cell-directed self-assembly in *Acetobacter xylinum*. In: Cellulose and other natural polymer systems. Plenum press, New York, p 273
153. Benziman M, Haigler CH, Brown RM jun., White AR (1980) Proc Nat Acad Sci USA 77: 6678
154. Ledbetter MC, Porter KR (1963) J Cell Biol 19: 239
155. Hepler PK, Newcomb EH (1964) J Cell Biol 20: 529
156. Newcomb EH (1969) Ann Rev Plant Physiol 20: 253
157. Dustin P (1978) Cell shape In: Microtubules. Springer, Berlin Heidelberg New York, p 226
158. Hogetsu T, Shibaoka H (1978) Planta 140: 7
159. Hogetsu T, Shibaoka H (1978) Planta 140: 15
160. Lawson VR, Weintraub RL (1975) Plant Physiol 55: 1062
161. Hardham AR, Gunning BES (1978) J Cell Biol 77: 14
162. Quader H, Robinson DG (1981) Ber Dt Bot Ges 94: 75
163. Schliwa M (1976) J Cell Biol 70: 527
164. Hertel C, Quarder H, Robinson DG, Marme D (1980) Planta 149: 336
165. Lloyd CW, Slabas AR, Powell AJ, Lowe B (1980) Planta 147: 500
166. Grimm J, Sachs H, Robinson DG (1976) Cytobiol 14: 61
167. Schnepf E, Stein U, Deichgraber G (1978) Protoplasma 97: 221
168. Heat JB (1974) J Theor Biol 48: 445
169. Montezinos D, Brown RM (1979) Cytobiol 23: 119
170. Shibaoka H (1974) Exp Cell Res 15: 255
171. Volfova A, Chvojka L, Hankovska J (1977) Biol Plant Acad Sci Bohemosl 19: 421
172. Volfova A, Chvojka L (1981) The role of microtubules in cell wall formation. In: Plant growth and differentiation. Nauka, Moscow, p 20
173. Northcote DH (1969) Essays Biochem 5: 90
174. Ryser U (1979) Protoplasma 98: 223
175. Muzzarelli RAA (1977) Chitin. Pergamon press, Oxford
176. Herth W, Schnepf EE (1982) Chitin-fibril formation in algae. In: Cellulose and other natural polymer systems. Plenum press, New York, London, p 185
177. Ruiz-Herrera J (1982) Synthesis of chitin microfibrils *in vitro*. In: Cellulose and other natural polymer systems. Plenum Press, New York, London, p 207
178. Bracker CE, Ruiz-Herrera J, Bartnicki-Garcia S (1976) Proc Nat Acad Sci USA 73: 4570
179. Calvo C (1981) Mechanisms de synteus de quitosana en Mucer rouxii. Thesis, Centro de investigation y avanjados, Instituto Politicuico Nacional, Mexico
180. Kurosaki F, Tashiro N, Nishi A (1988) Plant Cell Physiol 29: 527

181. Tarchevsky IA, Lozovaya VV, Gordon LKh, Akhmerova VA (1982) Physiologiya rastenii. (English transl.: Sov Plant Physiol) 29: 895
182. Vakhmistrov DB (1969) Accumulation of ions by plants and cell membranes (potassium as an example). Thesis, Inst Plant Physiol, USSR Acad Sci, Moscow
183. Malone C, Koeppe DE, Miller RJ (1974) Plant Physiol 53: 388
184. Gordon LKh (1976) Respiration and water-ions regime of plant tissue. Nauka, Moscow
185. Matus VK, Vorobei AV, Chernitsky EA (1977) Biophysika 22: 861
186. Papahadjopoulos D, Jacobson K, Poste J, Shepherd J (1975) Biochim Biophys Acta 394: 504
187. Vorobei AV, Chernitsky EA (1977) Doklady Akad Nauk Belorussian SSR 21: 756
188. Ordin L, Hall MA (1968) Plant Physiol 43: 473
189. Kursanov AL (1982) Scientist and his Audience. Nauka, Moscow
190. Heiniger U, Kulow C, Delmer DP (1974) Enzymatic studies of cellulose synthesis using a detached fiber assay. In: Plant research. MSU, Michigan, US, p 107
191. Heiniger U, Delmer DP (1974) Enzymatic studies on cellulose synthesis in isolated membrane preparation. In: Plant research. MSU, Michigan, US, p 109
192. Tsai CM, Hassid WZ (1973) Plant Physiol 51: 998
193. Delmer DP, Heiniger U, Kulow C (1977) Plant Physiol 59: 713
194. Tarchevsky IA, Lozovaya VV, Markova MN (1980) Doklady Akad Nauk SSSR 255: 510
195. Wagner GJ (1979) Plant Physiol 64: 88
196. Skazkin FD, Lovchinovskaya EI, Miller MS, Anikiev VV (1958) Practical works in plant physiology. Sovetskaya Nauka, Moscow
197. Kursanov AL, Vyskrebentseva EI (1953) Biokhimiya (English transl.: Sov Biochemistry) 18: 448
198. Usmanov KhU, Tushkevich TI, Tillaev RS (1955) Physiologiya rastenii. (English transl.: Sov Plant Physiol) 2: 358
199. Usmanov KhU, Mingazieva TSh (1956) Doklady Akad Nauk Uzbek SSR 3: 27
200. Pinkhasov YuI (1981) Izvest Akad Nauk SSSR, Ser biol (English transl.: Sov News of USSR Acad Sci) 1981: 402
201. Prokof'ev AA (1980) Izvest Akad Nauk SSSR, Ser biol (English transl.: Sov News of USSR Acad Sci) 1980: 214
202. Raven JA (1976) Division of labour between chloroplasts and cytoplasm. In: The Intact Chloroplast. Elsevier, Amsterdam, p 403
203. Walker DA (1976) CO_2 fixation by intact chloroplast: Photosynthetic induction and its relation to transport phenomena and control mechanisms. In: The intact chloroplast. Elsevier/North-Holland, Amsterdam, p 235
204. Krause GH, Heber U (1976) Energetics of intact chloroplasts. In: The Intact Chloroplast. Elsevier/North Holland, Amsterdam, p 171
205. Tarchevsky IA, Lozovaya VV, Vorob'eva LZ (1982) Physiologiya i biokhimiya kult rastenii. (English transl.: Sov Plant Physiol Biochem Agric Plants) 14: 45
206. Tarchevsky IA, Lozovaya VV, Ruseva LG, Markova MN (1980) Physiologiya rastenii. (English transl.: Sov Plant Physiol) 27: 1052
207. Morosova A (1976) Izvest Akad Nauk Tajik SSR. Ser biol (English transl.: Sov News of Tajik Academy of Sci) 1976: 9
208. Tarchevsky IA (1964) Photosynthesis and drought. Kazan Univ. Publishers, Kazan
209. Lavergne D, Bismuth E, Champigni M (1974) Plant Sci Lett, 3: 391
210. Evans ML (1974) Ann Rev Plant Physiol 25: 195

211. Thimann KV (1974) Plant Physiol 54: 450
212. Went FW (1974) Ann Rev Plant Physiol 25: 1
213. Chailakhyan MKh, Khryanin VN (1982) Plant sex and its hormonal regulation. Nauka, Moscow
214. Polevoy VV (1982) Phytohormones. LGU Publishers, Leningrad
215. Chailakhyan MKh (1937) Hormonal theory of plant development. Academy of Sciences Publishers, Moscow-Leningrad
216. Kholodny NG (1939) Phytohormones: Essays on Physiology of Hormonal Phenomena in Plant Organism. Academy of Sciences of Ukrainian SSR Publishers, Kiev
217. Nishitani K, Masuda J (1981) Physiol plant 52: 482
218. Jacobs M, Ray PM (1976) Plant Physiol 58: 203
219. Ray PM (1977) Plant Physiol 59: 594
220. Bates GW, Cleland RE (1979) Planta 145: 437
221. Bates GW, Cleland RE (1980) Planta 148: 429
222. Cline MG, Evans ML, Rehm MM (1979) Plant Physiol 64: 679
223. Evans ML, Vesper MJ (1980) Plant Physiol 66: 561
224. Evans ML, Mulkey TT, Vesper MJ (1980) Planta 148: 510
225. Cleland RE (1976) Plant Physiol 58: 210
226. Johnson KD, Daniels D, Dowler MJ, Rayle DI (1974) Plant Physiol 53: 224
227. Fan DF, Maclachlan GA (1966) Canad J Bot 44: 1025
228. Osborne D (1968) Hormonal mechanisms regulating senescence and abscission. In: Biochemistry and Physiology of Plant Growth substances. Ottawa, p 815
229. Ridge I, Osborne DJ (1969) Nature 223: 318
230. Ferrari TE, Arnison PG (1974) Plant Physiol 54: 487
231. Tepfer M, Taylor I, MacKay A, Bloom M (1980) Plant Physiol 65: 105
232. Labavitch JM, Ray M (1974) Plant Physiol 54: 499
233. Rees DA, Wight NB (1969) Biochem J 115: 431
234. Rees DA, Wight NG (1971) J Chem Soc B.: 1366
235. Yamamoto R, Sakurai N, Shibata K, Masuda G (1980) Plant and Cell Physiol 21: 373
236. Katsu N, Kamisaka S (1981) Plant and Cell Physiol 22: 327
237. Dowler MJ, Rayle DL, Cande WZ, Ray PM, Durand H, Zenk MH (1974) Plant Physiol 53: 229
238. Boysen-Yensen P (1938) Plant growth hormones. Biomedgiz, Moscow, Leningrad
239. Leopold A (1968) Plant growth and development. Mir, Moscow
240. Torrey JG (1973) Science 181: 1075
241. Kefeli VI (1974) Natural Growth Inhibitors and Phytohormones. Nauka, Moscow
242. Kefeli VI (1978) Primary Mechanism of Integration and Growth of Plant Organism. In: Plant Growth. Primary Mechanisms. Nauka, Moscow
243. Moore TC Biochemistry and physiology of plant hormones. Springer, Berlin, 1979, vol 12
244. Abdul-Baki AA, Ray PM (1971) Plant Physiol 47: 537
245. Ray PM (1973) Plant Physiol 51: 601
246. Ray PM (1973) Plant Physiol 51: 609
247. Montague MJ, Ikuma H (1978) Plant Physiol 62: 391
248. Hall MA, Ordin L (1968) Auxin-induced control of cellulose synthetase activity in Avena coleoptile sections. In: Biochemistry and physiology of plant growth substance. Proc. 6th Intern. conf. plant growth subst. Runge press, Ottawa, p 659

249. Spencer FS, Shore G, Ziola B, Maclachlan GA (1972) Arch Biochem and Biophys 152: 311
250. Bevan M, Northcote DH (1981) Planta 152: 32
251. Beasley CA, Ting TP (1973) Amer J Bot 60: 133
252. Beasley CA, Ting TP (1974) Amer J Bot 61: 188
253. Birnbaum EN, Beasley CA, Dugger WM (1974) Plant physiol 54: 931
254. Nishitani K, Masuda J (1982) Physiol plant 56: 38
255. Naithani SL, Rama Rao N, Singh YD (1982) Physiol plant 54: 225
256. Chrispeals MJ (1976) Annu Rev Plant Physiol 27: 19
257. Willison JHM, Klein AS (1982) Cell-wall regeneration by protoplasts isolated from higher plants. In: Cellulose and other natural polymer systems. Plenum press, New York, London, p 61
258. Enzmann-Becker G (1973) Zeitschr. Naturforsch. 28 C: 470
259. Edwards GE, Huber SC, Guttieres M (1976) Photosynthetic properties of plant protoplasts.—In: Microbial and plant protoplasts. London, p 299
260. Ravinder G, Rashid A, Maheshvari SC (1981) Protoplasma 106: 351
261. Hanke DE, Northcote DH (1974) J Cell Sci 14: 29
262. Ryabushkina NA (1979) Physiologiya Rastenii (English transl.: Sov Plant Physiol) 26: 1135
263. Takeuchi Y, Komamine A (1978) Planta 140: 227
264. Blaschek W, Haas D, Koehler H, Franz G (1981) Plant Sci Lett 22: 47
265. Lozovaya VV, Gorshkova TA, Raimanov IT, Salnikov VV, Khusainov MB, Tarchevsky IA (1982) Doklady Akad Nauk SSSR 266: 229
266. Tarchevsky IA, Lozovaya VV, Gorshkova TA et al. (1988) Physiologiya rastenii 35: 129
267. Willison JHM, Brown RM jun. (1978) Planta 141: 51
268. Roland JC, Vian B (1979) Intern Rev Cytol 61: 129
269. Burgess J, Linstead PJ, Bonsall VE (1978) Planta 139: 85
270. Roland JC, Prat R (1973) Colloq intern CNRS 212: 243
271. Willison JHM (1973) Colloq intern CNRS 212: 215
272. Burgess J, Fleming EN (1974) J Cell Sci 14: 439
273. Robinson DG, Schlosser UG (1978) Planta 141: 83
274. Morre DJ, Franke WW, Deumling B, Nyquist S, Oytracht L (1971) Golgi apparatus function in membrane flow and differentiation: Origin of plasma membrane from endoplasmic reticulum. In: Biomembranes. Proc Symp Membranes and coordination of cellular activities. Plenum press, New York, vol. 2, p 94
275. Morre DJ, Mollenhauer HH (1974) The endomembrane concept: A functional integration of endoplasmic reticulum and Golgi apparatus. In: Dynamic aspects of plant ultrastructure. McGraw-Hill, New York, p 85
276. Mollenhauer HH, Moore DJ (1976) Cytobiologia 13: 297
277. Sawano F, Fujita H (1981) Histochemistry 71: 335
278. Northcote DH (1974) Philos Trans Roy Soc London A 269: 119
279. Willison JHM, Cocking EC (1975) Protoplasma 84: 147
280. Wilkinson MJ, Northcote DH (1980) J Cell Sci 42: 389
281. Meyer G, Abel VO (1975) Planta 125: 1
282. Willison JHM (1976) Protoplasma 88: 187
283. Asamizu T, Tanaka K, Takebe J, Nishi A (1977) Physiol Plant 40: 215
284. Robenek H (1980) Biol Zentr.-Bl 99: 13
285. Hereward FV (1974) Exp Cell Res 87: 213

286. Weinecke K, Glas R, Robinson DG (1982) Planta 155: 58
287. Nagata T, Yamaki T (1973) Ztschr Pflanzenphysiol 70: 452
288. Herth W, Meyer I (1977) Biol Cell 30: 33
289. Takeuchi Y, Komamine A (1981) Plant and Cell Physiol 22: 1585
290. Takeuchi Y, Komamine A (1982) Plant and Cell Physiol 23: 249
291. Takeuchi Y, Komamine A (1980) Physiol Plant 48: 271
292. Takeuchi Y, Komamine A, Sato T, Watanabe K, Morikawa H (1980b) Physiol Plant 48: 536
293. Takeuchi Y, Komamine A (1980) Physiol Plant 50: 113
294. Burgess J, Watts JW, Fleming EN, King JM (1973) Planta 110: 291
295. Popova PYa (1975) Biology of Development and Technological Properties of Cotton Fiber. Fan, Tashkent
296. Usmanov AN (1972) On accumulation and assimilation of various forms of nitrogen by cotton plant. In: Problems of physiology and fertilization of cotton plants. Uzbekistan, Tashkent, p 59
297. Belousov MA (1973) Studies on physiology of cotton-plant root nutrition. In: Progress in cotton-growing in Uzbekistan in 50 years. Fan, Tashkent, p 180
298. Nazirov NN, Akbarov D, Koshaev TKh, Tillabaev A (1976) Response of several new cotton varieties to aqueous-nutritional conditions. In: Physiology, Biochemistry, and Radiobiology of cotton plant. Fan, Tashkent, p 8
299. Belousov MA, Isaev BM (1976) Physiological basis of optimization of cotton-plant mineral nutrition. In: Physiology and biochemistry of cotton plant. Proc SojuzNIKHI, Tashkent, p 3
300. Azimov RA (1973) Physiological role of Calcium in cotton-plant salt resistance. Fan, Tashkent
301. Willison JHM (1983) J Applied Polymer Sci 37: 91
302. Colvin JR (1980) Planta 149: 97
303. Grout BWW (1975) Planta 123: 275

Part II

Cellulose Structure

1 General Notes and Definitions

Many problems arising from the investigation of cellulose and products of its processing are related to their stereochemistry. Therefore one should know static and dynamic aspects of stereochemistry. Static stereochemistry takes into account the spatial disposition of the atoms in the molecule in the static state (constitution, configuration), while dynamic stereochemistry deals with the spatial relations between atoms and groups in a molecule involved directly in the course of chemical transformations which are due to the distinct effect of these relations on the equilibrium conditions and reaction rate (conformation, structure).

It is necessary in this connection to define the usage of some ordinary stereochemical terms such as "constitution", "configuration", "conformation" and "structure", since they are often used rather ambiguously for cellulose-type polymers [1–6].

The following definitions required for consideration of cellulose stereochemistry are used in this book.

1. *Constitution of a molecule* defines the nature and sequence of the linkages in a molecule, i.e. the architecture of the main chain of a molecule. Thus the basic stereochemical information involves the data about linearity or branching of the main chain, its carbon-chain or heterochain character, the presence of chiral carbon atoms, etc.
2. *Configuration of a molecule* specifies the spatial disposition of atoms or groups of atoms without consideration of possible rotation around one or several single covalent bonds and it is completely represented by a three dimensional model or three dimensional or projection formulae. Thus the configuration is the basic term of the static stereochemistry.
3. *Conformation* defines spatial disposition of atoms realized with regard to the certain limitations of rotation around single bonds, the position of which is determined according to the structure of the corresponding molecule. In this connection one should keep in mind that an infinite number of non-identical conformations is possible for even a not very complex system due to rotation around single bonds. However the consideration is usually limited by the definite conformations which correspond to the minimum (gauche conformations) or maximum (eclipsed conformations) of the potential energy. Thus, the conformation is the basic term of dynamic stereochemistry.
4. *Structure*—this is an integral term comprising constitution, configuration, and conformation. While the elucidation of the structure of simple molecules

requires the information about constitution and configuration, the structure of complex systems can be determined only if the constitution, configuration, and conformation are known.

These definitions are rather comprehensive in the case of simple molecules. But when we transfer them to polymers, in particular of rigid-chain types like cellulose, these terms become essentially more complex due to the overlapping effects of intra- and intermolecular interactions caused by the linear sizes of the polymeric chain, its rather random spatial disposition, and the presence of chemically different functional groups. These specific features of polymers introduce new meaning into the terms "configuration", "conformation", "structure".

The configuration in this case is the integral characteristic of an elongated chain which is composed from local configurations of separate chain fragments. The hierarchy of configuration comprises four basic levels; *the configuration of the unit*; *the short-range configuration order* i.e. the configuration of the addition of adjacent units; *the long-range configuration order*, i.e. the length and distribution of branches, the uniformity of the units distribution along the chain, their blocking etc.; *the configuration of the elongated chain as a whole* defined by mutual disposition of large sequences of the units, which may be fully characterized by long-range configuration order. Thus the configuration of the chain as a whole is the superposition of local configurations. The conformation in this case is a physical characteristic derived from the configuration. In general it is the variable distribution in space of the atoms and atomic groups which form the polymer molecule. When internal rotation is fully hindered the conformation coincides with configuration. The polymeric chain is usually able to assume the infinite number of "instant" conformations which arise continuously due to solely internal thermal motion. The conformation-like configuration is also characterized by a similar set of hierarchic levels comprising *the conformation of the unit*; *the short-range conformation order* governed by the interaction of neighbouring units; *the long-range conformation order* governed by the additional influence on the conformation of elongated chains by the spacial interaction of side substituents which happen to be positioned close to each other and because of this the simple rotation around C—C bonds prove to be insufficient to provide the minimum energy of interaction. The appearance of a spiral conformation is partially caused by this effect. The conformation of the chain as a whole is naturally the superposition of the lowest conformation levels.

The knowledge of the constitution, configuration and conformation of the polymer chain makes it possible to determine the molecular structure of the polymer. However, this concept does not characterize completely the polymer structure, since it neglects the effects of intermolecular (interchain) interactions. These effects result in the formation of associates from a great number of chains with such elements of morphological structure as fibrils, regions with various accessibility for some reagents, amorphous and crystalline fractions, and some other forms usually called supermolecular.

The morphological structure of the polymer comprises the set of spatially distributed discrete structural elements possessing the certain thermodynamic and kinetic stability that provides retention of their forms and dimensions upon some macroscopic transformations of the polymer [7]. One can deduce from this definition the stability and the autonomy of the elements of the morphological structure as well as their ability to make a transition from one level of supermolecular organization to another without substantial changes. The basic morphological form, which adequately indicates the orientation order of the polymer chains associated by interchain interaction forces, is the microfibril.

Depending on the conditions, the structure of microfibrils can be different but they are always complex polymolecular formations. The specific feature of microfibrils is the preferred orientation of macromolecules within them along the main axis. The presence of the distinct side borders is also substantial.

Thus, the microfibril is a lengthy aggregate of macromolecular chains stacked in a parallel manner due to intermolecular interaction and they have a real border and developed internal structure. The monocrystal is the most perfect morphological shape of the crystalline polymers. Usually it is a flat, plate-like lamella with a thickness of about 100 Å. The plate dimensions may attain several micrometers. It is important that polymeric monocrystals can only be obtained from very dilute solutions upon slow crystallization.

Since the thickness of the plate-like lamella is markedly (more than one order of magnitude) less than the macromolecule length, they must be multiply folded. This is the basic peculiarity of the polymer crystals distinguishing them from those of low molecular weight compounds. It is natural that the regions of folding cannot have a crystalline structure, therefore the plate-like monocrystal usually comprises the central part, where the macromolecules are packed into a crystalline lattice, and disordered at the border regions where the macromolecules are folded. It was mentioned above that polymer monocrystals can be obtained only during slow crystallization from diluted solutions. Enhancement of either the crystallization rate or the solution concentration results in the formation of more complex morphological forms—spiral formations and "twins" (two plates, bound at the crystallographic plate) as well as various dendrite forms.

Upon further increase of concentration and crystallization from the melt, spherulites are formed. They are the most common and general form of supermolecular formations in polymers. The dimensions of spherulites may vary from several micrometers up to several millimeters and in some cases even centimeters, depending on the crystallization conditions. It is substantiated that the degree of crystallinity of the spherulite, used to correlate that of the polymer as a whole, demonstrates the existence of regions with disordered amorphous structure within the spherulite. Thus, *the spherulite* is a morphological form—a polycrystalline formation with spherical symmetry comprising the separate crystallites. It is important that the polymeric monocrystal as a flat plate-like lamella can also comprise the separate crystallites. However their formation occurs, in this case (in contrast to spherulite formation), in conditions which are

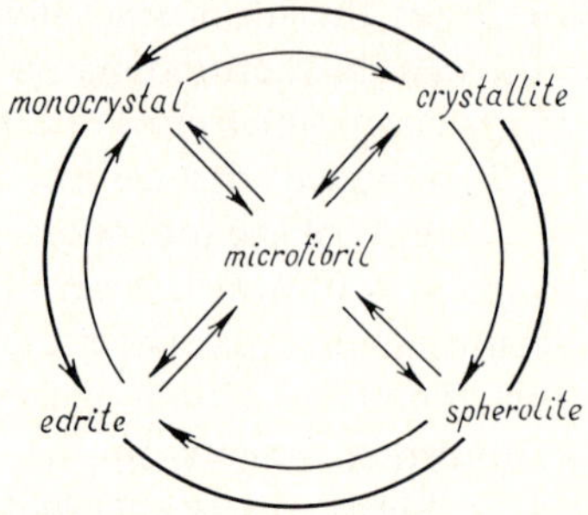

Fig. II.1. Interconversions of polymer morphological forms

more favorable thermodynamically due to creation of the more perfect orientation order.

Thus, the crystallites are the smallest discrete elements, which are able to exist separately, of any stable supermolecular organization in polymers. They represent the intermediate morphological form based on the hierarchy of supermolecular formations between spherulite and polymeric monocrystal. Like spherulites they are also comprised of structural elements with the chains stacked in parallel, but in contrast to polymeric monocrystals, the parallel stacking in one region of macromolecular chain aggregates may combine with distortions of parallel arrangement in other regions. This takes place, for example, upon chain folding. Among supermolecular formations under consideration which characterize the certain level of hierarchy of supermolecular organization, *the edrites* take a specific position [8]. Their specific features involve the structure, consisting of aggregates from the great number of lamellae, and are similar in shape to single crystals although they are essentially larger in dimensions (their cross-section is usually higher than 1 μm). Meanwhile, with reference to the complexity in the crystallographic sense, the edrites, like crystallites, take the intermediate position between monocrystals and spherulites.

Thus the supermolecular organization, as the general basis of the morphological structure, is defined as the required set of interactions between morphological forms and the certain structural elements of the polymacromolecular associates which produce macroscopic polymeric matter taking into consideration their internal structure and mutual spatial disposition.

According to the general definition of the morphological structure of polymers given at the beginning of the chapter, each morphological form has the certain thermodynamic and kinetic stability that predetermines its ability to interconversions without change of individual properties. The general scheme of such interconversions is presented in Fig. II. 1.

2 The Constitution of Cellulose

The constitution of cellulose has been studied up to now rather comprehensively using the common methods of carbohydrate chemistry (hydrolysis, methylation, periodate oxidation). Acid or enzymatic hydrolysis of polysaccharides results in the formation of free monosaccharides, the ratio of which indicates the monomer (sugar) composition of the initial polysaccharide. Partial acid or enzymatic hydrolysis results in the formation of oligosaccharides, the structure of which exhibit fragment composition of the initial polysaccharide. It is important that the oligosaccharides obtained after partial hydrolysis, although less complex objects for structural analysis, reflect rather strictly the architecture of the main chain of the initial molecule. At the same time, the determination of molecular weights of the initial polysaccharide and the oligosaccharide allows us to specify whether the corresponding oligosaccharide is formed due to hydrolysis of the glucosidic linkages of the main chain or due to cleavage of the hydrolytically unstable side or terminal substituents or groups. Upon acid hydrolysis of cellulose usually only D-glucose is formed [9].

The methylation method involved methylation of all hydroxyl groups of a polysaccharide followed by hydrolytic cleavage of the glycoside bonds in the methylated polysaccharide to give the corresponding methylated monosaccharides [10]. The nature and the ratio of methylated monosaccharides demonstrate the relative content of nonreducing terminal groups, degree of branching, the character of interchain links, and the type of predominant bonds in the polysaccharide. The hydrolysis of methylated linear glucan with β-1–4-bonds (**1**) is given as an example [11] (Fig. II. 2).

The nonreducing terminus affords 2,3,4,6-tetra-*O*-methyl-D-glucose (**2**), whereas reducing one and central units give 2,3,6-tri-*O*-methyl-D-glucose (**3**) with free hydroxyl group at C-4. The determination of the ratio between the amount of three- and tetramethyl glucose derivatives shows that in the cellulose molecule one terminal group corresponds to about 1500 glucose residues. Thus natural cellulose is the linear macromolecule with about 3000 glucose units bound mainly by 1–4 bonds.

The periodate cleavage of 1,2-diol and 1,2,3-triol groups is the widely used method for the determination of the nature of terminal groups and the type of the glycoside bonds [11]. A typical scheme of periodate oxidation is given in Fig. II. 3.

This reaction, usually carried out in aqueous solution, proceeds quantitatively. Depending on the type of linkage of glycopyranose residues in the polymer

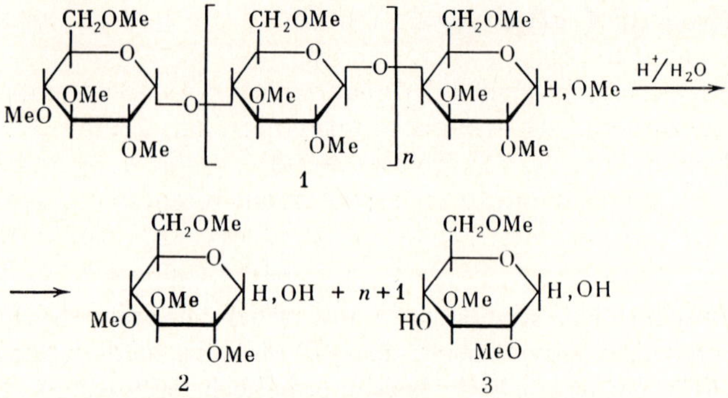

Fig. II.2. The hydrolysis of methylated cellulose

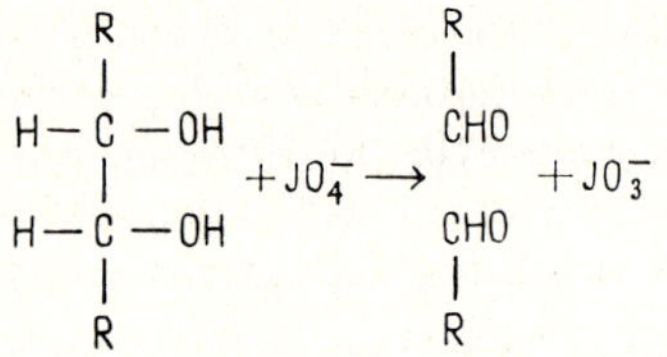

Fig. II.3. Periodate oxidation of polysaccharides

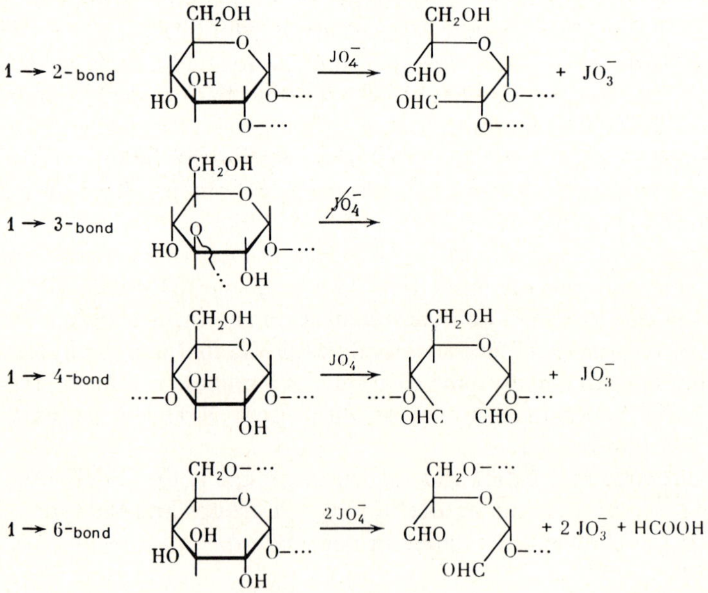

Fig. II.4. Influence of the type of linkage of glucopyranose residues in the polymer chain on the chemical nature of final products chain, the direction of oxidation and thus the character of the final products formed are significantly changed (Fig. II. 4).

Since every glucose residue undergoes individual oxidation the determination of periodate consumption and the amount of formic acid formed together with the study of the periodate-oxidized polysaccharide results in detailed information about the nature and the ratio of various bonds in initial polysaccharide. The behavior of terminal monoglucoside residues upon periodate oxidation is illustrated by an example of glucan with β-1–4-bonds (in Fig. II. 5).

The oxidation of nonreducing terminal glucose leads to consumption of 2 mole of periodate and formation of 1 mole of formic acid. The reducing terminal glucose consumes 1 mole of periodate with formation of 1 mole of formic acid and the acid ester. The oxidation of every central glucose residue leads to the consumption of 1 mole of periodate without formation of low molecular weight products. Thus the only side reaction resulting in non-selective oxidation is hydrolysis of formic acid ester (**2**) which is obtained in the terminal reducing residue in the course of the primary stage oxidation (**1**) (Fig. II. 6).

Upon hydrolysis of this ester, 1 mole of formic acid is released and 1,2-diol (**3**) is formed which is oxidized by 1 mole of periodate with the formation of 1 mole of formaldehyde from primary alcohol group and substituted derivative of malonic dialdehyde (**4**). The active hydrogen of the product (**4**) is oxidized by 1 mole of periodate with formation of hydroxyaldehyde (**5**) which is cleaved by 2 moles of periodate releasing 2 moles of formic acid and forming intermediate (**6**). This compound eliminates CO_2, giving rise to hydroxyaldehyde (**7**) which interacts with 1 mole of periodate and affords formic acid ester (**8**), similar to ester (**2**). Thus, the sequence of reactions 1–8 is repeated and results in stepwise degradation of the chain.

The above data allows us to consider the pure cellulose as the unbranched linear polymer of 1–4 bonded D-glucose (Fig. II. 7).

The additional data obtained by partial acid hydrolysis, enzymatic cleavage and acetolysis [3], show that probability of other bonds in the cellulose molecule does not exceed one bond per 1000 glucose units.

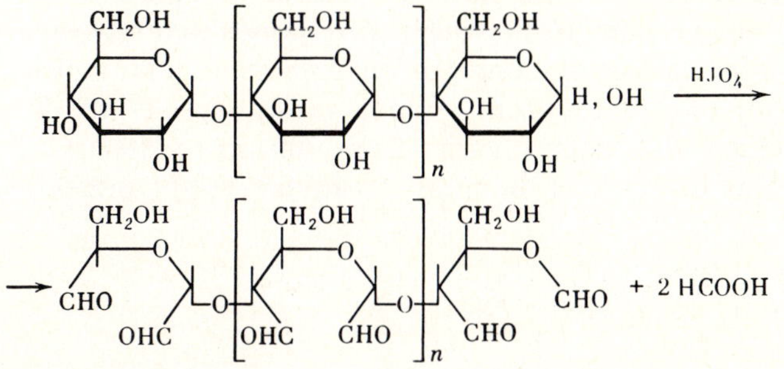

Fig. II.5. The behaviour of terminal monoglucoside residues upon periodate oxidation of cellulose

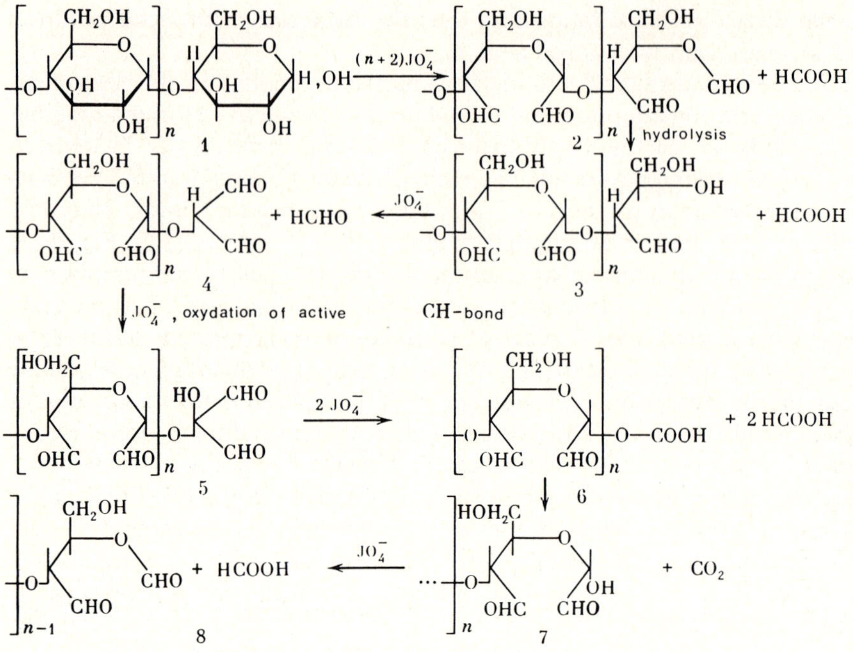

Fig. II.6. Configurational scheme of the oxidation of cellulose and products of the reaction

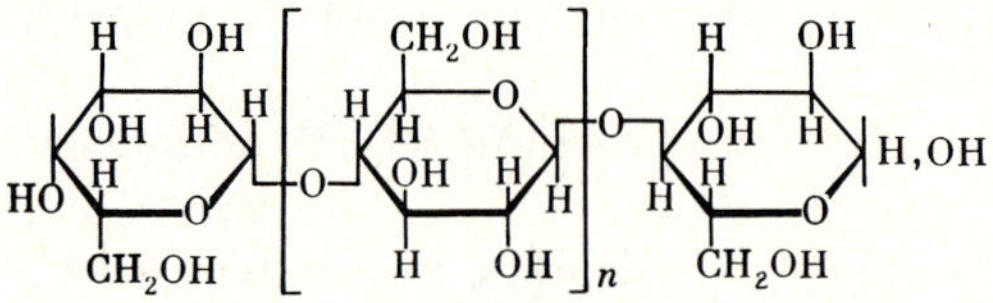

Fig. II.7. Structural formula of the cellulose macromolecule

The degree of polymerization (DP) of natural cellulose preparations obtained in mild conditions may exceed 10000 which corresponds to a molecular weight of about 10^6 [12]. In the case of more rigorous treatment, taking place, for example, upon isolation of cellulose from wood of various species, the DP value is markedly decreased and usually varies from 300 up to 5500 [13], resulting in a corresponding decrease of the mean molecular weight values.

3 Cellulose Configuration

According to optical rotation data of its solutions [2] glucose comprises the equilibrium mixture of α- and β-anomeric forms, the optical rotation of which varies from $[\alpha]_d^{20} + 111°$ up to $[\alpha]_d^{20} + 19.2°$ (Fig. II. 8).

Taking into account the configuration of aldo- or ketopyranoses the scheme of glucose anomerization can be presented as in Fig. II. 9.

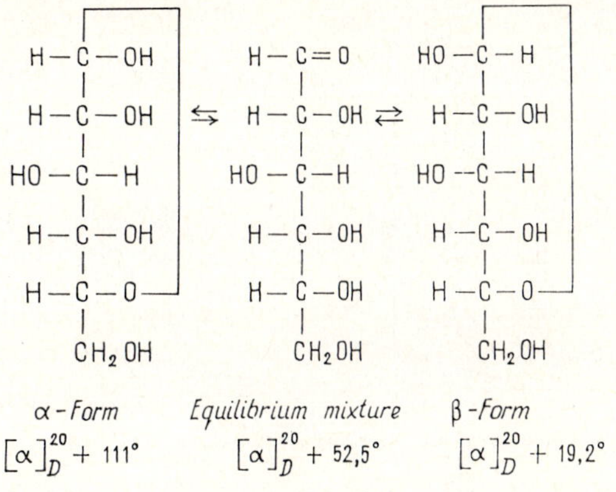

Fig. II.8. Configurational scheme of the equilibrium mixture of the glucose solution

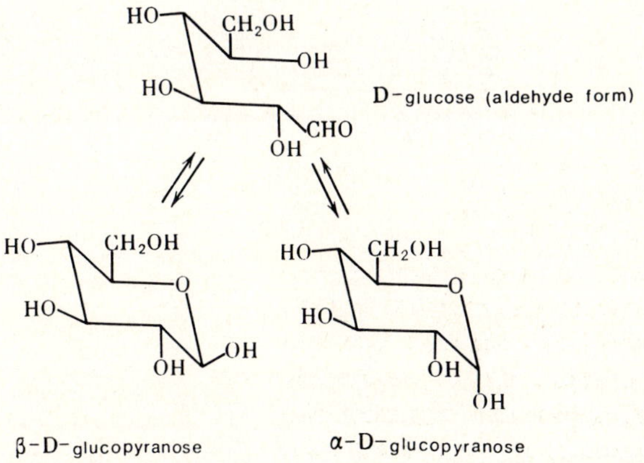

Fig. II.9. The anomerization of aldo- and ketopyranose solutions

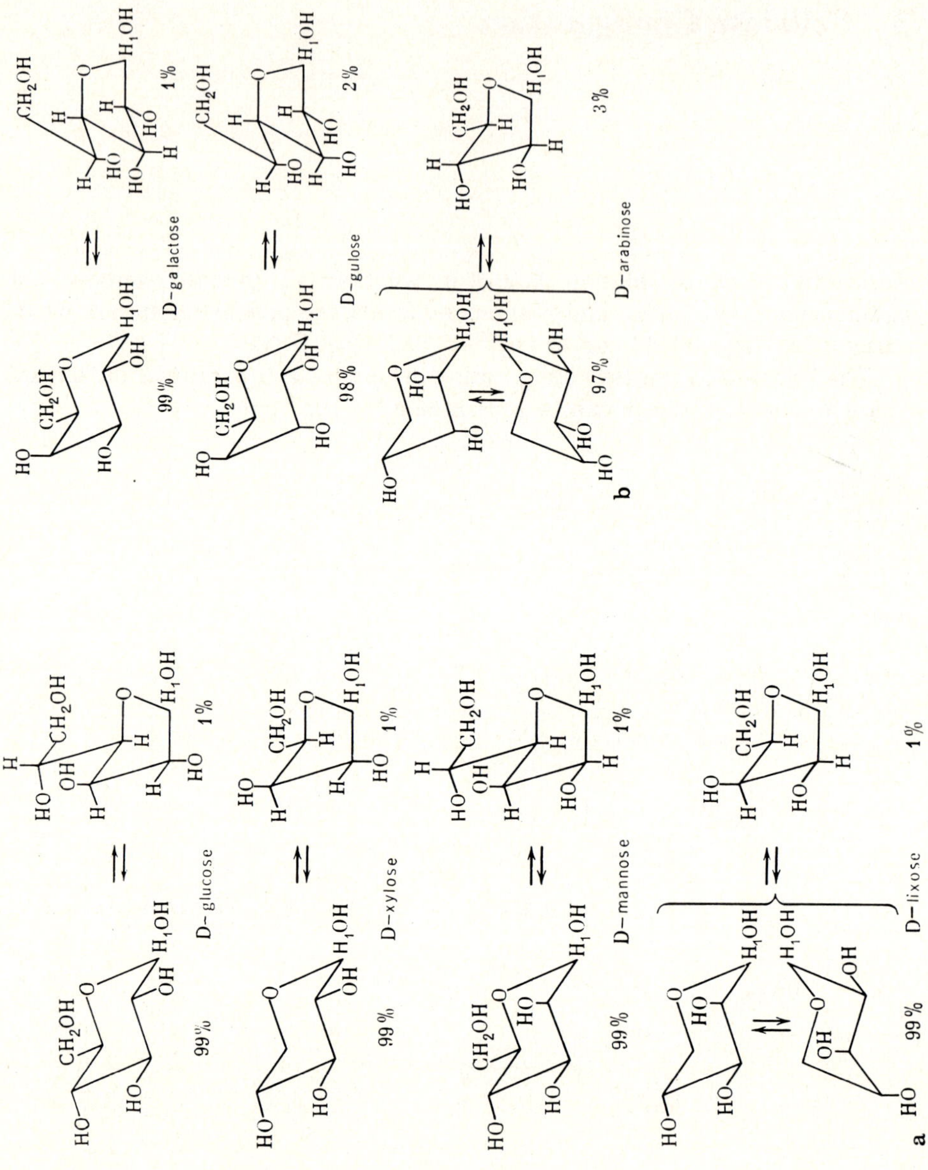
a
D-glucose
99%
1%
D-xylose
99%
1%
D-mannose
99%
1%
D-lixose
99%
1%
b
D-galactose
99%
1%
D-gulose
98%
2%
D-arabinose
97%
3%

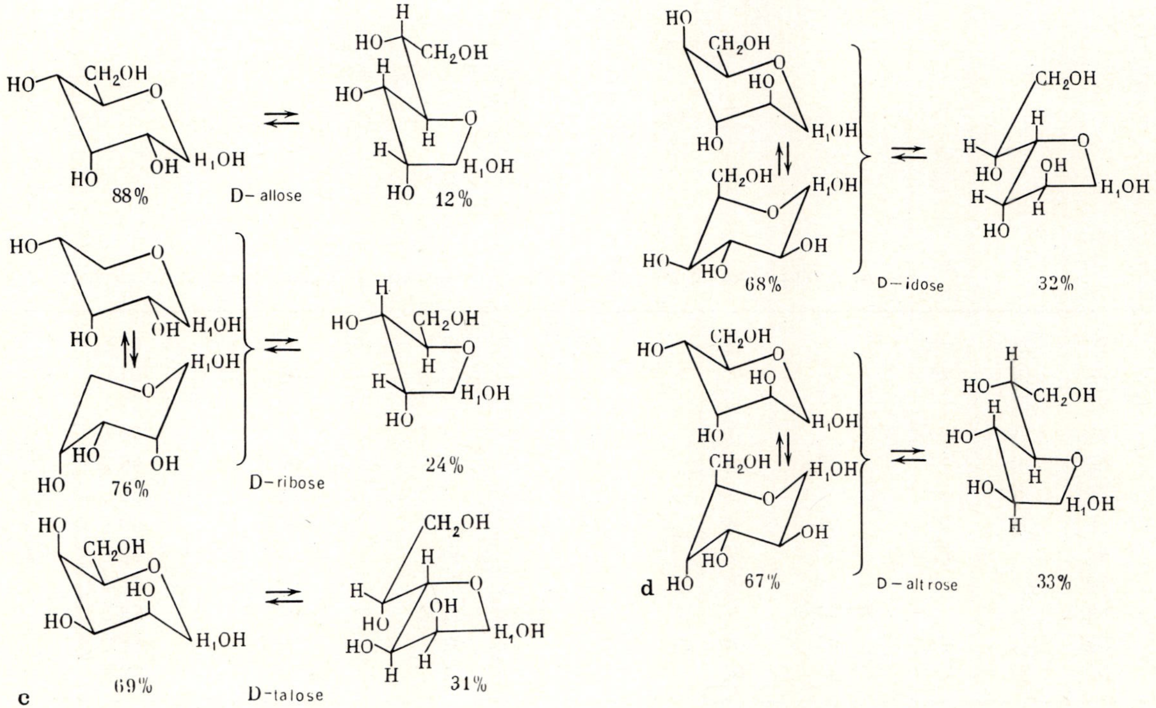

Fig. II.10a–d. Stereochemical formulas of D-monosaccharides and schemes of corresponding equilibria in aqueous solutions.

a D-glucose, D-xylose, D-mannose, D-lyxose;
b D-galactose, D-gulose, D-arabinose;
c D-allose, D-ribose, D-talose;
d D-idose, D-altrose

Table II.1. The relative free energies, content of α-anomer and equilibrium composition of the aqueous solutions of aldohexoses and aldopentoses (according to PMR-spectroscopy [15])

				Content of α-anomer (%)		Furanose (%)		Pyranose (%)	
Monosaccharides[a]	G^0_α (kcal/mole)	G^0_β (kcal/mole)	G^0pyranose (kcal/mole)	Calculated	Found	α	β	α	β
D-allose	3.85	2.95	2.85	18	20	5	7	18	70
D-altrose	3.35	3.35	2.95	50	40	20	13	27	40
D-galactose	2.85	2.50	2.25	36	27	traces		36	64
D-glucose	2.40	2.05	1.80	36	36	—		—	64
D-gulose	3.85	3.05	2.85	21	22		22		78
D-idose	3.65	4.00	3.40	64	46	16	16	31	37
D-mannose	2.50	2.95	2.25	68	67	—	—	68	32
D-talose	3.55	4.00	3.30	68	58	20	11	40	29
D-arabinose	1.95	2.20	1.65	60	63		3	63	34
D-lyxose	1.85	2.40	1.65	72	71	—	—	72	28
D-ribose	3.10	2.30	2.15	21	26	6	18	20	56
D-xylose	1.90	1.60	1.35	37	33	—	—	37	63

[a] Formulae of monosaccharides and schemes of corresponding equilibrium are given in Fig. II.10

This scheme of anomerization is generally accepted now [14]. It should be noted that a more complex mixture of configuration isomers is formed in aqueous solution (Table II.1, Fig. II.10 (1, 2, 3, 4)).

However, glucofuranoses are practically absent in glucose aqueous solutions. This fact is primarily caused by the level of relative free energy of corresponding aldoses, as it follows from the recent data about the equilibrium composition of aqueous solutions of variety of aldohexoses and aldopentoses

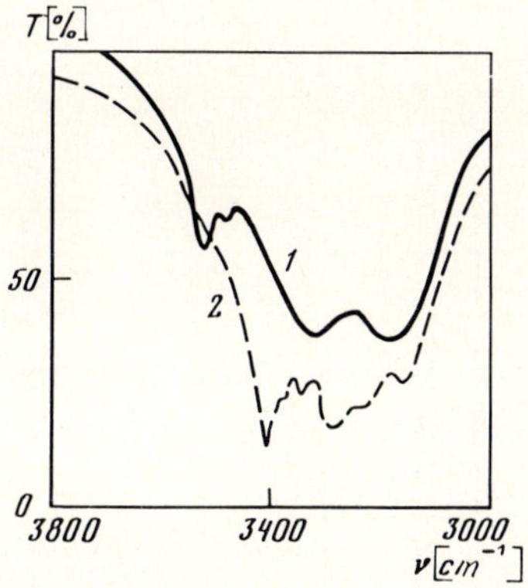

Fig. II.11. IR-spectra of β-D-glucose (*1*) and α-D-glucose (*2*) at liquid nitrogen temperature

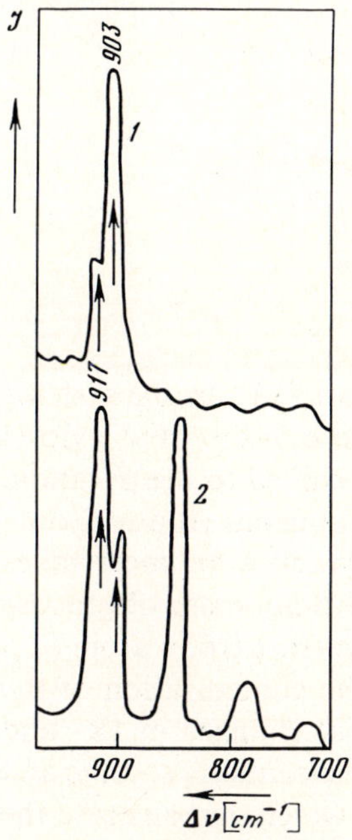

Fig. II.12. Raman spectra of *β*-D-glucose (*1*) and α-D-glucose (*2*)

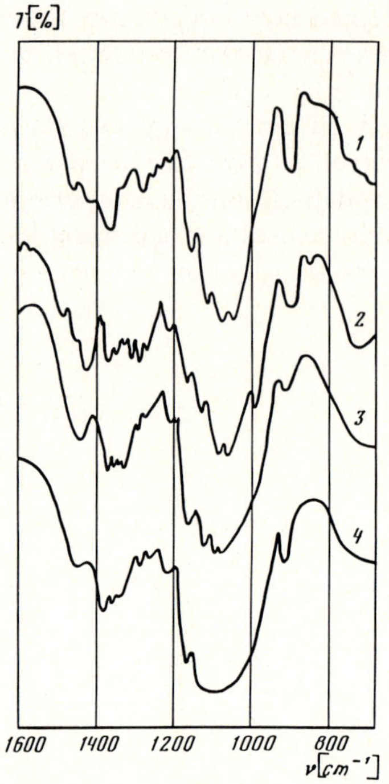

Fig. II.13. IR-spectra of β-D-glucose (*1*), cellobiose (*2*), cellulose (*3*) and hydrate cellulose (*4*)

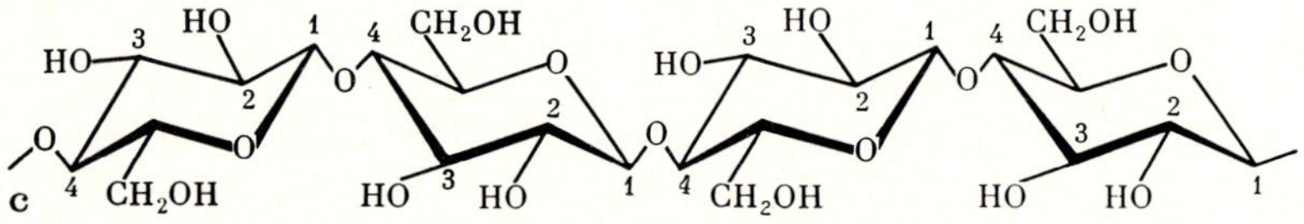

Fig. II.14. Configuration of the cellulose macromolecule

obtained by PMR-spectroscopy and reviewed by Stoddart [15]. In the systems where free energy is minimal (galactose, glucose, mannose, lyxose and xylose) the equilibrium pyranose–furanose is almost completely shifted to the pyranose anomers (see Table II.1). Thus, in the case of glucose in aqueous solution, only anomerization up to equilibrium between two glucopyranose anomers takes place. At the same time the difference between α- and β-anomers of glucose consists of the position of the hydroxyl group at the anomeric carbon atom (C_1) in relation to the plane of the ring. The alteration of the configuration of the OH-group at C_1 of glucopyranose results in substantial changes in IR- and Raman spectra [16, 17] (Fig. II.11, 12). These changes are essentially specific and they are distinctly seen in cellulose IR-spectra [18, 19], which demonstrate the β-configuration of its glucoside linkage (Fig. II.13). Taking into account these data as well as the axial and equatorial position of hydroxyl groups in the glucose residue [20], the configuration of the cellulose macromolecule can be represented as in Fig. II.14.

The specific features of cellulose are implied from a given configuration of its molecular chain, they govern mechanical and physico-chemical properties of this natural polymer. These features involve a linear structure of the chain, its rigidity and the possible realization of various associates, caused by the forces of intra- and intermolecular interactions.

4 Cellulose Conformation

4.1 Conformation of Glycosidic Centers

Like cyclohexane derivatives, the glucopyranose ring may have the chair or boat conformation [2]. It is generally accepted that the more stable conformation of cyclohexane is the chair one with D_{3d}-symmetry (Fig. II.15). The bonds parallel to the C_3-axis are usually called axial (a), those forming a tetrahedral angle with this axis—equatorial (e). The cyclohexane ring actually is not tetrahedral and the endocyclic angle C–C–C seems to be 111.5° (Fig. II.16). With regard to this value the torsion angle C–C–C (φ) happens to be 54.5°. The torsion angle between the adjacent axial and equatorial bonds is also 54.5°. Taking into account that the H–C–H angles in cyclohexane are less than tetrahedral the value of the projection angle H–C–H (ω) is accepted as being 118°.

With regard to these assumptions the angle between adjacent equatorial bonds is 63.5°. Due to these specific features, some flattening of the cyclohexane ring and deviation of axial bonds from the main C_3-axis of about 4° are observed. These calculations are confirmed by NMR data [15, 21]. It is essential that axial and equatorial bonds can transfer into each other due to confluent interconversion (Fig. II.17).

Such conversion involves pseudorotation-type transition [22] via some intermediate conformations with higher energy than that of the chair conformation. This process consists of the simultaneous and continuous changes of torsion angles in such a way that every ring carbon atom occupies sequentially every possible position in the ring. This is illustrated by the data given in Fig. II.18, where it follows that the chair conformation has no angular tensions. The stresses in the ring due to shielding of the bonds are also excluded and the potential energy (Fig. II.18, point *a*) is minimal. The transfer of conformation from chair to boat is accompanied by distortion of angles between valent bonds, since simultaneous rotation around several bonds should occur on this transition. In this connection, the molecule passes through a state with high potential energy (Fig. II.18, point *b*). Although the distorted boat conformation (Fig. II.18, point *c*) is less stable (the difference in energies is 5.5 kcal/mole) than the chair conformation, it nevertheless corresponds to the energy minimum on the pseudotransformation pathway. Thus the enantiomeric conformers as distorted boats (Fig. II.18, points *b* and *c*) are converted into each other via the boatshape intermediate, the energy of which exceeds that of the chair conformation by 6.4 kcal/mole. The situation on the pseudotransformation pathway at the points

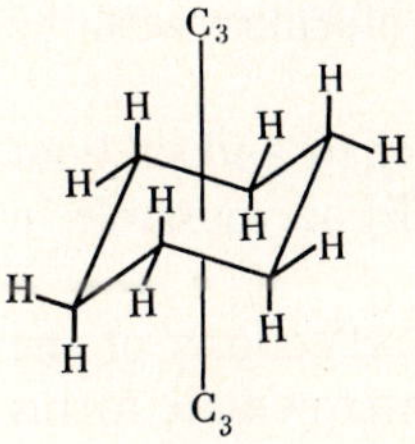

Fig. II.15. Chair configuration of the cyclohexane molecule with D_{3d}-symmetry

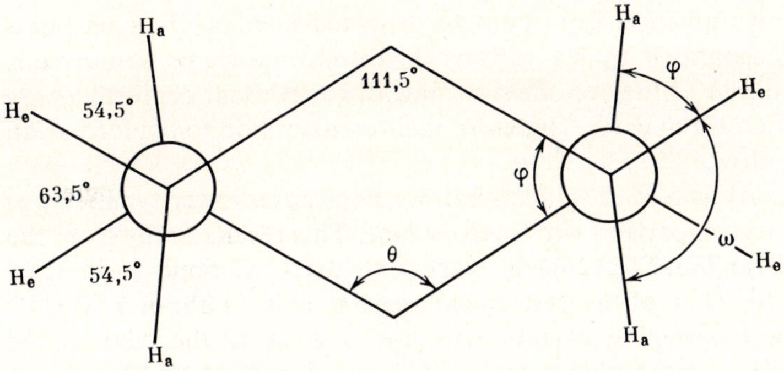

Fig. II.16. Newman projection of the chair conformer of cyclohexane
θ—endocyclic angle, *a*—axial bonds, *e*—equatorial bonds

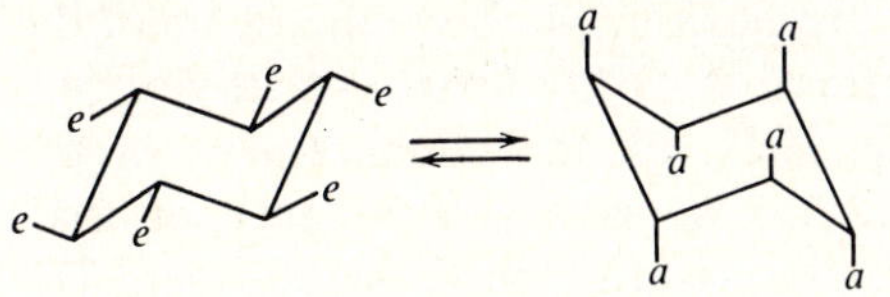

Fig. II.17. Interconversion of axial and equatorial bonds of the monomeric unit

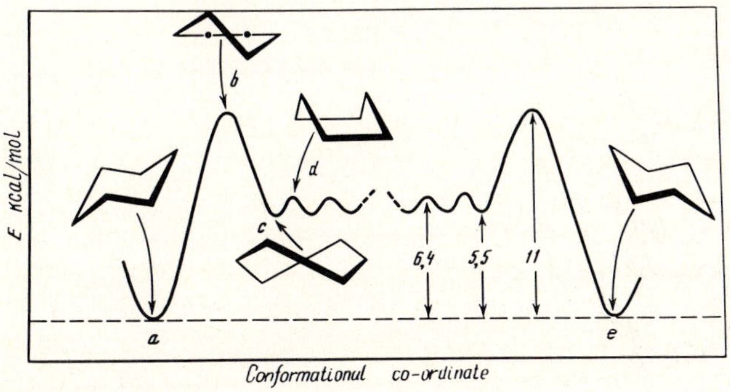

Fig. II.18. Potential energy of various conformations of cyclohexane [15]. E—kcal/mole; conformation coordinate

a and *d* reflects the confluent interconversion of chair-shape conformers with the transition of axial and equatorial linkages.

Since the glucopyranose residue has an oxygen atom, two chair and six boat conformations denoted as C1(1C) and B1(1B), B2(2B), B3(3B) are possible in principle (Fig. II.19).

The conformations 1C, 1B, 2B and 3B are the mirror reflections of the conformations C1, B1, B2 and B3. In some conditions the intermediate forms "semichair", "semiboat" etc. can be realized (Fig. II.20).

As follows from the data given in Fig. II.20 the cycle of interconversion of pyranose ring upon transition from boat to distorted boat involves six boat-shape conformers separated by six various distorted boat-shape conformers. However, as in the case of the cyclohexane ring, distorted boat conformers are much less stable than C1 and 1C. Therefore their contribution to conformation equilibrium is also low.

The presence of oxygen in the glucopyranose ring causes lower symmetry of the pyranose ring in comparison with cyclohexane. This results firstly from the lower length of C–O bond (1.42 Å) in respect to the C–C bond (1.54 Å) in glucopyranoses. The value of the endocyclic angle C–O–C is about 112–114°. Thus the significant flattening of the pyranose ring, as in the case of the cyclohexane ring, should take place and this is confirmed by NMR-spectroscopy of some glucopyranoses [23]. The significant destabilizing effect on the conformation properties of glucopyranoses is produced by axial hydroxy- and especially hydroxymethyl groups. Therefore the C1 conformer is the stable form of the majority of α-D-aldohexapyranoses with axial hydroxymethyl group (Fig. II.21, a). Provided two axial groups are positioned above the plane of the pyranose ring, then due to syn-axial interaction of the hydroxymethyl group with the anomeric hydroxyl group (Fig. II.21, b), the enhancement of the destabilizing effect (Hassel–Ottar's effect) takes place and because of that, the stable form of all β-D-aldohexapyranoses is also the C1-conformer. The

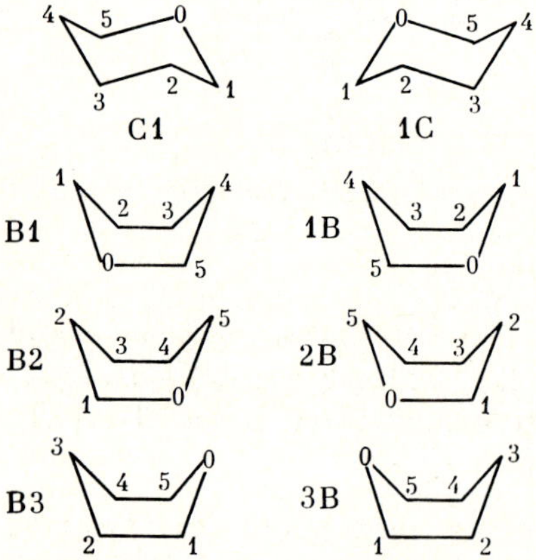

Fig. II.19. Conformations of the glucopyranose unit

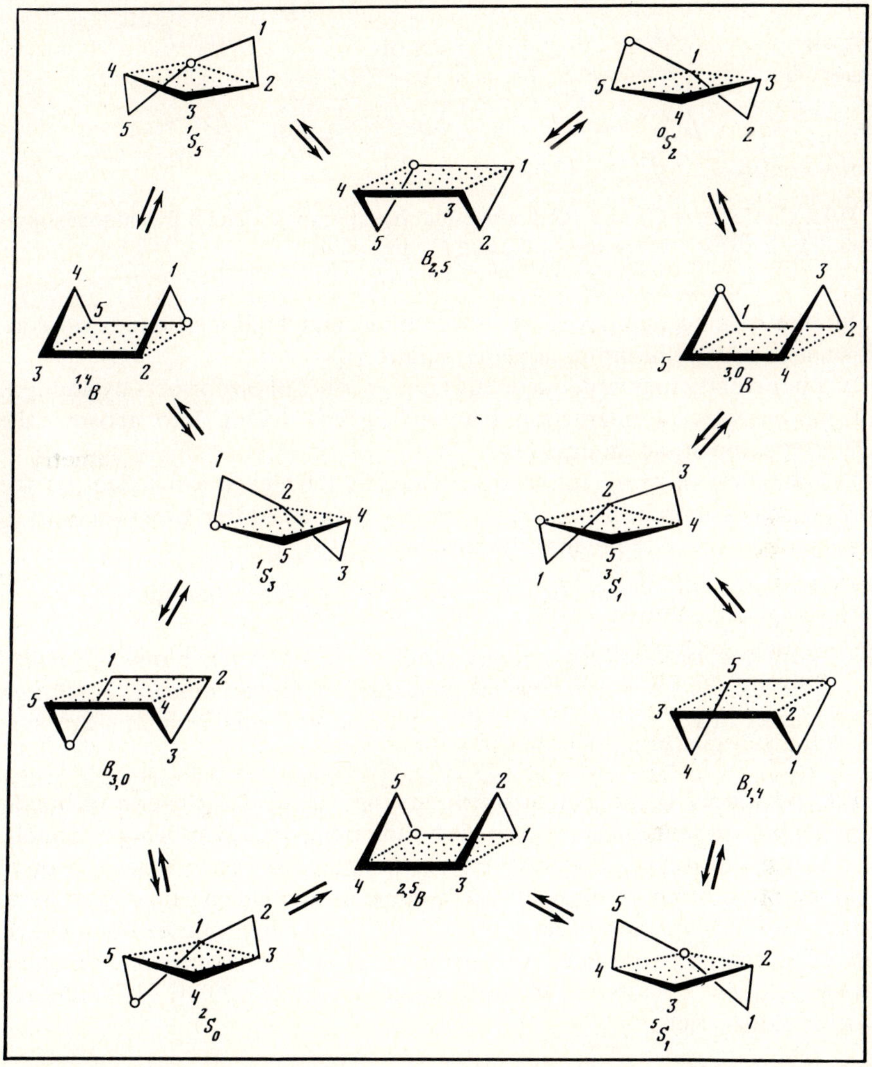

Fig. II.20. The cycle of pseudoconversion of the glucopyranose ring upon transition boat (*B*)—distorted boat (*S*)

semiquantitative approach to the calculation of the relative free energies of pyranoses in aqueous solution which are necessary for analysis and prediction of conformation properties of the corresponding derivatives is based on the following propositions [15]:

1. The equivalence of pyranose and cyclohexane rings geometry. The basic difference between them is connected with the lower length of C–O bonds in pyranose ring.

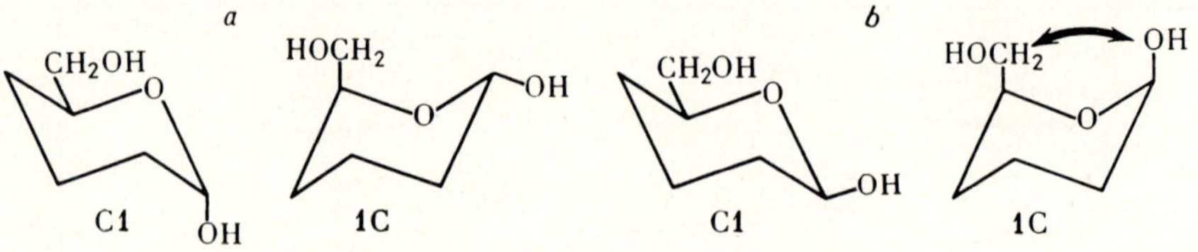

Fig. II.21. Conformers C1 and 1C of **a** α-D-aldohexopyranoses and **b** β-D-aldohexopyranoses. The hydroxy groups at C_2, C_3 and C_4 are omitted

2. The lack of a substantial effect of intermolecular hydrogen bonds on conformational equilibrium in aqueous solutions.
3. The opportunity to obtain the relative free energies of conformers by addition of the energies of interaction between ligands taking into account the electronic and steric interactions.
4. Consideration of every interaction separately, which is equivalent to the proposition about the lack of their mutual influence. At that time two types of interactions are of the highest importance:

   a) 1,3-diaxial interaction between syn-axial ligands other than hydrogen atoms $(X_a : Y_d)$;
   b) 1,2-interaction between ligands other than hydrogen atoms, provided these ligands are at adjacent carbon atoms in skewed position in relation to each other. Axial-equatorial and equatorial-axial positions are suggested as being equivalent $(X_1 : Y_1)$.

The differences in energies of strain, arising due to deformation of bonds, valent angles and rotation around single bonds are proposed as being negligible. The values of interaction of the above types were determined by investigation of the anomeric equilibrium of pyranoses because α- and β-anomers of pyranoses from an equilibrium mixture in aqueous solutions allow one to obtain their relative free energies experimentally (see Table II.1). The energies of interaction between pyranose substituents in aqueous solution at 295–298 K [15] are given below (in kcal/mole):

$(C_a{:}O_a)$	2.5[a]	$(O_a{:}H_a)$	0.45
$(O_a{:}O_a)$	1.5	$(C_1{:}O_2)$	0.45
$(C_a{:}H_a)$	0.9	$(O_1{:}O_2)$	0.35

[a] At 313 K

Together with the above interaction, the steric interaction between the axial hydroxyl at C_2 or C_4 and the syn-axial lone electron pair of the pyranose ring oxygen is of some interest. The study of equilibrium between α-D-glucose and α-D-galactose phosphates due to enzymatic epimerisation at C_4 (Fig. II.22) [24] allows one to estimate the difference in free energies as 0.6–0.8 Kcal/mole. Taking into account that α-D-galactose has an additional incoherent interaction $(O_a{:}H_a)$ with energy of 0.45 kcal/mole, one can postulate the steric interaction

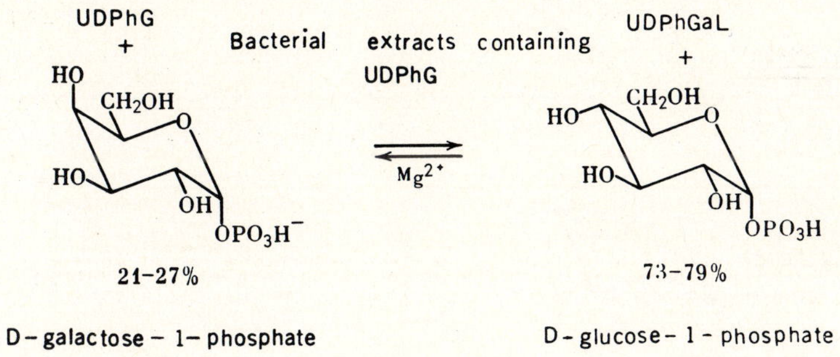

Fig. II.22. Enzymatic equilibrium in the system: α-D-galactose-1-phosphate, α-D-glucose-1-phosphate

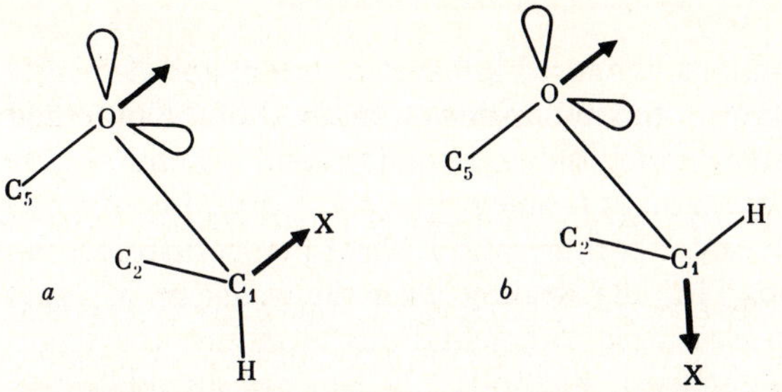

Fig. II.23. Relative disposition of dipoles for **a** equatorial- and **b** axial anomeric substituents—*X*

including the lone electron pair of pyranose cycle oxygen. Thus one should consider both steric and electronic interactions upon summation of interactions in glucopyranose conformers. However, the most significant electronic effect results from the interaction of the electron negative substituent at the anomeric carbon atom with ring oxygen. This interaction governs the axial orientation of such substituents. This is illustrated by the scheme given in Fig. II.23. It can be seen that orientation (*a*) with polar bond C_1—*X* which is hindered due to the presence of the lone electron pair of ring oxygen is destabilized in comparison with orientation (*b*) when this polar bond is skewed with respect to one lone pair and is in *trans* position to another one. This situation is usually considered as electrostatic repulsion between the carbon-substituent dipole and the resulting dipole of lone electron pairs of ring oxygen and it is called the anomeric effect.

The tendency of the substituent *X* to axial orientation is characterized [15, 25] by the difference between the free conformation energy ($-\Delta G_X^0$) of the equilibrium of the substituted pyranoses and that ($-\Delta G_X^0$) of the corresponding equilibrium of the substituted cyclohexanes in Fig. II.24.

The calculation of the geometrical parameters of α-D- and β-D-glucopyranose molecules [26] using Dashevsky's potentials; [27] and taking into

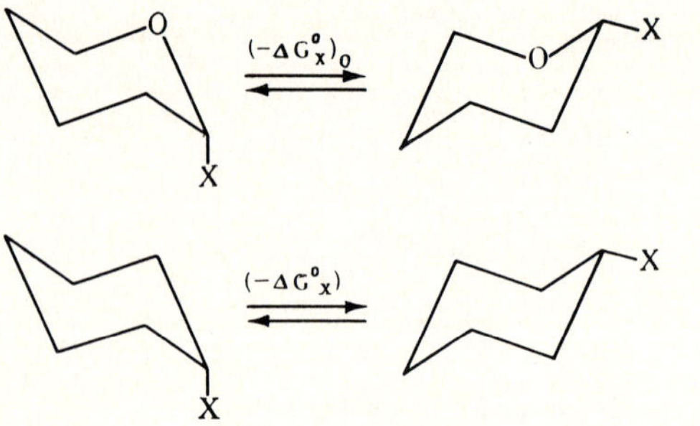

$$(0:X) = (-\Delta G^{\circ}_{X}) - (-\Delta G^{\circ}_{X})_0 \quad \text{or} \quad (0:X) = (\Delta G^{\circ}_{X})_0 - (\Delta G^{\circ}_{X})$$

Fig. II.24. The quantitative determination of the anomeric effect in the six-membered ring

account the electrostatic interactions, hindered internal rotation, and deformation of the valent angles by means of the minimization method of the potential function of independent geometric parameters (torsion and valent angles) demonstrates that the more stable form of D-glucopyranose is the C1 conformation regardless of the anomeric center configuration. The same conclusion can be drawn from detailed studies [16, 28] dealing with the problem of sugar conformation. The later papers dedicated to calculations of mono- and oligosaccharides conformations by empirical methods [29–31] are an additional confirmation of this conclusion. It was an important step when the above authors made some improvements to the calculation model which involves the potential function form and the increase of the number of degrees of freedom and is used for function minimization.

A distinct tendency has been recently evident in the wide application of various quantum mechanical methods for the investigation of mono- and oligosaccharides conformations [32–35]. In connection with conformational analysis of glucosides it is interesting to consider paper [36] dealing with various versions of the approximation of zero differential overlapping (CNDO with modified potential of the stem repulsion [37], CNDO/2 [38] and MINDO/3 [39, 40]). The comparative data obtained in this paper about the energies of different conformers of β-D-glucose are given in Table II.2.

It has been deduced from the data obtained that CNDO/2 gives the best results. The conformation 1C appears to be also the most stable one for β-D-glucose and the difference in energies $E_{1C}-E_{C1}$ is in close correlation with that in free energies of these conformers, obtained on the basis of the mass of experimental data [15].

Taking into account these data and calculation of the charge distribution in glucopyranose by CNDO method the conformation of β-D-glucopyranose can be represented as follows [41] in Fig. II.25.

The above authors applied semi-empirical SCF-MO-CNDO methods with Boyd and Whitehead parameters (CNDO/BW) [42–44]. The CNDO/BW

Table II.2. The calculated values of the energies of various β-D-glucose conformers [36]

	CNDO with modified repulsion potential		CNDO/2		MINDO/3	
Conformation[a]	Total energy (eV)	ΔE in relation to C1 (kcal/mol)	Total energy (eV)	ΔE in relation to C1 (kcal/mol)	Total energy (eV)	ΔE in relation to C1 (kcal/mol)
C1	– 2937.076	00.0	– 4429.353	0.0	– 2807.213	00.0
1C	– 2936.055	23.5	– 4429.250	2.4	– 2806.291	21.3
B1($^{1.4}$B)	– 2936.321	17.4	– 4429.122	5.4	– 2806.724	11.3
1B($B_{1.4}$)	– 2936.227	19.6	– 4429.227	2.9	– 2806.779	10.0
B2($^{2.5}$B)	– 2936.342	16.9	– 4428.894	10.6	– 2806.392	19.0
2B($B_{2.5}$)	– 2935.255	42.0	– 4429.044	7.1	– 2806.234	22.6
B3($^{3.0}$B)	– 2936.678	9.2	– 4429.181	4.1	– 2807.162	1.2
3B($B_{3.0}$)	– 2935.601	34.0	– 4428.924	9.9	– 2806.307	20.9
1S_3	– 2937.747	– 15.5	– 4428.667	15.9	– 2807.234	– 0.5
3S_1	– 2936.710	8.4	– 4429.265	2.0	– 2807.127	1.2

[a] The notations of the boat shape conformers according to [15] nomenclature are given in parentheses (see Fig. II.20)

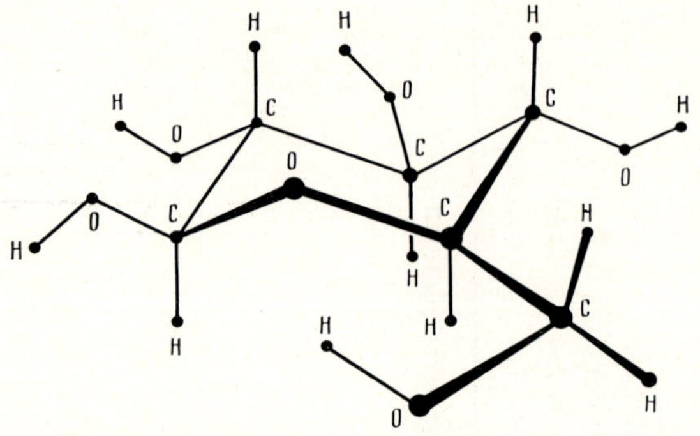

Fig. II.25. Conformation of β-D-glucopyranose determinated by semi-empirical methods [40]

method is superior to MINDO/3 for the description of the systems with a significant effect of the hydrogen bonds while it is inferior in precision for the calculation of the heat of formation and molecule geometry [45–47]. This superiority was of great importance in our case, since the comprehensive experimental data [4] demonstrated the substantial effect of the hydrogen bonds on the structure and reactivity of cellulose.

4.2 Conformation of the Isolated Cellulose Helix

The direct experimental determination of the conformation and structure of cellulose by diffraction methods is very problematic. Thus, for example, the typical X-ray diffraction pictures of even well-oriented cellulose fibers are rather vague and they used to comprise only several tens of reflections, which is totally insufficient for the determination of the coordinates of all atoms in elementary unit. The number of independent reflections of the electron-diffraction pattern of cellulose is somewhat higher, however these data can hardly be considered conclusive due to the strong dependence of the reflection intensities on the orientation of the reflecting planes. Some additional information can be obtained by spectroscopic methods, however their efficacy is limited, and the results can be interpreted ambiguously.

Due to the above reasons, the conformation problems can usually be resolved by drawing on of some stereochemical principles, in particular those about optimum packing of the molecules. Thus, one can suggest when optimizing the structural model that the number of intermolecular contacts and hydrogen bonds should be the maximum, and the deviation of the bond length, valent and torsion angles from their standard values should be the minimum.

The general approach to the selection of the most suitable structural models is illustrated by analysis of cellobiose—the dimer fragment of the cellulose chain

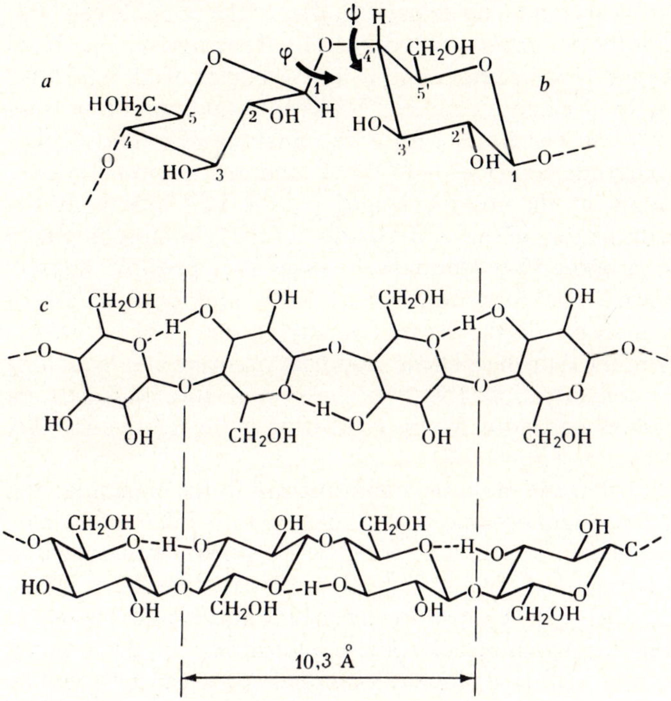

Fig. II.26. The conformation of **a**, **b** the cellulose fragment and based on it **c** Herman's conformation of cellulose in two projections

(Fig. II.26). In order to specify the conformation of the cellobiose residue when the angle of the glycosidic bond is close to 117° one should know the values of torsion angles between H_1–C_1–O'_4–C'_4 and H'_4–C'_4–O'_4–C_1 bonds, denoted in the scheme as φ and ψ respectively. The origin of the coordinates in every residue coincides with the glycosidic oxygen atom, and the axis X in residue a is superposed with the O'_4–C_1 bond and in residue b—with the O'_4–C'_4 bond. The planes XZ may be considered as fitting the planes passing through O'_4, C_1, H_1, and O'_4, C'_4, H'_4 respectively. In this case the Y axes are perpendicular to these planes. Then the coordinates of any conformation of cellobiose residue are determined by φ and ψ parameters and they can be obtained by standard methods of transfer and rotation of the series of axes. The new coordinates of the "b" residue defined with respect to the "a" residue axes can be acquired by gradual rotation of the axes of the "b" residue around the O'_4–C'_4 bond at ψ angle followed by rotation around the Y axis in the plane C_1, O'_4, C'_4 at the angle complementary to that at the glycosidic oxygen atom and then around O'_4–C_1 bond at φ angle. It is possible to determine from these data the interatomic distances and to estimate the possibility of the formation of the hydrogen bonds. Only those conformations are "permitted" where "shortened" contacts are absent, this means the contacts when the distance between two atoms is less than the sum of their van der Waals radii. Systematic studies of the possible cellulose conformations show that the conformation proposed by Hermans (see Fig. II.26) [48] is in better agreement. It corresponds to φ and ψ angles of 25° and 146° respectively for each pair of glucose residues. This conformation is free from "shortened" interactions and correlates with X-ray analysis of cellulose [49], where the macromolecule axis coincides with the second order symmetry axis, and the projection value of the residue height on the coil axis is 5.15 Å (Fig. II.26. c). The specific feature of the conformation of the cellobiose fragment of the cellulose chain proposed by Hermans involves two possible ways of intramolecular hydrogen bond formation: 05 . . . H03′ and 06′H . . . 01 or 06′H . . . 02 at tg orientation of the hydroxymethyl group. The structural models of the majority of the known cellulose modifications were created taking into account the above considerations [50–54]. The aforementioned papers are characterized by the lack of a criterion for the estimation of the relative stability of these conformations.

It is well known [55] that the molecule conformation in the crystal for the majority of low molecular weight organic solids is only slightly different from its conformation in the isolated state. The differences for long-chain molecules can be more significant. However, one can also suggest in this case that the molecule conformation in the crystal will be close to one of the local minima on the energetic surface of the isolated molecule. The conformation of the isolated cellulose molecule was analyzed taking into account these propositions [56, 57]. The energy of cellulose macromolecule has been calculated as the sum of the non-valent interactions, hydrogen bond potentials, the deformation energies of the valent and torsion angles. Dashevsky potential functions were applied for non-valent interactions [27].

The glucopyranose conformation was proposed in the course of calculation as being unchanged. The values of the bond lengths, valent and torsion angles within the cycle were taken from another paper [58]. Four torsion angles which govern the orientation of hydroxymethyl and three hydroxy groups (τ_2, τ_3, τ_4) were selected as independently varying parameters of the repeating units (Fig. II.27). The torsion angle τ_1 is defined by the atoms C(4)–C(5)–C(6)–O(6), and the angles τ_2, τ_3 and τ_4—by the atoms C(5)–C(6)–O(6)–H(61), C(1)–C(2)–O(2)–H(21) and C(2)–C(3)–O(3)–H(31) respectively. The angle τ_1 was proposed as being equal to zero, when the bonds A–B and C–D in sequence A–B–C–D were in *cis*-position. The counter-clockwise rotation of the C–D bond in the direction of the B–C bond was suggested as positive rotation.

The special algorithm described below was used for the creation of the chain from monomer units. The angles in conjunction with monomer units β, φ, ψ accepted in conformation analysis of polysaccharides appear to be dependent variables, defined by parameters of the helix.

The construction of the regular helix in the course of energy minimization requires special attention. In the majority of early papers this problem was not considered at all, i.e. the conformation of the macromolecule basic chain was suggested as suitable to spiral symmetry without variation in the course of

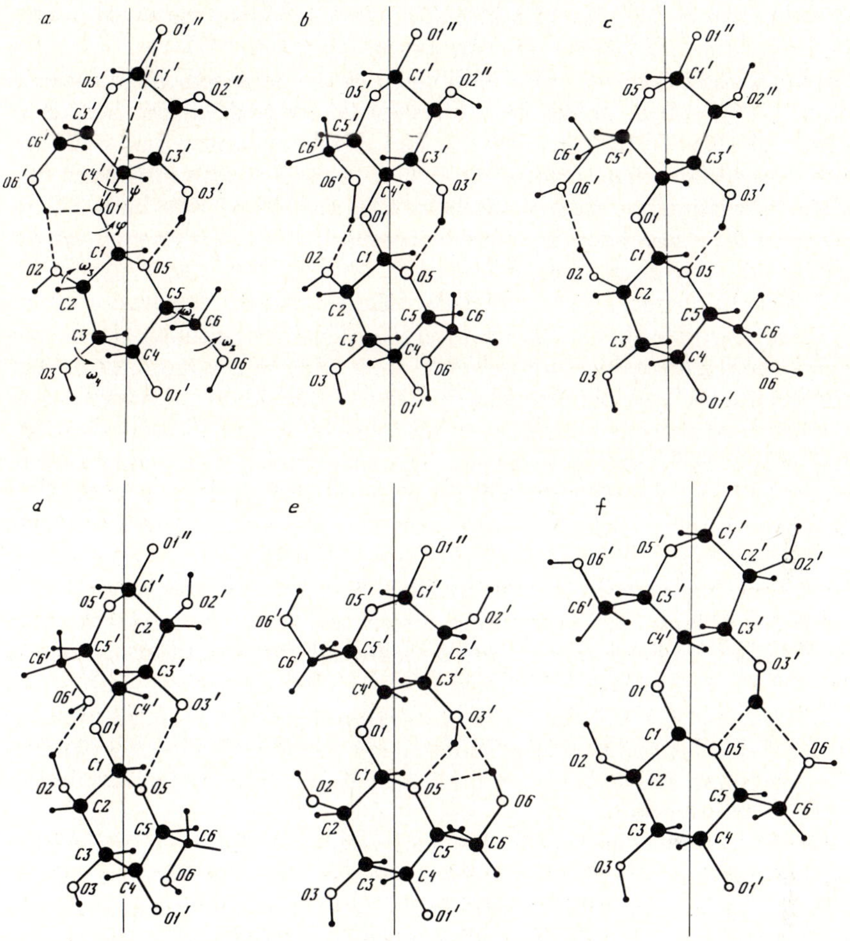

Fig. II.27. The projection of the conformations **a** A1, **b** A2, **c** B1, **d** B2, **e** C1, **f** C2 of the isolated cellulose coil

calculations. Four parameters were used in the paper [59] for definition of the special position of the monomer unit—that is the distance up to the coil axis and the three Euler's angles, which describe the monomer unit orientation in relation to the helix coordinates, although only two independent variables, described the conjunction of the monomer units, are required for the helix of a given type 2_1 in particular and given geometry of repeating unit like in cellulose. In the paper cited [59] the elimination of two dependent variables among the four mentioned above was carried out with the help of "penalty" functions, "drawing up" the monomer units on the coil, that is not quite satisfactory since it results in a sharp decrease of convergence of the minimization procedure.

Besides, the polymer chain in each step of minimization process is not a strict helix and thus does not possess the translation symmetry.

The mentioned drawbacks were eliminated by the development of a special algorithm [56], applied, unlike known analytical methods of elimination of dependent variables [60], to macromolecules of any symmetry and with any structure of the main chain (except the ladder type). It is also important that within this algorithm the step of the helix (the magnitude well known from experiment) is the independent parameter of conjunction that is very convenient for carrying out a given calculation.

The calculation procedure involves the following stages: Separation of the polymer chain into the chemically-identical periodically-repeating units in such a way that the separate units are joined to each other at a point (atom). Such resolution is always possible for linear polymers. The glucoside oxygen atoms (01, 01′ etc) can be chosen as coupling points for cellulose and its derivatives (see Fig. II.27). Let us propose that the elementary unit involves both coupling atoms (at the inlet and outlet of the unit). The fact that each coupling atom happens to be within two units can be easily taken into account upon calculation of the polymer chain energy.

Let us consider the point O_m on Fig. II.28 as one of the coupling point of initial elementary unit. Choose the unit coordinates so that the origin coincides with O_m. The coordinate axis X_m, Y_m and Z_m may be combined with the unit in many ways. It is convenient for cellulose to superpose the X_m axis in the direction along the O1–C1 bond and to propose Y_m axis as being within the plane O1–C1–C2. In the first stage, Cartesian coordinates of the atoms of initial unit are computerized in coordinates O_m, X_m, Y_m, Z_m. The input data in this case are 3N-6 independent natural coordinates—bond lengths, valent and torsion angles (N—the number of atoms in the unit, for cellulose $N = 22$). Four variable independent coordinates mentioned above are involved in natural coordinates. The addition of the following unit to the previous one is performed in the following way. One can find vector a which binds the coupling angles at

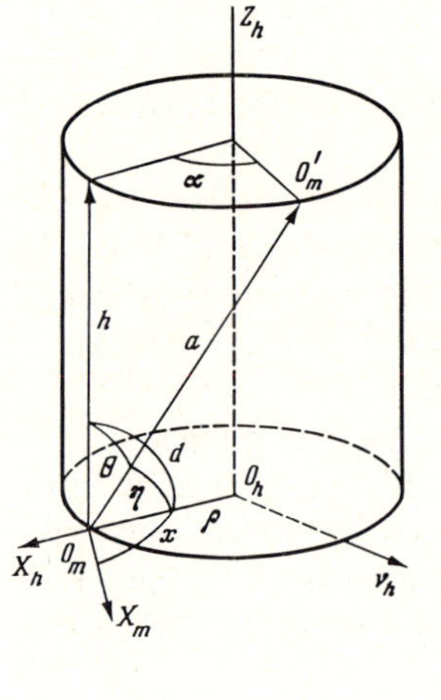

Fig. II.28. The coupling of repeating units into the helix

the beginning and the end of the initial unit (points O'_m and O_m in Fig. II.28 or atoms 01 or 01′ in Fig. II.27, *a*). Provided the helix step h is selected as independent coupling parameter, then the distance from O_m to helix axis (the position of which is not known yet) can be easily determined by the formula:

$$\rho = [(a^2 - h^2)/2(1 - \cos\alpha)]^{1/2}$$

where $a = |a|$, α—the twisting angle of the helix (180° for cellulose). Then the angle θ and η which are formed between the helix axis and radius-vector with vector a (see Fig. II.28) can be easily determined: $\cos\eta = \rho(1 - \cos\alpha)/a$, $\cos\theta = h/a$.

Let us introduce the second parameter of coupling—the angle χ which vector ρ forms with the X_m axis. Now according to the standard method of vector calculation, based on two known vectors and two angles which the vector under study forms with them, one can easily find vector ρ (from vectors a, X_m and angles η and χ) and vector h (from vectors a, ρ and angles θ, $\alpha = 90°$).

The application of the χ angle as one of the independent parameters upon minimization of helix energy as a function of geometric parameters is not very convenient. It can be easily shown that the variation area of parameter χ is not always singly associated. The more useful parameter appears to be the dihedral angle δ formed by two planes where the vectors a, X_m and a, ρ are situated respectively. The angle δ indicates the rotation of the monomer unit around vector a and it may vary from $-180°$ up to 180°. The correlation between δ and the earlier defined angle χ is the following:

$$\cos\delta = (\cos\chi - \cos\eta\cdot\cos\varkappa)/\sin\eta\cdot\sin\varkappa$$

where $\varkappa$—the angle between vectors a and X_m. Note, that the angle δ defined in the above way, coincides with the angle φ [61] for the description of mutual orientation of glucopyranose cycles. The angle δ is equal to O when all three vectors ρ, a and X_m are situated in one plane. The values δ are positive when three vectors ρ, a and X_m are right-handed.

Thus, the position of the helix axis in the coordinates $\{O_m, X_m, Y_m, Z_m\}$, which is unequivocally defined by vectors ρ and h, can be easily found for any a on the basis of two independent h and δ. One should remember that the helix type which is defined by the twisting angle α is supposed to be known and unchanged in the course of calculation.

When the helix axis has been found, the coordinates of the monomer unit are transferred into the cylindrical coordinate system denoted in Fig. II.28 as O_h, X_h, Y_h and Z_h. The plotting of the subsequent elementary units was performed by the rotation of the initial unit coordinates around the Z_h-axis on the angle α and by the displacement along the same axis on the step h.

The helix energy, U_{hel} is calculated as the sum of two components: $U_{hel} = U_{mon} + U_{conj.}$, where U_{mon}—the intrinsic energy of the monomer unit, $U_{conj.}$—the conjunction energy. The value of U_{mon} is identical for all monomer units and it depends solely on (3N-6) independent coordinates. Only four from these coordinates are varied in this paper and thus $U_{mon} = U_{mon}(\tau_1, \tau_2, \tau_3, \tau_4)$.

The conjunction energy involves the energy of interaction of the monomer units (the possible hydrogen bonds are included) and the deformation energy of the valent and torsion angles in the conjunction; at the same time $U_{conj.}$ also depends on the independent helix parameters δ and *h*. Thus the total energy of the helix is the function of the six variables: $U_{hel.} = U_{hel}(\tau_1, \tau_2, \tau_3, \tau_4, \delta, h)$. The stable conformations of the helix were found by minimization of $U_{hel.}$ in the space of six variables. Fletchir's method was used for minimization [62].

The preliminary calculation where the helix step was considered as constant and the angle δ was varied from $-180°$ up to $180°$ with steps of $10°$ was performed for determination of an area of accepted conformations of the main chain correlated with the experimental helix period. The dependence of the angle β at the glycoside oxygen atom on the angle δ for $h = 0.51$ nm is presented in Fig. II.29. It can be seen that the reasonable values of β angles ($114° < \beta < 118°$) are present in two relatively narrow regions: $44° < \beta < 58°$ and $-70° < \beta < -56°$. The dependence of the distance 02 . . . C6′ on the angle β is also given in Fig. II.29. It is evident that the region $-70° < \delta < -56°$ can be excluded from consideration due to the presence of contacts 02 . . . C6′ which are too short. This simplifies considerably the problem of finding the optimal cellulose helix. The following calculations involve minimization of $U_{hel.}$ in all six independent parameters. The initial values for τ were taken from the range $44° < \delta < 58°$, the initial value for *h* was assumed to be 0.51 nm. The initial values for variables τ_1, τ_2, τ_3 and τ_4 for minimization were chosen randomly with the help of a random numbers generator.

The calculations resulted in six stable conformations, denoted as A1, A2, B1, B2, C1, C2 in Tables II.3 and II.4. The values of independent variables corresponding to local energy minimum U_{hel} are given in the first six columns of Table II.3. Three dependent parameters of coupling are mentioned further—they are the β-angle of glucoside bond as well as φ and ψ angles which

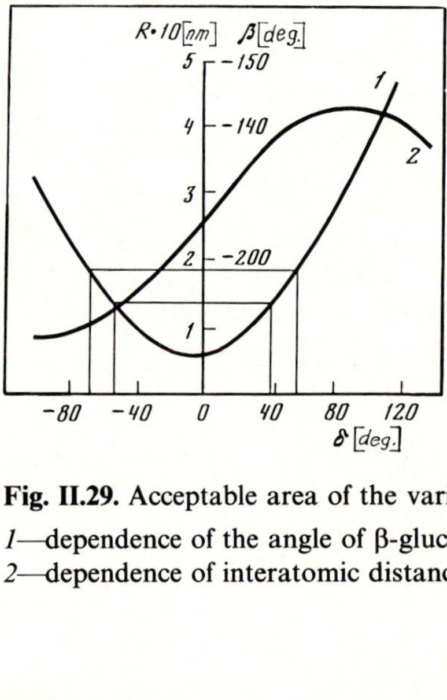

Fig. II.29. Acceptable area of the variation of angle δ

1—dependence of the angle of β-glucoside bond on the angle δ

2—dependence of interatomic distance 02 . . . 06′ on the angle δ

Table II.3. Structural and energetic parameters of stable conformations of the cellulose helix

	Independent variables					
Conformation	τ_1	τ_2	τ_3	τ_4	δ	h
A1	64	− 52	174	170	44	5.16
A2	− 20	79	179	180	36	5.17
B1	68	167	− 53	171	44	5.16
B2	− 17	184	− 38	180	36	5.17
C1	174	47	− 56	− 122	53	5.11
C2	− 168	179	− 56	156	51	5.13

	Dependent parameters					
Conformation	β	φ	ψ	U_{mon}	U_{conj}	U_{hel}
A1	117.3	28.1	− 29.5	− 15.1	− 4.2	− 19.3
A2	116.1	23.7	− 24.9	4.6	8.4	13.0
B1	117.5	28.0	− 29.4	− 9.6	10.9	1.3
B2	115.9	23.7	− 24.9	− 6.7	4.2	− 2.5
C1	117.2	34.4	− 36.0	− 10.0	9.6	− 0.4
C2	117.5	32.4	− 34.0	− 7.7	18.8	11.1

Table II.4. Parameters of hydrogen bonds, present in stable conformations of cellulose helix

Conformation	Bond type	$R_{O \ldots O^{10}}$ nm	$R_{O \ldots H^{10}}$ nm	Bond energy KJ/mole
A1	05 . . . H03′	2.60	1.81	− 16.7
	02 . . . H06′	2.69	1.85	− 16.4
	01 . . . H06′	2.71	2.06	− 11.8
A2	05 . . . H03′	2.70	1.86	− 16.2
	02 . . . H06′	2.70	1.88	− 16.0
B1	05 . . . H03′	2.60	1.81	− 16.7
	02H . . . 06′	2.60	1.82	− 16.7
B2	05 . . . H03	2.70	1.86	− 16.3
	02H . . . 06′	2.75	1.93	− 15.0
C1	05 . . . H03′	2.41	1.89	− 15.7
	06H . . . 03′	2.86	1.98	− 13.9
C2	05 . . . H03′	2.60	1.80	− 16.7
	06 . . . H03′	2.87	2.11	− 10.5

are usually used for depiction of orientation of cycles in polysaccharides [51, 53]. Some energetic parameters of stable conformations are given in the subsequent columns of Table II.3. Table II.4 summarizes the parameters of the hydrogen bonds formed in each conformation, the energy of bonding, as well as distances 0 . . . 0 and 0 . . . H. All values having the dimension of energy are given in Kilojoule per 1 mole of monomer units, distance—in nanometer, and angles—in degrees.

As it can be seen from Table II.4, all stable cellulose conformations are characterized by the hydrogen bond 05 . . . 03′ with energy 16.7 Kj/mole. The presence of such a bond in cellulose is confirmed by various experimental data and it is an indisputable fact. A1, A2, A3 and B2 conformations comprise also 02 . . . 06′ bonds and C1 and C2 conformations—06 . . . 03′ bonds (as for 06′ . . . 01 bond in A1 conformation see below). As can be seen from Table II.3, the orientation of the hydroxymethyl group in the A1 and B1 conformations is close to the tg-position ($\tau_1 = 60°$) and in C1 and C2 conformation—to the gt-position. The group orientation in A2 and B2 conformations is close to the gg-position, although deviations of τ_1 angles from the ideal value corresponding to the gg-position are rather high. The difference of the A-type conformation from the B type consists of the way of formation of the 02 . . . 06′ bond. The bond 02 . . . H06′ is present in A type conformations, the bond 02H . . . 06′—in B conformations, the bond 06H . . . 03′ in the C1 conformation, and the bond 06 . . . H03′—in the C2 conformation. All three types of hydrogen bond have been widely discussed in the literature [11, 58, 63]. It is important to mention the following. It was proposed for type A and B conformations that the formation of the 02 . . . 06′ bond is possible only at the tg-position of hydroxymethyl group [11, 51]. The calculations show that there is another position of this group ($\tau_1 \approx -20°$) when formation of the 02 . . . 06′ bond is possible. And this is valid both for hydrogen bond 02H . . . 06′ and for the 02 . . . H06′ bond. The C1 conformation (gt-orientation) can be additionally stabilized by the intermolecular hydrogen bond 06H . . . 05. The analysis of the conformation parameters given in Table II.3 demonstrates that all six stable conformations of the cellulose macromolecule have an equilibrium value of helix step *h*, which is in good correlation with the experimental period of identity (1.03 nm). The equilibrium values of the angle at the glucoside oxygen atom are also close to those values of this angle usually observed in carbohydrates and polysaccharides.

As can be seen from Table II.3, the conformation A1 is the most preferable one, since its energy is lower at 18.8 Kj/mole than that of the B1, B2 and C1 conformations and at 32.2 Kj/mole than that of the A2 conformation (the projections of A1, A2, B1, B2, C1 and C2 conformations on the plane $X_hO_hZ_h$ are depicted in Fig. II.27). It should be mentioned, however, that the A1 conformation has a three center (bifurcate) hydrogen bond

01
06′H ⋰⋱
02

which involves one donor (06′) and two acceptor oxygen atoms (01 and 02). The bonds of this type as proposed in Refs [64, 65] were denoted as biacceptor in contrast to bidonor hydrogen bonds

```
     H—O
   ·
O
   ·
     H—O
```

The energetic preference of the biacceptor bond in comparison to linear OH . . . O has been recently demonstrated by a non-empiric calculation of the water trimer in basis 4-31G [66]. The authors found that the appearance of the second acceptor of protons results in stabilization of the system at 4.2–8.4 Kj/mole as compared with the linear system. According to [64, 65] the energetic difference between the biacceptor and the ordinary bond is about 9.5 Kj/mole for the isolated chain and as we shall see, it lies in the range of 4.2–9.5 Kj/mole for the crystalline regions of cellulose, that is in good agreement with non-empirical data [66]. The presence of a biacceptor bond in cellulose can be also confirmed by the abundance of these bonds in the crystals of some low molecular weight carbohydrates and their derivatives [67]. The neutron diffraction analysis of these carbohydrates demonstrated that biacceptor hydrogen bonds were present in 25 from 100 cases and in one case even a four center (triacceptor) hydrogen bond was observed.

The calculations of conformation energy of the single cellulose macromolecule performed at the fixed geometry of glucopyranose cycle demonstrated six the most stable conformations, characterized by the presence of the 05 . . . 03′ hydrogen bond as well as the bonds 02 . . . 06′ (type A and B conformation) and 06 . . . 03′ (conformation C1 and C2). All six stable conformers are characterized by the intramolecular hydrogen bond 05 . . . H03′, the presence of which should be considered as indisputable. The formation of intramolecular hydrogen bonds 02H . . . 06′ and 02 . . . H06′ is possible both at tg- and gg-orientations of hydroxymethyl group. This corresponds to the values of the dihedral angle τ, which defines position of the CH_2OH group with respect to the ring plane, as 60° and about −20° respectively.

There are interesting data demonstrating the existence of two bifurcate (biacceptor) hydrogen bonds at tg and gt-orientations of the hydroxymethyl group and the type 06H . . . 03′ hydrogen bond. The results of the determination of the geometric and energetic parameters of intramolecular hydrogen bonds are of interest too. These results together with the determination of the number of points of potential contacts for formation of such hydrogen bonds dictate the significant differences in packing of the single chains that is confirmed by polymorphism of cells.

5 Equilibrium and Kinetic Rigidity of the Cellulose Macromolecular Chain and Some of its Derivatives in Solution

Cellulose and its derivatives are the classic examples of macromolecules characterized by enhanced skeleton rigidity. This property is directly deduced from the configuration of its molecular chain presented in Fig. II.30. In order to calculate the statistic dimensions of the chain, its monomer unit can be substituted by an equivalent "effective" unit [68] consisting of two parallel bonds $b = 2.7$ Å around which the rotation takes place and a single bond $d = 1.45$ Å perpendicular to the two previous ones where rotation is excluded. It has been shown, using this scheme, [68] that, for a rather long chain on full freedom of rotation around O–C_1 and O–C_4 bonds, the mean square distance between its ends h_f^2 is equal to

$$h_f^2 = P[d^2 + (2b)^2 + (1 + \cos v)/(1 - \cos v)] \tag{5.1}$$

where P is the number of monomer units in the chain, $(\pi - v)$ is the valent angle between adjacent bonds O–C_1 and O–C_4. Substituting b and d values and assuming $v = 70°$ we obtain $h_f^2 = 62P \times 10^{-6}$ and for a segment length of cellulose chain with full freedom of rotation.

$$A_f = h_f^2/P\lambda = h_f^2/P(2b\cos v/2 + \sin v/2) = 12 \text{ Å}$$

This value corresponds to the number of glucoside units in the segment $S_f = 2.5$. In the case of the carbon chain polymer free of cycles in the chain, a similar calculation [69] affords $A_f = 3.8$ Å provided full freedom of rotation. Thus, even with full freedom of rotation the length of the segment of cellulose chain exceeds by more than 3 times that of flexible carbon chain (aliphatic) polymers, demonstrating the increased skeleton rigidity of cellulose derivatives.

The rigidity of polymer molecules is usually defined on the basis of their hydrodynamic properties in the solutions. It follows from analysis of dependence of coefficients of sedimentation (S), diffusion D and intrinsic viscosity $[\eta]$ with the molecular weight M. The most widespread forms of these dependences are the empirical Mark–Kuhn expressions:

$$[\eta] = K \times M^{\alpha} \tag{5.2}$$

$$D = K_D \times M^{-\beta} \tag{5.3}$$

$$[S] = K_s \times M^{1-\beta} \tag{5.4}$$

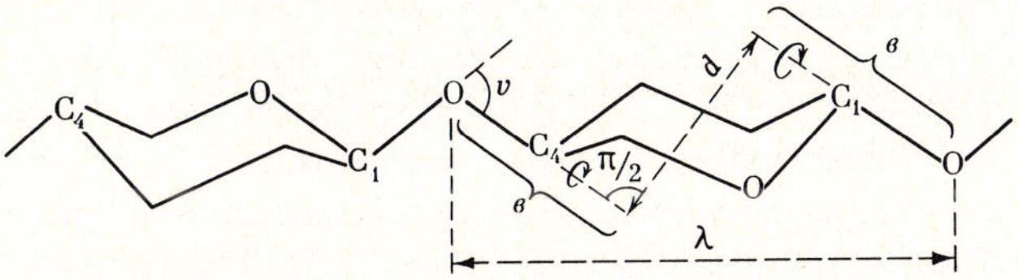

Fig. II.30. Configuration of the cellulose chain fragment. λ—the projection of glucoside unit on the chain direction; b = 2.7 Å—the length of the "effective" bond where rotation is possible; d = 1.45 Å—the length of the "effective" bond where rotation is impossible

When measurements are carried out in θ-conditions i.e. without excluded volume effect, the high α and β-values demonstrate the significant influence of the permeability effect and thus may show the enhanced rigidity of the molecular chain. It is important that in many cases the excluded volume effect for rigid chain molecules does not play a significant role and it can be neglected in the first approximation [70]. The theory of hydrodynamic properties of persistent (wormlike) chains can be applied for evaluation of chain rigidity [71, 72]. In the area of rather high molecular weight when a molecule can be represented by a Gauss chain, this theory leads to the expressions

$$M/[\eta] = \Phi^{-1}(M/LA)^{3/2}M^{1/2} + 0.93[\ln(A/d) - 1.43] \times (M/L)^2 A^{-1} \quad (5.5)$$

$$DM/RT = (1.8433\pi\eta_0 N_A)(M/LA)^{1/2}M^{1/2} + (3\pi\eta_0 N_A)^{-1} \times (M/L)[\ln A/d - 1.43] \quad (5.6)$$

Apart from L (contour chain length) and A (length of statistic Kuhn segment) values, which characterize conformation and rigidity of the chain, the above formulas also contain the diameter d of persistent chain, which should be taken into account for the proper description of its hydrodynamic properties.

One can see considering the left side of Eqs. (5.5) and (5.6) as function $M^{1/2}$ that the corresponding dependence is linear and one can determine the M/L-value on the basis of the curve slope—i.e. estimate the chain rigidity and A/d value on the basis of segments intercepting at ordinate and thus chain diameter. In the case of small M and significant chain rigidity, which is typical for cellulose type polymers, the role of the second term in the right side of Eqs. (5.5) and (5.6) is increased and this results in an increase of value α to about a unit in expressions for intrinsic viscosity.

5.1 Hydrodynamic Properties of Cellulose Derivatives

Since cellulose itself is soluble only in complex solvents (see for example, [73]) and these solutions are not true ones, information about the structure of cellulose macromolecules in the solution has been mainly obtained by analysis of its derivatives, esters in particular. The most complete and systematic studies of the hydrodynamic properties of solutions of cellulose derivatives were performed for nitrocellulose solutions [74, 75]. The typical dependences of the dispersion coefficient $\sigma^2 = 2Dt$ of diffusion curves and position X of the sedimentation border on the time for solutions of different fractions of nitrocellulose (degree of substitution $n = 1.9$) are presented in Fig. II.31. The values of diffusion D and sedimentation S coefficients at the corresponding concentrations were determined from these data. It is important that the $S(c)$ and $D(c)$ values measured in the concentration range $C = 0.05$–0.1% are practically independent of C and can be proposed as equal to their limiting values S_0 and D_0 at $C \to 0$. The molecular weights of fractions M_{SD} were determined according to the Svedberg equation using specific partial volume $V = 0.54$ (methylethylketone). The result of hydrodynamic studies and molecular weights obtained are given in Table II.5.

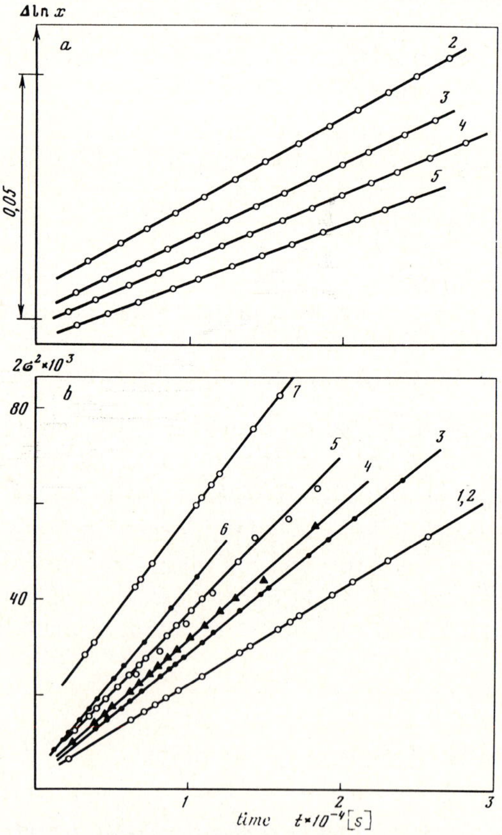

Fig. II.31. Dependence of **a** Δlnx and **b** $2\sigma^2$ on the time for various nitrocellulose fractions. The curve numbers correspond to those of the fractions in Table II.5

Table II.5. Hydrodynamic properties of nitrocellulose fractions (degree of substitution 1.9) in methylethylketone (MEK) at 299 K and in dioxane at 294 K

Fraction N	$D \times 10^7$ cm^2/s	$S \times 10^{10}$ s	$[\eta] \times 10^{-2}$ cm^3/g in MEK	in dioxane	$M_{SD} \times 10^{-3}$	$A_0 \times 10^{10}$ erg/grad
1[a]	5.0	8.8	1.74	2.00	76.9	3.2
2	5.0	8.9	1.60	1.80	78.3	3.2
3	6.4	7.3	1.38	1.72	51.0	3.3
4	6.7	7.0	1.00	1.40	46.0	3.0
5	8.2	6.4	0.84	0.97	34.1	3.2
6[a]	9.2	5.8	0.66	0.78	27.8	3.1
7[a]	12.0	4.6	0.54	0.60	15.9	3.1
non-fractionated sample	5.7	8.3	1.72	2.20	65.0	3.5

[a] Molecular weights are determined according to Eq. (5.9)

The dependences S_0, D_0 and intrinsic viscosity

$$[\eta] = \lim_{C \to 0} [(\eta - \eta_0)/\eta_0 C]$$

(η and η_0—solution and solvent viscosity) on molecular weight in Fig. II.32 in a studied range of molecular weights are described by the following equations

$$[\eta] = 1.4 \times 10^{-2}\, \mathrm{M}^{0.86} \quad \text{(in dioxane)} \tag{5.7}$$

$$[\eta] = 1.2 \times 10^{-2}\, \mathrm{M}^{0.86} \quad \text{(in methylethylketone)} \tag{5.8}$$

$$D_0 = 4.3 \times 10^{-5}\, \mathrm{M}^{-0.6} \quad \text{(in methylethylketone)} \tag{5.9}$$

$$S_0 = 9.6 \times 10^{-15}\, M^{-0.4} \quad \text{(in methylethylketone)} \tag{5.10}$$

Using the data shown in Table II.5, the values of hydrodynamic constant $A_0 = \eta_0 D_0 (M[\eta])^{1/3} \times T^{-1}$ were determined for all fractions, its mean value equal to $(3.3 \pm 0.2) \times 10^{-10}$ erg/grad is close to A_0 values determined for other cellulose esters [76, 77]. Quantitative estimation of the equilibrium rigidity of the nitrocellulose molecular chain has been performed within the scope of the theory of the translational friction of wormlike chains [72, 78]. According to this theory the following relationship is realized for the chain with the length $L > 2.27$.

$$DM/RT = 1.84(M_0/\lambda A)^{1/2} M^{1/2}/(3\pi N_A \eta_0) + M_0[\ln A/d - \gamma]/(3\pi N_A \eta_0 \lambda) \tag{5.11}$$

where R—gas constant; M_0—molecular weight of the monomer unit; N_A—Avogadro number; d—hydrodynamic chain diameter; A—length of Kuhn segment; $\lambda = 5.15$ Å—the length of monomer unit projection on the chain direction; γ—coefficient equal to 1.43 [72] or to 1.05 [79]. The graphic

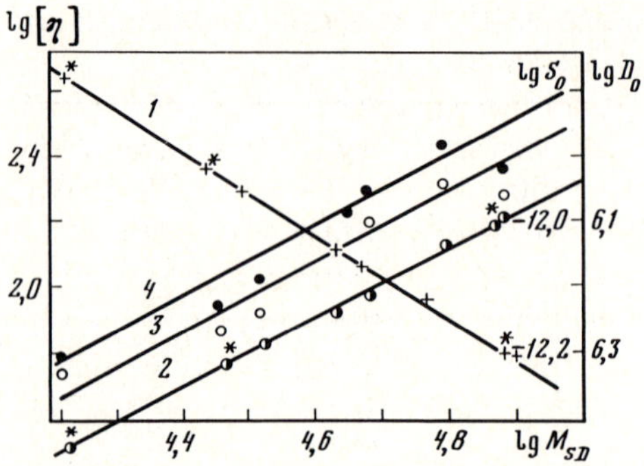

Fig. II.32. Dependence of D_0(*1*), S_0(*2*) and [η] (*3, 4*) on the molecular weight of nitrocellulose fractions in methyl ethyl ketone (*1–3*) and dioxane (*4*). The points marked with *asterisks* correspond to the fractions the molecular weights of which were determined by Eqs (5.9) and (5.10)

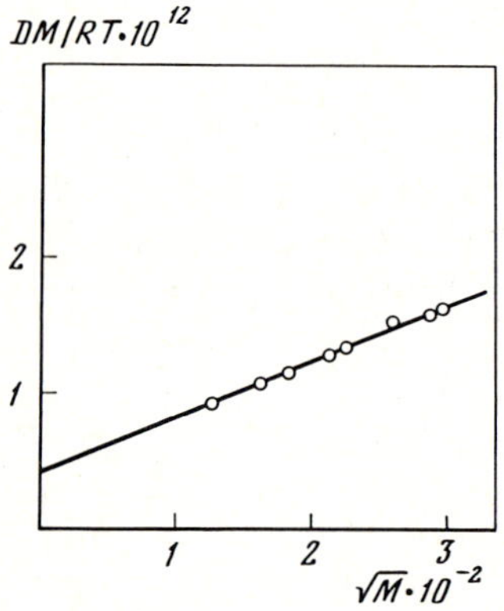

Fig. II.33. Dependence of DM/RT on $M^{1/2}$ for different nitrocellulose fractions in methyl ethyl ketone

interpretation of the corresponding experimental data is given in Fig. II.33. It follows from these results that the length of the Kuhn segment $A = (150 \pm 30)$ Å and diameter of the wormlike chain $d = (10 \pm 4)$ and (14 ± 5) Å for the values 1.43 and 1.07 respectively of numerical coefficient γ in Eq. (5.11). Note that the values of A and d found for nitrocellulose with the above degree of substitution happen to be close to those obtained for some other polysaccharides [76, 80]. The acquired magnitudes of the translational diffusion coefficient D and the intrinsic viscosity [η] were used for the determination of the equilibrium rigidity of the molecules with the application of the theory of translational friction [78] and the viscosity of wormlike chains [79]:

$$KT/D = P \times \eta_0 \times M^{1/2}(\lambda A/M_0)^{1/2} \tag{5.12}$$

$$[\eta] = \Phi M^{1/2}(\lambda A/M_0)^{3/2} \tag{5.13}$$

where P and Φ—the functions L/A and d/A; L and d—contour length and hydrodynamic chain diameter; λ—the projection of the monomer unit length on the chain direction.

Specifying the structural characteristics of the chain $\lambda = 5.17$, $M_0 = 247$ (at a given degree of substitution equal to 1.9) and varying d/A parameter, one obtained (according to viscometry and diffusion) the length of the Kuhn segment and the hydrodynamic chain diameter, shown in Table II.6. It can be seen from these data that the proposition about the two-fold change of the hydrodynamic chain diameter result in the Kuhn segment length taken from viscometry data differing by less than 20%. This fact makes possible the determination of the equilibrium rigidity of molecular chains of nitrocellulose, in particular from hydrodynamic data, using theory [78, 79].

Averaging the results given in Table II.6 affords the most probable A values from viscometric measurements: $A = (209 \pm 14)$ Å at $d = (6.8 \pm 1.8)$ Å in dioxane; $A = (192 \pm 13)$ Å at $d = (6.3 \pm 1.7)$ Å in cyclohexanone; from translational friction data: $A = (250 \pm 33)$ Å at $d = (8.0 \pm 1.6)$ Å in methylethylketone.

The hydrodynamic properties of nitrocellulose are markedly affected by the degree of substitution of hydroxyl groups [75].

The results of viscometric studies of some fractions of highly substituted cellulose nitrate (degree of substitution is 2.7 in ethyl acetate) of the above authors are shown in Fig. II.34. In this paper the determination of molecular weight was carried out with the Schultz equation (5.14) [81] which combines the intrinsic viscosity $[\eta]_{13.8}$ with the degree of polymerization $P = M/M_0$ for nitrates with nitrogen composition 13.8%.

$$\begin{aligned} P &= 0.95[\eta]_{13.8} && \text{at } P < 1000 \\ \lg P &= 1.32 \lg [\eta]_{13.8} - 1.0 && \text{at } P > 1000. \end{aligned} \tag{5.14}$$

Table II.6. The values of the length of Kuhn segment (A) and chain hydrodynamic diameter d(Å) for nitrocellulose molecules depending on parameter d/A according to diffusion and viscometric data

Nitrocellulose solution	d(Å)					
	0.02		0.03		0.05	
	A	d	A	d	A	d
In methylethylketone ($D = 5.2 \times 10^{-7}$)	300	6	250	7.5	200	10
In dioxane ($[\eta] = 2.08 \times 10^2$)	230	4.6	206	6.20	190	9.5
In cyclohexanone ($[\eta] = 1.9 \times 10^2$)	210	4.2	190	5.80	177	8.9

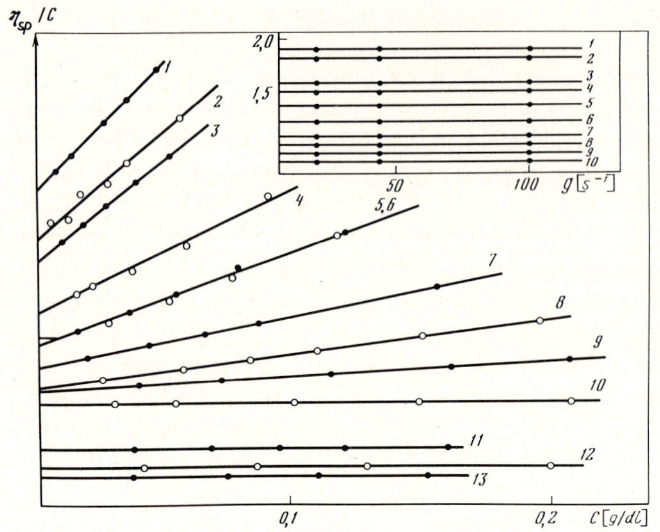

Fig. II.34. Dependence of η_{cp}/C values on the concentration C for various nitrocellulose fractions in ethyl acetate and the dependence of relative viscosity τ/τ_0 on the velocity gradient for fractions I in cyclohexanone. The numbers on the curves correspond to the number of the fractions presented in Table II.7, for fractions *1* to *10*—the upper scale of concentration C, for fractions *11* to *13*—the lower one. Concentration (in g/ml): *1*—0.04; *2*—0.038; *3*—0.028; *4*—0.025; *5*—0.021; *6*—0.015; *7*—0.011; *8*—0.01; ·*9*—0.007; *10*—0.005

In order to obtain experimental data about intrinsic viscosity in ethyl acetate, $[\eta]_{N=13.4}$ values were recalculated to $[\eta]_{N=13.8}$ using formulas:

$$[\eta]_N = [\eta]_{13.8}[1 - \varkappa(13.8 - N)] \tag{5.15}$$

where $\varkappa = 0.32 + 5.0 \times 10^{-5}[\eta]_{13.8}$.

The magnitudes of intrinsic viscosity $[\eta]_{13.8}$ and molecular weights of the corresponding fractions as well as their weight portions P_i in milligrams are given in Table II.7. Molecular-weight distribution of the initial sample in integral $[\omega(M)]$ and differential $(d\omega/dM)$ forms (Fig. II.35) was calculated on the assumption that the polydispersity of this sample was governed only by the difference in molecular weight. The results of diffusion studies as the curves $\sigma^2 = 2Dt$ from t are given in Fig. II.36. The coefficients of diffusion $D(C)$ at corresponding concentrations are determined on the basis of the curves' slopes. It is important that the $D(C)$ value is practically independent of C within the concentration range $C = 0.03$–0.1% and it may be considered as equal to the limit value at $C \rightarrow 0$. The values of diffusion coefficient D and hydrodynamic constant $A_0 = (3.5 \pm 0.15) \times 10^{10}$ erg/grad were close to the corresponding values for a low substituted sample of nitrocellulose (see Table II.5). The experimental dependence of DM/RT on $M^{1/2}$ (Fig. II.37) for the highly substituted cellulose nitrate under consideration results in the length of the Kuhn

Table II.7. Hydrodynamic characteristics of nitrocellulose fraction (degree of substitution 2.7) in ethylacetate at 293 K

Fraction number	P_i (mg)	$[\eta]$ (dl/g)	N, %	$[\eta]_{13.8}$ (dl/g)	$M \times 10^{-3}$	$D \times 10^7$ (cm²/sec)	$A_0 \times 10^{10}$ (erg/grad)
1	102.9	20.0	13.42	22.8	770	0.95	3.5
2	153.4	16.2	13.35	18.7	590	1.05	3.3
3	252.2	15.6	13.34	18.2	570	1.10	3.4
4	216.0	11.9	13.60	14.0	380	1.40	3.5
5	173.3	9.5	13.56	11.3	304	1.60	3.5
6	132.8	9.5	13.36	11.3	304	1.80	3.5
7	119.7	7.6	13.50	9.1	245	2.05	3.8
8	115.2	6.0	13.69	7.2	194	2.13	3.4
9	101.5	5.6	13.92	6.8	180	2.40	3.6
10	80.4	4.8	12.87	5.9	160	2.80	3.9
11	87.5	3.1	13.02	3.7	100	3.40	3.5
12	51.2	2.0	12.34	2.5	70	4.40	3.5
13	45.5	1.6	12.28	1.9	50	5.10	3.4
14	19.4	1.3	—	1.5	40	5.60	3.2

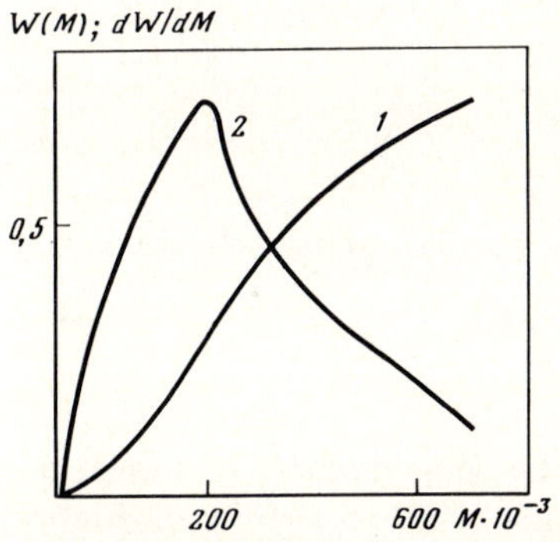

Fig. II.35. Integral ω(M)(*1*) and differential dω/dM (*2*) molecular weight distribution of the initial nitrocellulose sample

segment $A = (300 \pm 50)$ Å at $\lambda = 5.15$ Å and in the diameter of the wormlike chain $d = (22 \pm 7)$ Å and $d = (15 \pm 5)$ Å for the values of coefficient v 1.43 and 1.07 respectively in Eq. (5.11). The value $A = 300$ Å obtained for the highly substituted cellulose nitrate in ethyl acetate markedly exceeds that of $A = 160$ Å determined by the same method for low substituted nitrate which indicates the increase of equilibrium rigidity of nitrocellulose molecules with the enhancement of degree of substitution.

The increased kinetic rigidity of highly substituted cellulose nitrate molecules can be illustrated by the dependence of the Δn-value (flow

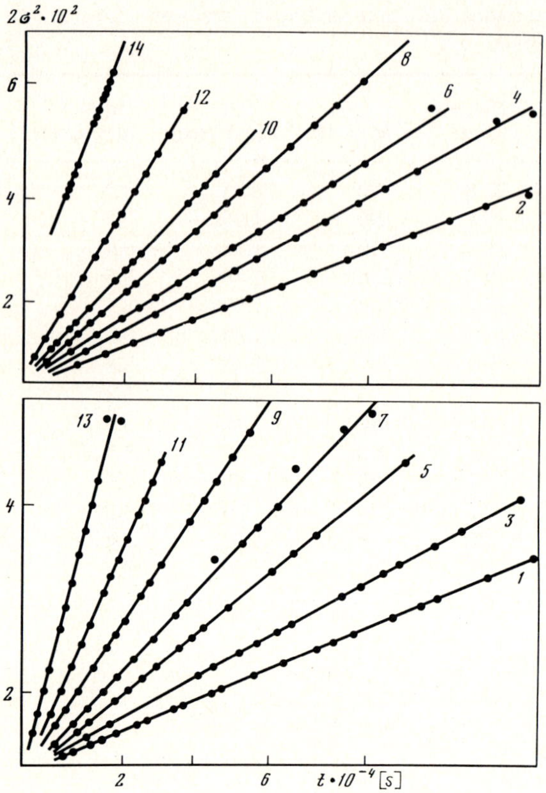

Fig. II.36. Dependence of $2\sigma^2$ on time t for various fractions (*1* to *14*) of a highly substituted nitrocellulose sample in ethyl acetate

birefringence—FB) on the shear stress $\Delta\tau$. This dependence is shown in Fig. II.38 for solutions of high molecular weight ($M = 5 \times 10^5$) of highly substituted cellulose nitrate in cyclohexanone. The dependence observed (negative deviation from linearity) is typical for kinetically rigid asymmetric particles (for example ellipsoids) in the solution of which FB is the result of their kinematic orientation without change of shape [69].

The characteristic value $\Delta n/\Delta\tau$ determined on the basis of the initial slope of the curve Fig. II.38, is equal to

$$(\Delta n/\Delta\tau) = -38 \times 10^{-10} \quad \Delta\tau \to 0$$

The absolute value of this negative magnitude markedly exceeds that of low substituted nitrocellulose. This variation is determined not only by the difference in molecular weights of the polymers but by the higher degree of substitution, since the magnitude and the sign of optical anisotropy of nitrocellulose molecules are responsive to the degree of esterification [82].

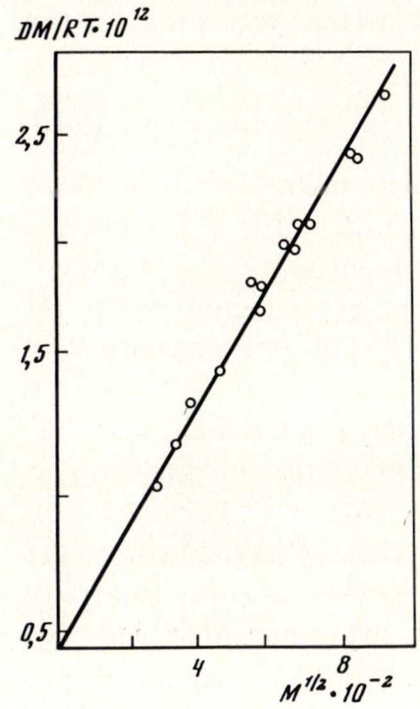

Fig. II.37. Dependence of DM/RT on $M^{1/2}$ for various fractions of a highly substituted nitrocellulose sample in ethyl acetate

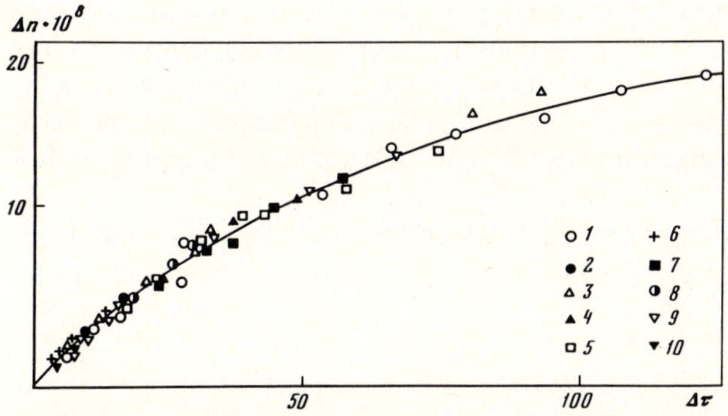

Fig. II.38. Dependence of Δn on the shear stress τ for a high molecular weight nitrocellulose sample in cyclohexanone. Concentration (%): *1*—0.155; *2*—0.107; *3*—0.101; *4*—0.064; *5*—0.0605; *6*—0.044; *7*—0.043; *8*—0.0302; *9*—0.0272; *10*—0.0214

The information about the effect of the cellulose origin (cotton, flax) on the conformation characteristics of its nitroesters is of some interest. The preparations studied were obtained from purified and bleached normal (NLC-1) and short staple (NLC-2) flax cellulose and the standard nitrating mixture.

The molecular weights (MW) values calculated on the basis of the Svedberg equation

$$M_{SD} = (RT/(1 - v\rho_0)) \times (S_0/D_0)$$

and those of hydrodynamic invariant A_0 [69] estimated according to the equation:

$$A_0 = \eta_0 D_0 (M[\eta]/100)^{1/3} T^{-1} \tag{5.16}$$

for some fractions and starting samples (Table II.8) demonstrate that the mean value for fractions of NLC-1 $A_0 = (3.21 \pm 0.22) \times 10^{-16}$ J $\times$ grad^{-1} $\times$ kmol$^{-1/3}$ and that of NLC-2 $A_0 = (3.30 \pm 0.36) \times 10^{-16}$ J $\times$ grad^{-1} $\times$ kmol$^{-1/3}$ are in good correlation with the mean experimental value $A_0 = (3.32 \pm 0.32) \times 10^{-16} \times$ J $\times$ grad^{-1} $\times$ kmol$^{-1/3}$ [84] for polysaccharides of different structure in the organic solvent.

Using M_{SD} values and weight portions of separate fractions $\omega_i = P_i/\Sigma P_i$ (P_i—fraction weight) the molecular weight distributions of the starting samples were calculated on the basis of the formulae $W(M) = \sum_{i=1}^{k=1} \omega_i + 1/2\omega_k$ (Fig. II.39). The average parameters M_n, M_w and M_z defined as the ratio of the corresponding moments of weight distribution, as well as polydispersion M_z/M_w and M_w/M_n for NLC-1 and NLC-2 are given in Table II.8 and Fig. II.39. The latter data demonstrate significant polymolecularity of nitrates prepared from flax cellulose which exceeds that of nitrates obtained from cotton cellulose (NCC).

The coefficients of diffusion and sedimentation at infinite dilution D_0 and S_0 presented in Table II.9 were determined by extrapolation of D and S to zero concentration of solution $C \to 0$. In the region of small polymer concentrations $C = (0.03\text{–}0.3) \times 10^{-2}$ g/cm^3 the concentration dependence of diffusion coefficient $D(C)$ is practically absent. The concentration dependences of the sedimentation coefficient $S(C)$ given in Fig. II.40 are related to the linear equation $S^{-1} = S_0^{-1}(1 + K_s C)$.

Fig. II.41 demonstrates the experimental dependences of $[\eta]$, S_0 and D_0 from the degree of polymerization Z, which can be approximated by equations (for D

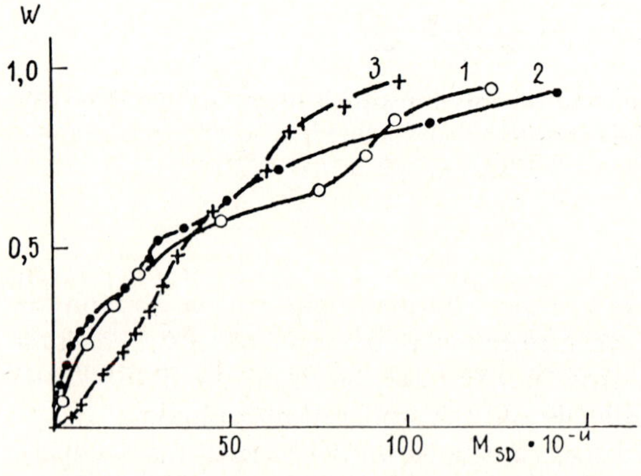

Fig. II.39. Weight MWD of reprecipitated NC in integral form according to fractional precipitation: NLC–1 (*1*), NLC-2 (*2*), NCC according to papers [83], [84] (*3*)

Table II.8. Polydispersity parameters of NC samples according to fraction precipitation

Sample	$[\eta] \times 10^{-2}$ (cm^3 g)	N (%)	$M_n \times 10^{-3}$	$M_w \times 10^{-3}$	$M_z \times 10^{-3}$	M_w/M_n	M_z/M_w
NLC-1[a]	10.5	12.9	108	490	844	4.54	1.72
NLC-2[a]	11.2	12.7	56	489	1014	6.75	2.07
NCC	10.0	13.4	256	427	582	1.67	1.35

[a] Reprecipitated samples

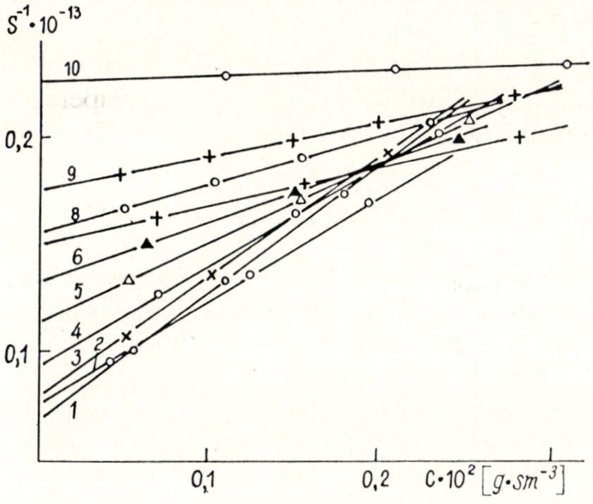

Fig. II.40. Concentration dependences for sedimentation coefficients S for NLC fractions in ethyl acetate. The number at the curves correspond the following fractions in Table II.10: *1*—1(1); *2*—1(4); *3*—2(4); *4*—1(5); *5*—1(6); *6*—2(10); *7*—1(8); *8*—2(11); *9*—2(13); *10*—1(9)

and S_0 within Z-values range from 21 up to 6300 and for $[\eta]$ Z-values range up to 3160) for both homologues series of NLC

$$
\begin{aligned}
[\eta] &= 1.46\, Z^{0.90} (\mathrm{cm^3\, g^{-1}}) \\
D &= 1.43 \times 10^{-5}\, Z^{-0.63} (\mathrm{cm^2\, s^{-1}}) \\
S_0 &= 7.33 \times 10^{-14}\, Z^{0.37}\, (\mathrm{s})
\end{aligned}
\tag{5.17}
$$

The equations obtained correlate with similar dependences for NCC samples with close nitrogen content [83], [84]. The high values of the exponent in Eqs. (5.17) demonstrate the significant role of permeability effects in NLC solutions and are typical for the majority of polysaccharides.

The estimation of equilibrium rigidity of NLC-1 and NLC-2 molecules on the basis of measurements of the translational diffusion D has been carried out based on Eq. (5.18) from the theory of translational friction of wormlike chains

Table II.9. Hydrodynamic properties of NLC-1 and NLC-2 samples and their fractions in ethylacetate at 298 K

Fraction	$P_i \times 10^2$ (g)	$[N]$ (%)	$[\eta] \times 10^{-2}$ ($cm^3 g^{-1}$)	k	$D_0 \times 10^7$ ($cm^2 s^{-1}$)	$S_0 \times 10^{13}$ (s)	$M_{SD} \times 10^{-3}$	$K_s \times 10^{-2}$ ($cm^2 g^{-1}$)	v	$A_o \times 10^{16}$ J $grad^{-1}$ $kmol^{-1/3}$
1	2	3	4	5	6	7	8	9	10	11
					Sample NLC-1					
1	0.2127	12.5	18.0	0.52	0.62	14.5	1242.0	8.50	0.47	3.06
2	0.1058	13.2	23.5	—	0.76	13.7	957.0	8.00	0.34	3.15
3	0.2874	13.2	16.4	0.53	0.75	12.5	885.0	6.90	0.38	2.77
4	0.0626	12.9	16.0	0.76	0.94	13.2	746.0	6.30	0.38	3.18
5	0.2934	13.0	12.3	0.40	1.19	10.5	468.5	4.80	0.39	3.14
6	0.2578	13.2	8.0	0.65	1.90	8.7	243.0	3.20	0.40	3.48
7	0.1084	12.7	5.0	0.72	2.40	7.5	166.0	2.10	0.42	3.32
8	0.2986	12.9	3.3	0.41	3.50	6.7	102.0	1.20	0.36	3.58
9	0.2928	11.7	0.7	0.22	9.40	4.4	24.8	0.14	0.20	3.58
NLC-1[a]	—	13.0	10.5	0.68	1.45	10.4	384.0	2.90	0.26	3.39

Table II.9 *(continued)*

1	2	3	4	5	6	7	8	9	10	11
					Sample NLC-2					
1	0.2394	12.58	22.50	0.53	0.50	16.1	1726.0	9.2	0.41	2.50
2	0.1196	13.04	20.00	0.46	0.60	14.3	1277.0	8.2	0.41	2.60
3	0.1645	12.05	13.40	0.52	1.04	12.5	638.0	6.0	0.45	3.13
4	0.0628	—	13.00	—	1.37	12.5	485.0	7.3	0.56	3.72
5	0.0534	13.00	11.80	0.38	1.46	10.4	378.0	—	—	3.53
6	0.0784	12.20	7.80	—	1.80	9.8	289.0	2.7	0.35	3.47
7	0.0475	—	7.00	0.37	1.84	—	284.0[b]	—	—	—
8	0.0536	12.80	8.00	0.44	2.00	9.2	244.0	2.7	0.34	3.67
9	0.0761	12.30	5.70	0.45	2.20	8.7	210.0	—	—	3.43
10	0.1077	13.70	4.50	0.36	2.70	7.6	150.0	2.5	0.56	3.48
11	0.0527	12.80	3.34	0.30	3.40	6.4	100.0	1.4	0.39	3.47
12	0.0391	—	3.00	0.20	6.10	—	80.0[(b)]	—	—	—
13	0.0344	12.60	1.80	0.41	5.00	5.7	60.6	1.0	0.56	3.51
14	0.1274	12.30	1.40	—	6.25	—	36.2	—	—	—
15	0.0988	12.40	0.96	—	6.46	4.3	35.4	—	—	3.08
16	0.0916	9.60	0.22	—	16.80	1.8	5.7	—	—	2.66
NLC-2	—	—	11.20	0.77	1.15	11.4	370.0	3.98	0.36	3.90

[a] Non-fractionated reprecipitated sample
[b] MW were obtained using mean experimental A value

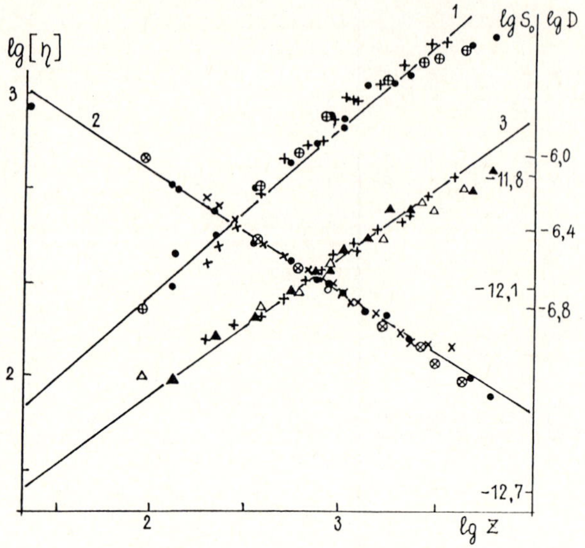

Fig. II.41. Plots of $[\eta]$ (*1*), d (*2*) and S (*3*) as a function of Z for fractions of NLC-1 (○---○), NLC-2 (●---●) and NLC (× --- ×) according to [84]

with permeation.

$$D_z/RT = (P\eta_0 N_A)^{-1}(A\lambda)^{-1/2\,z^{1/2}} + (3\pi\eta_0 N_A\lambda)^{-1}(\ln A/\alpha - Q) \quad \text{at} \quad L/A \geq 2.3 \tag{5.18}$$

where A—the length of Kuhn segment; d—hydrodynamic diameter of the molecule; $\lambda = 5.15$ Å—the length of projection of the cellulose monomer unit in the basic direction; $\eta_0 = 0.430 \times 10^{-2}$ g × cm^{-1} × s—solvent viscosity at 298 K; N_A—Avogadro number; R—gas constant; P and Q theoretical constants [72, 78]. From the slope of the curve D_z/RT versus $z^{1/2}$ (Fig. II.42, curve I) one can find $(P \times \eta_0 \times N_A)^{-1}$ $(A\lambda)^{-1/2} = 0.1677 \times 10^{-15}$, and from the intercept of the curve *I* on ordinate—$(3\pi\eta_0 N_A\lambda)^{-1} \times (\ln(A/d) - Q) = 1.6 \times 10^{-15}$. These data result in Kuhn segment values $A_D = (380 \pm 50)$ Å $(P = 5.11)$ [72, 78], and the hydrodynamic diameter of the molecule $d_D = (17 \pm 5)$ Å $(Q = 1.056)$ [78] and 12 ± 3 Å $(Q = 1.43)$ [72].

The presence of low molecular weight fractions for NLC-1 and NLC-2 in contrast to NCC [83, 84] allows one to use the model of the long direct rod [85] for analysis of the conformation of NLC molecules on the basis of diffusion data, where the following equation is valid

$$D_z/RT = (3\pi\eta_0 N_A\lambda)^{-1}(\ln z - \ln(d/A) + 0.38) \tag{5.19}$$

The relationship of D_z/RT to $\ln z$ (curve 2) is presented in Fig. II.42. The slope of the tangent curve plot in the low molecular weight region (Fig. II.42, curve 3) $\Delta(D_z/RT)/\Delta\lg z = 0.178 \times 10^{-14}$, was calculated according to Eq. (5.19) using the value $\lambda = 5.15$ Å. The free term of linear function of D_z/RT

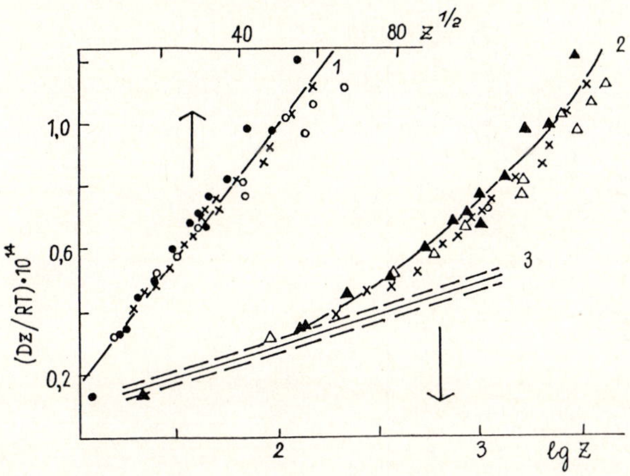

Fig. II.42. Plots of D_2/RT as a function of $Z^{1/2}$ and lg z (*2*) for fractions of NLC-1 (○ --- ○), NLC-2 (● --- ●) and NCC (× --- ×) according to [83, 84]; (*3*)—tangent to the curve 2 at small z

from $\ln z(3\pi\eta_0 N_A\lambda)^{-1}(\ln(A/d) - 0.38) = (0.58 \pm 0.18) \times 10^{-15}$ found from the curve 3 position on Fig. II.42 leads to the value of the hydrodynamic diameter $d_D = 16 \pm 5$ Å.

The relationship of $z/M_0[\eta]$ to $z^{1/2}$ is nonlinear (Fig. II.43, curve 1), which indicates both deviation of NLC molecule size from that of the Gauss coil and enhanced permeation effects. For rigid chain polymers the transfer to the linear part of the relationship described by Eq. 5.20 of the theory of the intrinsic viscosity of the weakly permeated Gauss coil [79, 85].

$$Z/M_0[\eta] = \Phi^{-1}(A)^{-3/2 \times z^{1/2}} + K(A\lambda^2)^{-1} \times [\ln(A/d) - Q] \text{ at } L/A \gg 1 \tag{5.20}$$

(Φ, K and Q—theoretical constants) proceeds at a L/A value about several tens of segments [85]. At $A/d = 22.4$ for NLC the area of the linear function related to Eq. (5.20) is $L/A \geq 70$ [85] and 24 [86] (Fig. II.43 vertical line 1). On the basis of the curve 1 – 1 slope (Fig. II.43) in the region $z^{1/2} \geq 42$ ($L/A \geq 24$) we can determine from Eq. (5.20) the value $A_n = (197 \pm 100)$ Å ($\Phi = 2.86 \times 10^{23}$ mol^{-1}) [79].

It is worthwhile plotting $(z^2/M_0[\eta]^{1/3})$ as a function of $z^{1/2}$ in this case [87] at $L/A \geq 2.3$

$$(z^2/M_0[\eta])^{1/3} = \Phi^{-1/3}(A\lambda)^{-1/2} Z^{1/2} + R[\ln(A/d) - 1.056] \div (3\pi\eta_0 N_A \lambda A_0) \tag{5.21}$$

The value $A_\eta = (250 \pm 30)$ Å was found from the slope of the curve 2 (Fig. II.43) in the region $L/A \geq 2.3$ (vertical line II), equal to $\Phi^{-1/3}(A\lambda)^{-1/2} = 0.046$ using the theoretical value $\Phi = 2.86 \times 10^{23}$ mol^{-1}. On the basis of the

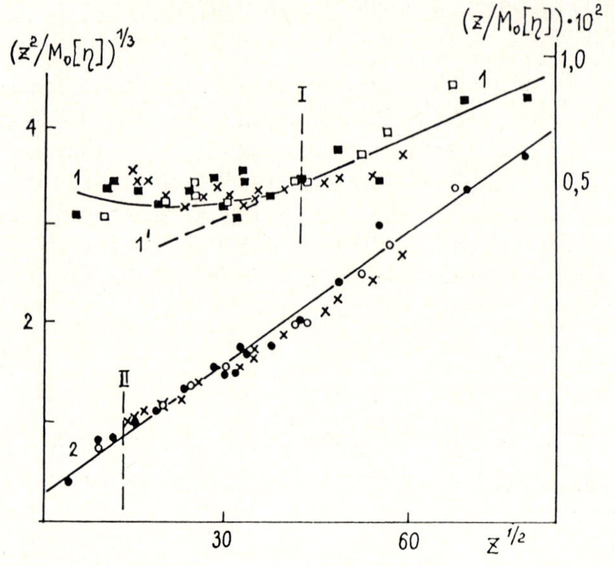

Fig. II.43. $z/M_0[\eta]$ (*1*) and $(z^2/M_0[\eta])^{0.33}$ (*2*) versus $z^{0.5}$ for fractions of NLC-1 (○ --- ○), NLC-2 (● --- ●) and NCC (× --- ×) according to [83, 84]. The curve 1–1′—linear approximation of the function $z/M_0[\eta]$ on $z^{1/2}$ in the region $z^{1/2} \geq 42$

intercept curve 2 at ordinate axis equal to $R \times (\ln A/d - 1.056)/(3\pi\eta_0 N_A A_0 \lambda) = 0.3$ we determine the hydrodynamic diameter of the molecule $d_n = (17 \pm 4)$ Å using the mean value $(A_0)_m = 3.3 \times 10^{-16}$ J × grad^{-1} × kmole$^{-1/3}$ for NLC-1 and NLC-2 fractions. Discrepancy between the A_D and A_n values correlates qualitatively with that of experimental and theoretical values of hydrodynamic invariant A_0 and it is related to the mutual inconsistency of the theoretical values of hydrodynamic constants P and Φ.

Thus, although points on the experimental curves (Fig. II.41–43) are positioned with significant dispersion (that might be connected with specific features of the polymer system under study) the obtained values of equilibrium rigidity and hydrodynamic diameter of the chain correlate with the corresponding molecular parameters of NLC with a similar degree of substitution.

It follows from the above data, that cellulose nitrates with a similar degree of substitution 2.6 ± 0.1 obtained from cotton and flax exhibit similar hydrodynamic and conformation properties typical for rigid chain polymers with high equilibrium chain rigidity $A \approx 400$ Å. At the same time NLC samples studied are characterized by a broader MWD, that is, the sample contains significant portions of both low and high molecular weight fractions. The parameters of polydispersity of nitroesters obtained may characterize in some extent the MWD of the initial flax and cotton celluloses, which is governed by the degree of destruction upon nitration by the above methods.

5.2 Dynamooptic Properties of Cellulose Derivatives

The length of Kuhn segment A or persistent length $a = A/2$ being a measure of equilibrium rigidity of the main chain of macromolecule allow us, to some extent, to evaluate the orientation order in the molecule. However the direct measure of the orientation order of the polymer molecule is optical anisotropy, defined as the difference between two basic polarizabilities $\gamma_1 - \gamma_2$ in the direction parallel to vector h connecting the chain ends and perpendicular to it. The optical anisotropy in the equilibrium conformation for the Gauss chain [71] does not depend on the molecular weight and close to anisotropy of the segment $\alpha_1 - \alpha_2$:

$$\gamma_1 - \gamma_2 = 3/5(\alpha_1 - \alpha_2) \tag{5.22}$$

The difference in polarizabilities of the segment $\alpha_1 - \alpha_2$ depends on the anisotropy of the segment $a_{\parallel} - a_{\perp}$ and the number of monomer units S involved in the segment:

$$\alpha_1 - \alpha_2 = S(a_{\parallel} - a_{\perp}) \tag{5.23}$$

This suggests that high optical anisotropy is inherent to rigid chain molecules characterized by a high S value. It is substantial that in contrast to hydrodynamic properties markedly dependent on the swelling effect in good solvents, optical anisotropy remains almost constant in this case [70]. Therefore optical anisotropy is the measure of the true skeletal rigidity of the polymer chain regardless of the excluded volume effect. It is important that anisotropy of the monomer unit and the polymer molecule as a whole in relation to the degree of regulation of valent bonds of the molecular unit with respect to the main chain may have both positive and negative values. This allows us to consider it as a quantitative feature of the axis orientation order of structural elements of both main chain and side substituents.

The relation between the anisotropy of the molecule and the degree of orientation order Q is governed by relationship

$$Q = (\gamma_1 - \gamma_2)/P(a_{\parallel} - a_{\perp}) \tag{5.24}$$

where P—degree of polymerization.

Optical anisotropy of the polymer molecule is usually determined experimentally by the measurement of flow birefringence of the studied polymer solution with respect to the Kuhn relationship [70]:

$$\begin{aligned}\Delta n/\Delta\tau = \Delta n/g(\eta - \eta_0) = [n]/[\eta] = [4\pi(n_S^2 + 2)^2/45\, KT_{ns}] \\ \times (\alpha_1 - \alpha_2)\end{aligned} \tag{5.25}$$

where $\tau = g(\eta - \eta_0)$—efficient shear stress in the flow with velocity gradient g; $[n]$—specific birefringence of the solution: $[n] = (\Delta n/g_0 C)$, $C \to 0$, where C—concentration.

Provided refractive index of the polymer n_k differs from that of the solvent n_s, additional positive contribution to anisotropy of the molecule appears which is governed by the mutual interaction of the chain segments—"shape effect" [88].

At that time the value $Q \approx 1$ is inherent to the systems with a rather perfectly regulated structure, which occurs in ideal space monocrystals. Less perfect, but rather a high orientation order is typical for mesomorphic systems. So, the Q value for the nematic mesophase is in the range 0.4–0.7 [89].

It is important that the above specific features of the optical properties of rigid chain molecules are independent of both the excluded volume effect and the hydrodynamic interaction. In this connection their application for characteristics of rigidity and orientation order of polymer chains does not require evaluation of the thermodynamic quality of the solvent and the permeation effect.

The results of investigation of flow birefringence (FB) of diluted solutions of non-fractionated low-substituted cellulose nitrate (degree of substitution 1.9) in dioxane and cyclohexanone [90] are presented in Fig. II.44. As follows from the above data, flow birefringence n_g of the studied solutions (with subtraction of the solvent effect) appears to be high, has a positive sign in dioxane, negative in cyclohexanone and it is proportional to the flow rate gradient at all concentrations studied. The slopes of the curves $\Delta n_g = f(g)$ (Fig. II.44) afford the value of reduced optical anisotropy $\Delta n/\Delta\tau$, where $\Delta\tau = g(\eta - \eta_0)$ is the shear stress. It is interesting to compare these results with similar data on birefringence in an

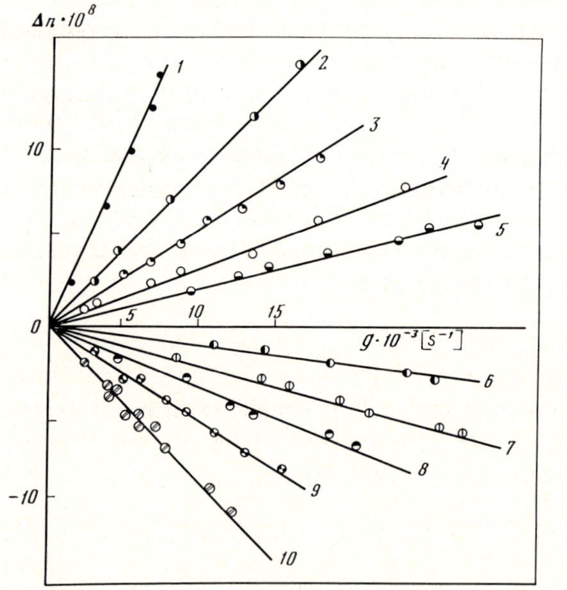

Fig. II.44. Plots of birefringence Δn as a function of flow rate gradient for nitrocellulose solutions in dioxane. Nitrocellulose concentration in dioxane (%) *1*—0.533; *2*—0.37; *3*—0.206; *4*—0.104; *5*—0.081; in cyclohexanone *6*—0.06; *7*—0.107; *8*—0.174; *9*—0.22; *10*—0.34

electric field (BE). Fig. II.45 demonstrates extrapolations of $\Delta n/\Delta\tau$ and K-values to zero concentration, while dynamo and electrooptical characteristics of low-substituted cellulose nitrate solutions are given below

$([n]/[\eta]) \times 10^{10}$	11.7 -5.0^a	$\mu_0(D)$	0.9
		$D_r \times 10^{-4}$ (FB)	0.8
$K \times 10^{10}$	7.1		
$\mu \times (D)$	95	$D_r \times 10^{-4}$ (BE)	6.6

[a] Nitrate solution in cyclohexanone, in other cases—in dioxane.

The effect of the configuration is the sum of two effects. The first one (the effect of "macroconfiguration") is caused by the interaction of rather remote regions of the chain (optical long-range action) and it is governed by the asymmetry of the macromolecular coil as a whole. In the case of the Gauss chain the reduced birefringence of the macromolecule is determined by

$$([n]/[\eta])_f = [0.058\Phi(n_S^2 + 2)^2(n_K^2 - n_S^2)]/\pi\rho^2 N_A RTn_S^3(M/[\eta]) \quad (5.26)$$

where Φ—Flory coefficient, P—polymer density.

The second effect (effect of "microconfiguration") is caused by the interaction of the adjacent regions of the chain. In contrast to the effect of macroconfiguration, the reduced birefringence for the Gauss chain in the case of the microconfiguration effect does not depend on the polymer molecular weight but it is proportional to the segment molecular weight $M_S = A(M/L)$ and thus it is increased with enhancement of the chain rigidity:

$$([n]/[\eta])_{f_S} = [(n_S^2 + 2)^2(n_K^2 - n_S^2)^2/180\pi RTn_S^2\rho](L_2 - L_1)_S M_S \quad (5.27)$$

where $L_2 - L_1$ is the asymmetry coefficient of the segment configuration. In total, the experimentally observed configuration effect, the anisotropy of the microconfiguration, is the most important for rigid chain polymers. Therefore the experimental measurement of the microconfiguration effect for rigid chain molecules is the independent method for the determination of skeletal rigidity of the molecular chain.

The difference in the main polarizabilities $(\gamma_1 - \gamma_2)$ of the molecule with L length for wormlike chains is determined according to the theory of persistent anisotropy [91] leading to the relationship:

$$\begin{aligned}(\gamma_1 - \gamma_2)_L &= 3/5(\alpha_1 - \alpha_2)(1 - e^{-5X/6}) \\ &= \alpha\beta_0 X[1 - 5X/12 + 25X^2/216 - 125X^3/5184 + \ldots]\end{aligned} \quad (5.28)$$

where $X = L/a$; $a = A/2$, $\beta_0 = (\alpha_1 - \alpha_2)/A$—anisotropy of the unit of chain length.

In contrast to Gauss chains, the anisotropy of wormlike chain increases with enhancement of its length, approaching $X \to \infty$ to the limit value $3/5(\alpha_1 - \alpha_2)$ which corresponds to the anisotropy of Gauss chain. According to Eq. (5.28) the

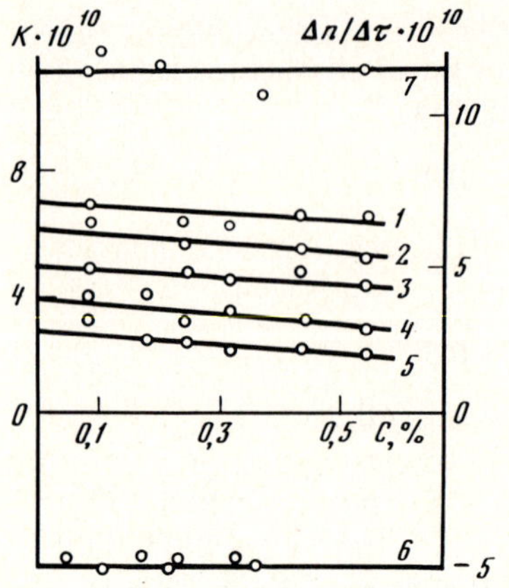

Fig. II.45. Relationship of K(*1–5*) and $\Delta n/\Delta\tau$-values (*6, 7*) to the concentration in dioxane at frequencies 0.15(*1*), 1(*2*), 4(*3*), 15(*4*) and 44 K Hertz (*5*) in cyclohexanone (*6*) and in dioxane (*7*)

reduced flow birefringence $([n]/[\eta])_L$ must grow in solutions of persistent chains with increase in molecular weight. Within small X-values ($X < 3$) this enhancement is defined by the expression

$$([n]/[\eta])_L = (4\pi/45\,KTn_S)(\gamma_1 - \gamma_2)_L \times F(X)(n_S^2 + 2)^2 \tag{5.29}$$

where $F(X)$ is the tabulated [92] function showing the change of geometric asymmetry of the molecule with increase of molecular weight; the value $(\gamma_1 - \gamma_2)_L$ is determined from Eq. (5.28). The comparison of experimental dependence of birefringence on the molecular weight with that of calculated from Eq. (5.27) allows us to determine the anisotropy of the length unit β_0 and persistent length $a = A/2$ of the molecular chain.

Equation (5.28) also permits us to estimate the change in the degree of intramolecular orientation order Q with increase in molecular weight. Thus, the joint consideration of Eqs. (5.23), (5.24) and (5.28), taking into account that $P/S = X/2$, results in the relationship

$$Q = (6/5X) \times (1 - e^{-5X/6}). \tag{5.30}$$

The relationship of Q to X according to Eq. (5.30) is presented in Fig. II.46 [70]. It follows from these data that the increase of X results in the decrease of ordering degree from about a unit up to values close to zero. It is also important that the coincidence of the signs of FB and BE is observed in nitrocellulose solutions in one solvent. This basic property indicates that there are (on average) three main directions in the macromolecule of rather high molecular weight: the directions of its longest geometric length, orientation-axial, and orientation-polar order. As it is demonstrated by comprehensive experimental data, this feature is typical of rather long chains of all rigid chain polymers, which can be approximated by Gauss coils [70, 93].

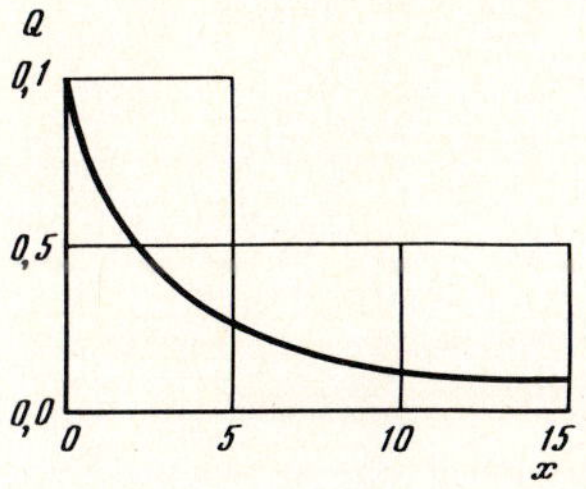

Fig. II.46. Relationship of the degree of high molecular weight orientation order Q to the parameter $X = L/a$ for the wormlike chain

Flow birefringence of the fraction of highly substituted nitrocellulose samples was studied in cyclohexanone that practically excluded the effect of configuration in observed FB [75]. The excessive birefringence was found to be significant in magnitude, negative in sign and proportional to the gradient of shear rate and concentration C (Fig. II.47). The values of reduced birefringence $[n]/[\eta]$ and reduced optical anisotropy $\Delta n/g(\eta - \eta_0) = \Delta n/\Delta\tau$ found from these data are presented in Table II.10.

As follows from these results, the values of $[n]/[\eta]$ are increased with enhancement of molecular weight, ($[n]/[\eta]$) is changed by 1.6 on a 20-fold increase of molecular weight, which is typical for all rigid chain polymers. In the case of high molecular weights where optical properties of nitrocellulose molecules follow Gauss statistics, the values $[n]/[\eta]$ attain the limit of $[n]/[\eta] = -44 \times 10^{10}\,\mathrm{cm} \times \mathrm{s}^2 \times \mathrm{r}^{-1}$. This allows us to use the following expressions for reduced birefringence

$$[n]/[\eta] = (4\pi/45\,KTn_S)(n_S^2 + 2)^2\,\Delta aS \qquad (5.31)$$

where Δa—optical anisotropy of monomer unit; S—the number of monomer units in the Kuhn segment; n_S—refraction index of the solution; K—Boltzmann constant. The calculation with this formula affords the value of segment optical anisotropy

$$(\alpha_1 - \alpha_2) = a \times S = -550 \times 10^{-25}\,\mathrm{cm}^3,$$

that corresponds to the $\alpha_1 - \alpha_2$ value obtained earlier for nitrocellulose with a close degree of substitution [88] and exceeds on almost an order of magnitude the value of $\alpha_1 - \alpha_2 = -62 \times 10^{-25}\,\mathrm{cm}^3$ determined for a low-substituted nitrocellulose sample [90]. The latter fact demonstrates that the negative optical anisotropy of nitrocellulose molecular chains is defined on the whole by the degree of substitution, and its absolute magnitude is markedly increased upon enhancement of the contents of the anisotrope side nitrate groups. The use of equilibrium rigidity value $S = A/\lambda = 60$ determined (see previous chapter) from hydrodynamic data at $\lambda = 5.15$ Å affords the magnitude of anisotropy of the monomer unit of studied nitrocellulose molecules in cyclohexanone

$$a = (\alpha_1 - \alpha_2)/S = -9 \times 10^{-25}\,\mathrm{cm}^3.$$

The experimental curves of relative value $(x) = ([n]/[\eta])/([n]/[\eta])_\infty$ versus parameter $X = 2M/M_0S$ at $M_0 = 282$ and $S = 40$ and 60 are presented in

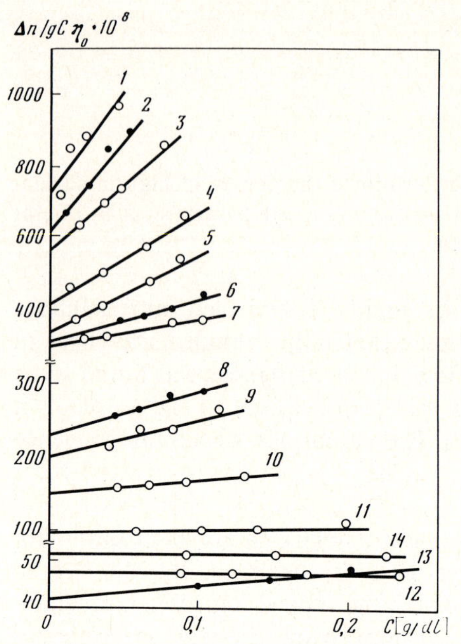

Fig. II.47. Plots of $\Delta n/gc\eta_0$—values as a function of the concentration C for various fractions (*1–14*) of nitrocellulose in cyclohexanone

Table II.10. Dynamooptic properties of nitrocellulose solutions in cyclohexanone at 293 K

Fraction number	$M \times 10^{-3}$	$[\eta]$ (dl/g)	$[n]/[\eta] \times 10^{10}$ ($cm\,s^2\,g^{-1}$)	$-\Delta n/\Delta\tau \times 10^{10}$ ($cm\,s^2\,g^{-1}$)	$[\lambda/g] \times 10^{10}$ (s)	G
1	770	16.5	− 43.0	− 43.0	53.00	0.26
2	590	14.6	− 42.5	− 43.0	26.00	0.33
3	570	13.5	− 43.0	− 43.0	22.00	0.32
4	380	10.2	− 43.9	− 42.8	12.00	0.34
5	304	8.5	− 42.3	− 42.5	8.20	0.35
6	304	7.4	− 42.6	− 42.5	6.50	0.32
7	245	7.4	− 42.6	− 41.0	5.90	0.36
8	194	5.9	− 40.6	− 40.7	4.00	0.39
9	180	5.2	− 39.1	− 40.6	3.50	0.41
10	160	4.1	− 37.8	− 37.6	2.00	0.33
11	100	3.0	− 34.5	− 34.2	1.30	0.49
12	70	1.7	− 29.5	− 27.0	0.65	0.59
13	50	1.5	− 26.6	− 27.0	0.40	0.55
14	40	1.6	− 32.0	− 35.5	0.60	0.71

Fig. II.48. The solid curves 1 and 2 depict the theoretical relationships for the kinetically flexible chain taking into account local rigidity [94] and for the kinetically rigid wormlike chain [95] respectively. The general type of experimental relationship presented in Fig. II.48 is markedly different from the theoretical curve 2 [95] but it correlates sufficiently with theoretical curve 1 [94], which is characterized by smooth dependence in the area of small X and by the slow approach to the limit in the area of large X. This means that FB in nitrocellulose solutions is connected with the kinetic flexibility of the macromolecular chain. The best correlation of experimental points and theoretical curve 1 within medium and large X is attained at $S = 40$ (black circles). Thus, the determined length of Kuhn segment $A = (250 \pm 50)$ Å is a qualitative estimation of the equilibrium rigidity which agrees with the results of hydrodynamic studies.

The application of photoelectric registration and high magnitude of $\Delta n/g$ allows us to measure [75] the orientation angles α in the region of rather low concentrations C, where the dependence of α on C is negligible. The dependence of $x = (45 - \alpha)$ for solutions of various nitrocellulose fractions (see Table II.10) is presented in Fig. II.49.

The values $(x/g)_{g \to 0} = \lim(\pi/4 - \alpha)/g$ are determined on the basis of the initial slope of the curves $x = f(g)$. The concentration relationships $(x/g)_{g \to 0}$ are shown in Fig. II.49, their extrapolation to $C \to 0$ affords the values of specific

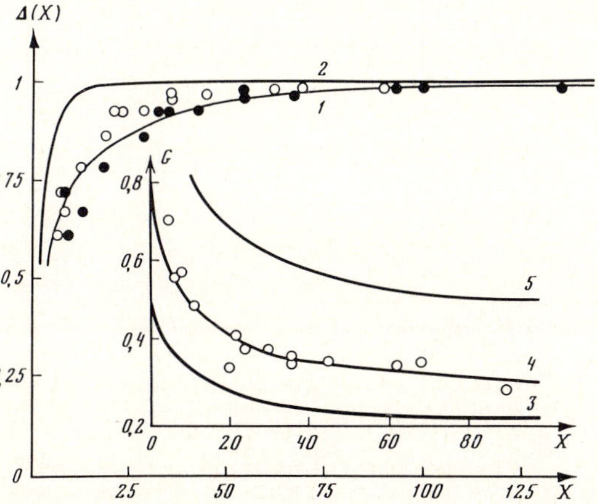

Fig. II.48. Relative values of $\Delta x = ([n]/[\eta])/([n]/[\eta])_\infty$ and coefficient G versus parameter X for the nitrocellulose fraction in cyclohexanone.

1—according to the Noda-Hearst theory; *2*—according to the Tsvetkov theory; *3*—according to the Noda-Hearst theory for a monodisperse sample; *4*, *5*—corrected using polydispersity for $M_w/M_n = 1.2$ and 1.6 respectively. *Black circles* correspond to experimental values of Δx at $S = 40$, *open circles*—at $S = 60$

angles $[X/Y] = \lim(X/g)$ given in Table II.10. The magnitudes of X/g obtained are comparable to those of $[\eta]$ in cyclohexanone and with molecular weights found by equations:

$$[X/g] = G(M[\eta]\eta_0/RT) \tag{5.32}$$

The values of coefficient G calculated by this way and dependent on conformation and hydrodynamic properties of nitrocellulose fractions studied are given in Table II.10 and in Fig. II.48 in relation to parameter x at $S = 60$. Curve 3 in Fig. II.48 shows the Noda–Hearst theoretical dependence according to which the coefficient G decreases from 0.5 up to 0.1 with an increase of compactness of molecular structure. The general character of experimental dependence corresponds to curve 3, with a 20-fold increase of molecular weight, the G-value decreases by 2.7 times. However the experimental G-values somewhat exceed the theoretical ones in absolute value. The basic reason of this seems to be polydispersity in the molecular weight which is not considered by

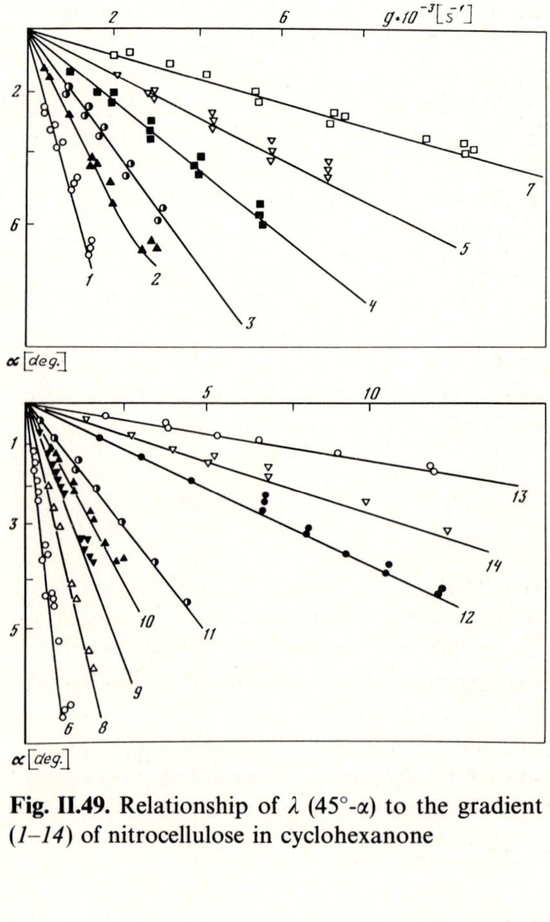

Fig. II.49. Relationship of λ (45°-α) to the gradient of low rate g for various fractions (*1–14*) of nitrocellulose in cyclohexanone

the Noda–Hearst theory. Equation (5.32) for a polydisperse sample should be transformed [96, 97] into expression

$$[x/g] = G_w(M[\eta]\eta_0/RT) \tag{5.33}$$

where $G_w = (P/w)G$, P—factor of polydispersity depending on the b-value in the Mark–Kuhn equation $[\eta] = K \cdot M^b$, on the width and type of molecular-weight distribution of the sample under study. Schultz–Zimm type distribution P/W is expressed by the following expression [96, 97]:

$$P/W = (n!(n+2b+1)!)/((n+1)[(n+b)!]^2) \tag{5.34}$$

where $n = 1/w - 1$; $w = M_w/M_n$.

According to some papers [81, 90] $b = 1$ for nitrocellulose in a wide molecular weight range.

Provided Schultz–Zimm type molecular-weight distribution is typical for studied nitrocellulose fractions, the values $G_w = (P/W)G$ were determined upon variation of the Schultz parameter W. Thus corrected with respect to polydispersity, plots of G_w as a function of X according to Noda–Hearst are shown in Fig. II.48 (curves 4 and 5). Experimental points correspond to curve 4 (see Fig. II.48) for which the Schultz parameter $W = 1.2$, which is markedly lower than the polydispersity of the non-fractionated sample ($W = 1.6$) (see Fig. II.48 curve 5). Thus the obtained experimental data of X/g values (Fig. II.50) and their molecular weight dependences for nitrocellulose can be interpreted within existing theories considering both local rigidity of the chain [94] and polydispersity in molecular weight [96, 97].

The results of the investigation of flow birefringence of nonfractionated nitrocellulose sample are given in Table II.11. They involve experimental values of reduced birefringence $[n]/[\eta]_{exp}$ and corresponding (Eq. 5.31) values of segment anisotropy $(\alpha_1 - \alpha_2)$ and monomer unit a_{exp}. For the calculation of a_{exp}, the parameter of equilibrium rigidity $S = 60$ was used, and this was obtained from hydrodynamic data (see previous chapter). The different values $[n]/[\eta]_{exp}$ obtained for nitrocellulose in various solvents can be explained by the effect of the microconfiguration [69], that is illustrated qualitatively by theoretical dependence:

$$[n]/[\eta] = [n]/[\eta]_i + [(n_S^2 + 2)^2 \times (n_K^2 - n_S^2)^2/180\pi RTn_S^3 \times \rho] \times M_0 S \times 2\pi \tag{5.35}$$

The reasonable agreement of experimental data with theoretical dependence, Eq. (5.35), is observed when $S = 40$, the refractive index of nitrocellulose solution $n_K = 1.49$, and the value of the inherent effect $[n]/[\eta]_i = -48 \times 10^{10}$ sm $\times$ s^2 $\times$ g^{-1}. It should be stressed that the limited solubility of cellulose nitrate (it is related to cellulose itself to an even greater extent) makes it difficult to investigate the flow birefringence with a wide variation of the solvent refraction index, n_S and thus set limits of application of the microconfiguration effect for quantitative estimation of the inherent anisotropy and rigidity of the molecular chain.

Table II.11. Dynamooptical properties of non-fractionated nitrocellulose sample solutions as a function of solvent nature

Solvent	$([n]/[\eta]_{exp}) \times 10^{10}$ (cm s^2 g^{-1})	$(\alpha_1 - \alpha_2)_{exp} \times 10^{25}$ (cm^3)	$a_{exp} \times 10^{25}$ (cm^3)	n_s
Acetone	– 8.4	– 114	– 1.9	1.357
Ethylacetate	– 10.4	– 138	– 2.3	1.372
Butylacetate	– 23.0	– 300	– 5.0	1.390
Amylacetate	– 24.8	– 324	– 5.4	1.402
Cyclohexanone	– 43.4	– 540	– 9.0	1.450

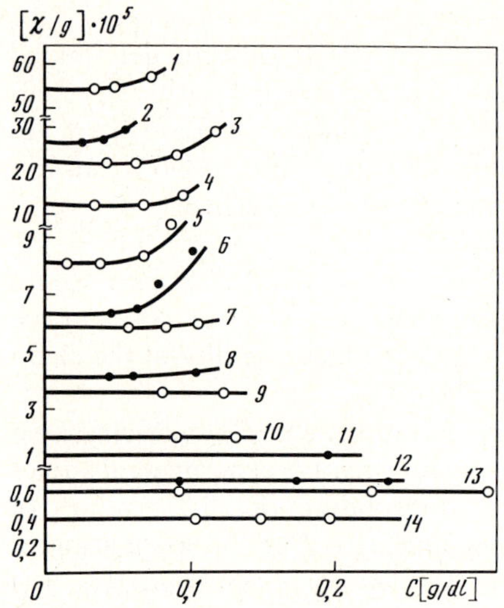

Fig. II.50. Plots of (x/g)-values as a function of solution concentration C of various nitrocellulose fractions (*1–14*) in cyclohexanone

5.3 Electrooptical Properties of Cellulose Derivatives

When considering polymer system order, one used to take into account both the orientation axis order and the orientation dipole order caused by the presence of the coaxial order in the mutual orientation of polar groups or bonds in the macromolecule, which should be characterized by a higher total dipole moment.

Due to the rigidity of the molecular structure in the case of rigid-chain molecules the orientations of the polar groups of the molecule in the electric field are not independent of each other, but they are intercorrelated according to the molecular chain or its segments. Therefore provided the vector sum of local group moments is not equal to zero, it indicates the presence of orientation polar order of the molecule then it can be characterized by the significant dipole moment and rotated as a rigid particle in the electric field [70].

The high birefringence appears in solutions of such polar rigid-chain polymers in an electric field (Kerr effect). The value of n and sign of birefringence are in this case also dependent on the value and sign of optical anisotropy of the molecule $\gamma_1 - \gamma_2$ as well as on the value of its dipole moment μ and angle v, formed by the dipole axis with the optical axis of the molecule:

$$\lim_{\substack{C\to 0\\ E\to 0}} (\Delta n/CE^2) = (\pi N_A/15n)((\varepsilon^2 + 2)/3)^2((n^2 + 2)/3)^2((v_1 - v_2)/M) \times (\mu/KT)^2(3\cos^2 v - 1) \tag{5.36}$$

where C is the concentration, n—the refractive index; ε—the solvent dielectric constant.

Experimental investigation of this effect and the comparison of the results with those of flow birefringence allows us to obtain information about the value and direction of the molecular dipole which affords additional information about the rigidity, the degree of order of the polymer molecule and thus about its conformation.

For the wormlike model of semirigid chains, the value of the dipole moment is connected [98] with the contour length $L = ax$ by the relationship

$$\bar{\mu}^2 = (\mu_0 S)^2(X - 1 + \exp(-X))/2 \quad \text{or} \tag{5.37}$$

$$\bar{\mu}^2/L = (\mu_S^2/A)[1 - (1 - \exp(-X))/X] \tag{5.38}$$

where μ_0 is the component of the dipole moment of the monomer unit along the molecule axis, S is the number of monomer units in its segment; μ_S is the segment dipole moment.

The experimental study of the dipole moments dependence on the molecular mass M allows us to determine the persistent chain length a and to estimate the rigidity of the molecule on the basis of electrooptical data.

The rigid chain polar molecules in solution in a periodical electric field demonstrate frequency dependence on the Kerr effect proving the presence of relaxation effects in the course of dipole orientation. The analysis of the frequency dependence produces information about the rotation mobility of the molecules which is also important from the point of view of ideas about their conformation.

The vast experimental data about electrooptical properties of nitrocellulose solutions have been obtained from the paper by Tsvetkov et al. [75, 99], where flow birefringence (FB) in an electric field was studied for solutions of nitrocellulose of various molecular weights in different solvents. The results obtained demonstrate that FB of nitrocellulose solutions are positive and proportional to the square of the field intensity E^2, and that the Kerr effect is valid at all studied concentrations C and frequencies v (Fig. II.51).

Kerr specific constants $K = \Delta n/cE^2$ were determined on the basis of the slope of the curves $\Delta n_E = f(E^2)$. Concentration dependence of the K-value for nitrocellulose samples of various molecular weights is presented in Fig. II.52 and corresponding extrapolated values are given in Table II.12.

With the growth of molecular weight, K-values increase accordingly (Fig. II.53) approaching the limit value $K_\infty = 6 \times 10^{-10}\, g^{-1}\, cm^5 (B/300)^{-2}$ in high molecular weight regions, where optical properties of nitrocellulose follow Gauss statistics. The type of dependence observed corresponds to the electrical properties of rigid chain polymers. According to this, the following equation has been used for the determination of the parallel component of the monomer unit dipole moment:

$$K_\infty = [2\pi N_A (n^2 + 2)^2 (\varepsilon + 2)^2 (\alpha_1 - \alpha_2) \mu_0^2 \times S \times \cos^2 v]/1215n \times (KT)^2 \times M_0 \quad (5.39)$$

Using this formula and experimentally found values of K_∞, $\alpha_1 - \alpha_2$, and S, we can determine the required value as $\mu_{0\|} = \mu_0 \times \cos v = 1D$. This value is close to that of $\mu_{0\|} = 1.15D$ found by a similar method for cellulose carbanilate [76]. The experimental dependences of K/K_∞ on parameter $x = 2M/M_0 \times S$ at $S = 30$ and $M = 247$ (see above) are presented in Fig. II.54. The theoretical plots of K/K_∞ for kinetically rigid wormlike chains are shown in Fig. II.54 (curves 4–7) [91]:

$$K/K_\infty = B[1 - 0.6\, \mathrm{tg}^2 v \times A]^{3/5} [\langle h^4 \rangle / \langle h^2 \rangle^2], \quad \text{where} \quad (5.40)$$

$$A = (X - 1 + \exp(-x))/(X^2 - 0.8(x - 1 + \exp(-x)))$$

$$B = (X - 1 + \exp(-x))^2/(X^2 - 0.8(x - 1 + \exp(-x)))$$

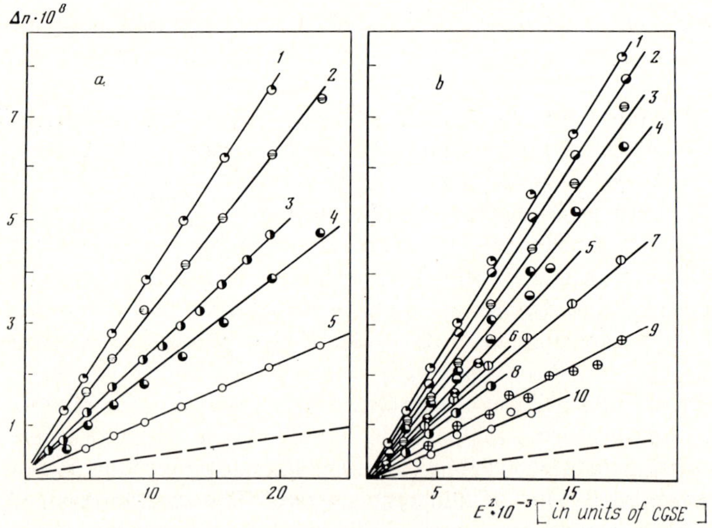

Fig. II.51. Plots of Δn-values as a function of E^2 for nitrocellulose solutions in dioxane. **a**—concentration (%): *1*—0.54; *2*—0.44; *3*—0.34; *4*—0.24; *5*—0.084; at $\gamma = 0.15$ (kHz); **b**—frequency (kHz): *1*—0.15; *2*—0.3; *3*—0.5; *4*—1; *5*—3; *6*—5; *7*—10; *8*—20; *9*—49; *10*—700 at C = 0.5%. *Dashed line*—pure dioxane

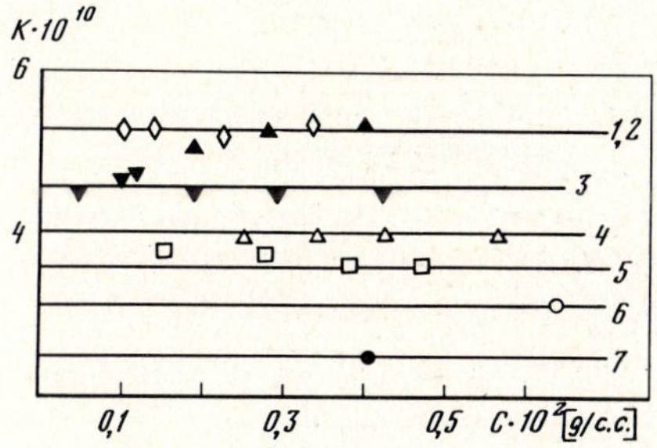

Fig. II.52. Relationship of K-value to the concentration for solutions of various nitrocellulose fractions (*1–7*) in dioxane

Table II.12. Electrooptical characteristics of nitrocellulose macromolecules in dioxane at 294 K

X	$K \times 10^{10}$ cm^5 g^{-1} $(B/300)^{-2}$	$\tau \times 10^6$ s	G	X	$K \times 10^{10}$ cm^5 g^{-1} $(B/300)^{-2}$	$\tau \times 10^6$ s	G
20.8	5.3	6.6	0.8	9.2	3.8	0.9	0.5
21.2	5.3	5.5	0.8	7.5	3.3	—	—
13.8	4.6	2.2	0.5	4.3	2.5	—	—
12.4	4.0	1.6	0.5	—	—	—	—

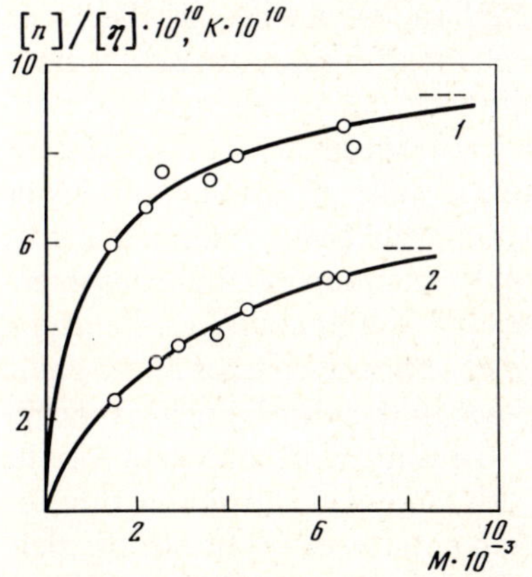

Fig. II.53. Experimental dependence of $[n]/[\eta]$ (*1*) and $K(2)$ on M for nitrocellulose fractions in dioxane. The limit values of $[n]/[\eta]$ and K are shown by *dashed lines*

where $\langle h^2 \rangle$ and $\langle h^4 \rangle$—are the second and fourth moments of the chain groups distribution upon h-values; K_∞ is the limit value, which is attained in the Gauss region and determined according to Eq. (5.39). These curves (see Fig. II.54) are related to various values of angle v formed by the dipole of the monomer unit μ_0

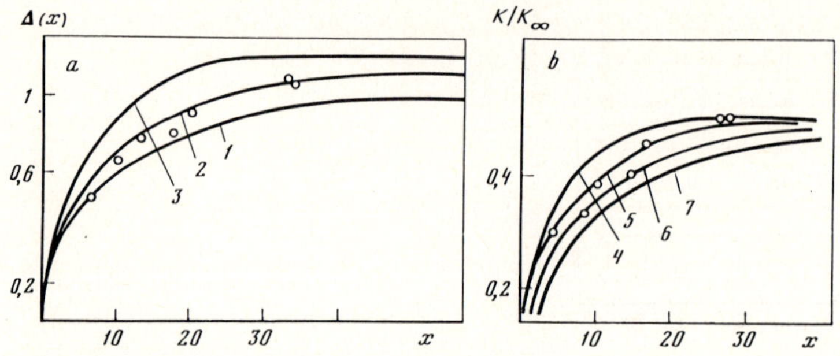

Fig. II.54. Plots of (**a**) $\Delta x = [n]/[\eta]/([n]/[\eta])_\infty$ and (**b**) K/K_∞ as a function of X for nitrocellulose fractions in dioxane.

1—according to the Noda-Hearst theory; *2*, *3*—according to the Tsvetkov theory without (*2*) and with (*3*) consideration of the dependence of shape asymmetry of the wormlike chain on its length; *4*–*7*—according to the Tsvetkov theory for Kerr constant of persistent chains at values of angle $\gamma = 0(4)$; $45(5)$, $60(6)$ and $65°(7)$

in the direction of the chain. The experimental values are in the range of K/K_∞ values corresponding to the range of angles $v = 45$–$60°$ which results in the magnitude of the dipole moment of the monomer unit of nitrocellulose molecules

$$\mu_0 = \mu_{0\|}/\cos v \approx (1.7 \pm 0.3)D$$

Thus in contrast to dynamooptic characteristics the electrooptical properties of nitrocellulose may be adequately interpreted in the frames of theory of kinetically rigid wormlike chains. This difference in the evaluation of kinetic flexibility of the polymer chains in their dynamooptic and electrooptic properties can be understood by taking into account the specificity of the "electric structure" of the polar chain molecule. If monomer units of the chain molecule have the dipole component uniformly directed along the basic chain and rigidly connected with it, then these components afford the summarized dipole moment of the molecule coinciding in direction with vector h which combines the chain ends. This is formally equivalent to the presence of opposite charges at the chain ends. Therefore reorientation of monomer dipoles in the electric field is inevitably connected with the change of spacial disposition of one end of the molecular chain with respect to the other, i.e. with large scale motion within the molecule. The cooperative character of the internal motions in the electric field in such a molecule can be carried out, not only due to rigidity of its structure but because of purely geometrical aspects of the dipole structure of the molecular chain. This clearly "polar" effect which favours the rotation of a molecule as a whole in the electric field is absent in the mechanical shear field of laminar flow. Therefore the same molecule, due to its kinetic characteristics, may be more flexible in a mechanical field than in an electrical one.

The significant deviation of angle v from 0 demonstrates that the side polar groups of the polymer chain make an important contribution to the dipole moment of the nitrocellulose monomer unit. The role of the polar side group in the observed Kerr effect (it follows from Eq. (5.40)) becomes more important the lower the polymer molecular weight. In contrast, the longitudinal component of the dipole moment of the monomer unit $\mu_{0\parallel}$ (Eq. (5.39)) fully defines the Kerr effect in the Gauss region, therefore $\mu_{0\parallel}$—values which are experimentally obtained from the Kerr effect are practically the same for all cellulose esters studied [76, 90, 100].

In nitrocellulose solutions, the molecular weight dependent dispersion of the Kerr effect was found in the radio frequency region [74] (Fig. II.55). The values $\log \nu$ at which the Kerr effect was twice reduced ($K_\nu/K = 0.5$) were determined on the basis of dependence shown in Fig. II.55. Here K and K_ν are specific Kerr constants in the area of the equilibrium effect and the dispersion at frequency ν respectively. The mean dipole relaxation times were calculated by expression $\tau = 1/2\pi\nu_{0.5}$ [93] using values of frequencies $\nu_{0.5}$ extrapolated to zero concentration. Values $\tau_s \to 0$ thus obtained are presented in Table II.12. The experimental values τ are related to the intrinsic viscosity and molecular weight by the following equation:

$$\tau = GM[\eta]\eta_0/RT \tag{5.41}$$

Calculated by means of this equation, coefficients G, dependent on the conformation properties of the molecule, are given in Table II.12.

It is important that experimental G-values are within the limits of theoretical G-values related to the motion of the molecules in the first Zimm mode [93, 101, 102]. This demonstrates that the basic motion of dissolved nitrocellulose molecules in an electric field is their rotation as a whole, this is valid for kinetically rigid chains.

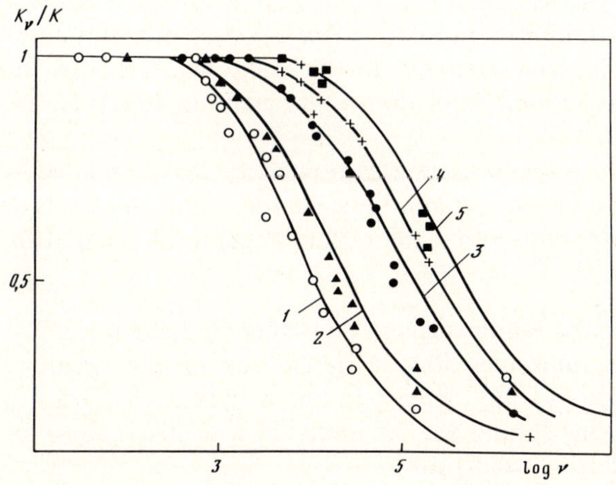

Fig. II.55. Plots of relative Kerr constant K_ν/K to the frequency of electric field V for various fractions (*1–5*) of nitrocellulose in dioxane

5.4 Light Scattering of Solutions of Cellulose Derivatives and Equilibrium Rigidity of their Molecules

Light scattering is usually applied in the physical chemistry of polymers for the determination of molecular weight, radius of gyration, and the second virial coefficient of dissolved molecules [103, 104]. The study of diluted solutions with the following extrapolation to zero concentration allows us to evaluate the properties of the isolated macromolecule. The detailed theoretical consideration of the method and a rather complete summary of experimental results dealing with solutions of cellulose and its derivatives is presented in a paper by Zandelev [105]. However this information is considered without any connection to hydrodynamic, dynamooptic and electrooptic properties which decreases its value.

The study of light scattering of nitrocellulose solutions in ethyl acetate is of interest [106] since the authors had investigated earlier the hydrodynamic, dynamooptic and electrooptic properties of the same nitrocellulose samples. In the paper cited the average molecular weight $\bar{M}_w$ the mean radius of gyration of the molecule $\langle R_z \rangle^{1/2}$, and the second virial coefficient A_2 for a wide variety of nitrocellulose fractions were determined after double extrapolation of light scattering data [103] using the equation

$$CH/R_v(\theta, C) = (1/M)\{1 + (16\pi^2 \langle R_z^2 \rangle / 3(\lambda/n_0)^2) \times \sin^2(\theta/2)\} + 2A_2 C + \ldots, \tag{5.42}$$

where $H = (4\pi^2 n_0^2 / N_A \times \lambda^4) \times (dn/dC)^2$, n_0 is the refractive index of ethyl acetate ($n_0 = 1.372$), N_A—Avogadro number, λ—wave length in vacuo ($\lambda = 54.6 \times 10^{-8}$ M).

When calculating the $\bar{M}_w$, the correction to depolarization Δv was not applied since the value $\Delta\omega$ was less than 1% even for the lowest molecular weight fractions. The plots of double extrapolation for 1, 3, 5, 7 and 9 fractions are presented in Fig. II.56. The characteristics of the fractions as well as the values of $\bar{M}_\omega$, $\langle R_z^2 \rangle^{1/2}$ and A_2 found from the results given in Fig. II.56 are presented in Table II.13.

The radius of gyration $\langle R \rangle$ is the molecular characteristic closely related to the equilibrium rigidity of the molecular chain. Thus for the chain with contour length L in the Gaussian coil conformation the radius of gyration is equal to

$$\langle R_z^2 \rangle = LA/6 \tag{5.43}$$

where A denotes the length of the Kuhn segment characterizing chain rigidity.

The properties of polymer molecules may, as in the case of nitrocellulose, deviate from Gaussian ones due to the increase in chain rigidity. In order to describe their conformations, one should use the model of wormlike chains for which the following relationship is valid [107].

$$\langle R^2/M \rangle = (\lambda_0 A / 6M_0)\{1 - 3/X + (6/X^2)[1 - (1/X) \times (1 - \exp(-x))]\} \tag{5.44}$$

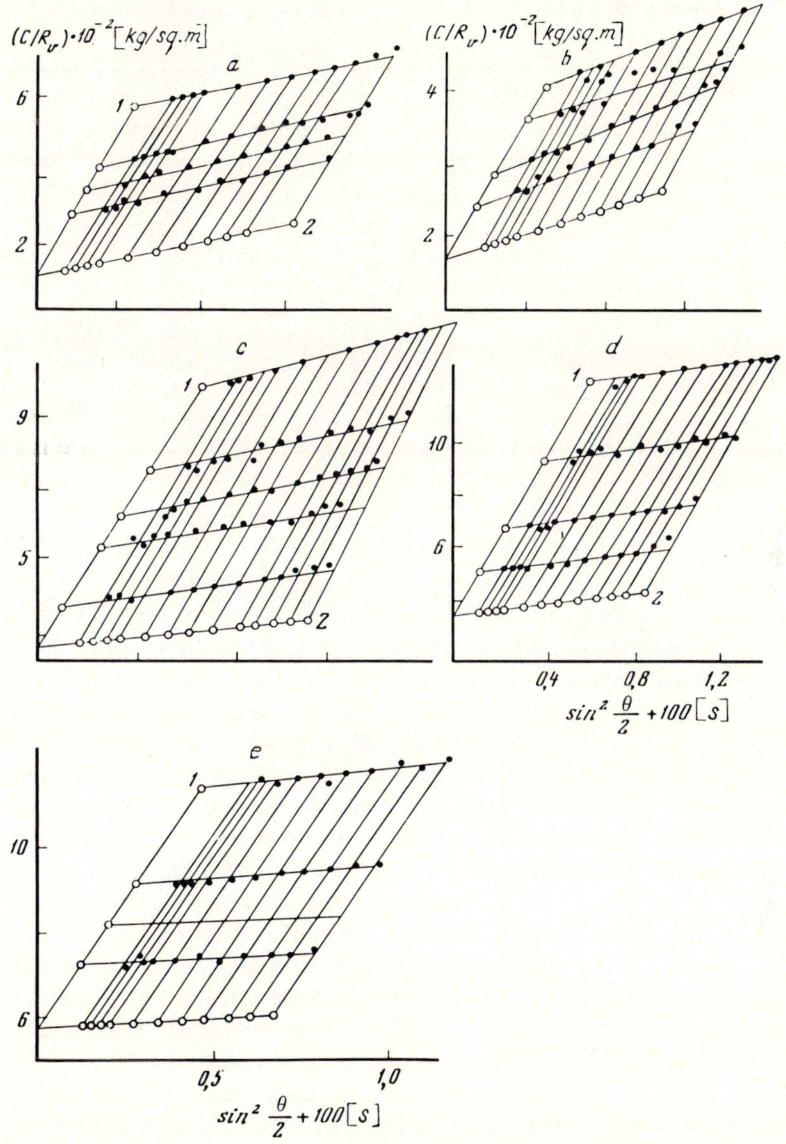

Fig. II.56. Extrapolation plots of light-scattering data for fraction 1 (**a**), 3 (**b**), 5 (**c**), 7 (**d**) and 9 (**e**) of nitrocellulose in ethyl acetate. The *circles* indicate extrapolation to $\theta = 0$ (1) and C = 0 (2)

where λ_0 denotes the projection of the monomer unit in the direction of the molecular chain (for nitrocellulose $\lambda_0 = 5.15 \times 10^{-10}$ M), M_0—the molecular weight of the monomer unit (for studied fractions with a nitrogen content equal to 13.24% it is 0.282 kg/mole). Parameter $X = 2L/A = 2M\lambda_0/M_0A$ describes

Table II.13. The values of $\bar{M}_w \langle R_z^2 \rangle^{0.5}$ and A_2 for various nitrocellulose fractions in ethylacetate

Fraction N	$\bar{M}_w \times 10^{-2}$ (kg/mole)	$\langle R_z^2 \rangle^{0.5m} \times 10^{10}$ (m)	$A_2 \times 10^4$ (m^3 mole/kg^2)	$\langle R_z^2 \rangle / \bar{M}_w \times 10^{17}$ (m^2 mole/kg)
1	5.95	674	7.3	0.76
2	5.40	626	8.8	0.72
3	4.33	561	4.4	0.73
4	2.86	458	7.4	0.73
5	2.74	391	9.1	0.56
6	2.30	411	6.7	0.73
7	1.98	268	7.8	0.36
8	1.74	250	8.0	0.36
9	1.31	214	7.9	0.35

the relative chain length. In the range of high molecular weights ($X \to \infty$, Gauss region), Eq. (5.44) is transformed into (5.43) and $\langle R^2 \rangle / M$ attaining the limit value $\lambda_0 M / 6 M_0$, is independent on M. On the other hand in the range of small X, $\langle R^2 \rangle / M$ is proportional to M according to Eq. (5.44). The expression (5.44) is valid in the whole range of molecular weights provided volume effects are absent. For nitrocellulose molecules [108] as well as for all rigid chain polymers [109] the coefficient of osmotic swelling α is close to unity even for the highest molecular weight fractions in a thermodynamically good solvent. Therefore the dimensions $\langle R_z^2 \rangle^{1/2}$ found in ethyl acetate are close to being unperturbed and the application of Eq. (5.44) is quite correct. The left part of Eq. (5.44) comprises equally averaged magnitudes $\langle R^2 \rangle$ and M whereas the experimentally found $\langle R \rangle$ and $\bar{M}_w$ are defined by different averaging, therefore the value $\{\langle R_z^2 \rangle / \bar{M}_w\} \bar{M}_{w \to \infty}$ affords the upper estimation of the magnitude of segment A. However, according to Pogodina and colleagues [75], the resulting inaccuracy of results is small since the polydispersity of the fractions under study is also small ($\bar{M}_z / \bar{M}_w = 1.2$).

The points in Fig. II.57 represent the experimental values of $\langle R_z^2 \rangle / \bar{M}_w$ (according to data given in Table II.13) as a function of $\bar{M}_w$. The curves illustrate the theoretical relationships plotted according to Eq. (5.44) for different chain rigidity. The general character of the experimental dependence of $\langle R_z^2 \rangle / M_w$ on $\bar{M}$ corresponds to the theoretical one which predicts the enhancement of the $\langle R_z^2 \rangle / M_w$ value for the rigid chain polymer with increase in molecular weight. However the errors in experimental results are too high for their application in the range of small molecular weights (for the lowest molecular weight fractions the error of $\langle R_z^2 \rangle / \bar{M}_w$ determination is 100%). In the range of high molecular weights the value $\langle R_z^2 \rangle / \bar{M}_w$ is constant within the accuracy limits and, thus, in this case the conformation of the nitrocellulose molecular chain can be described

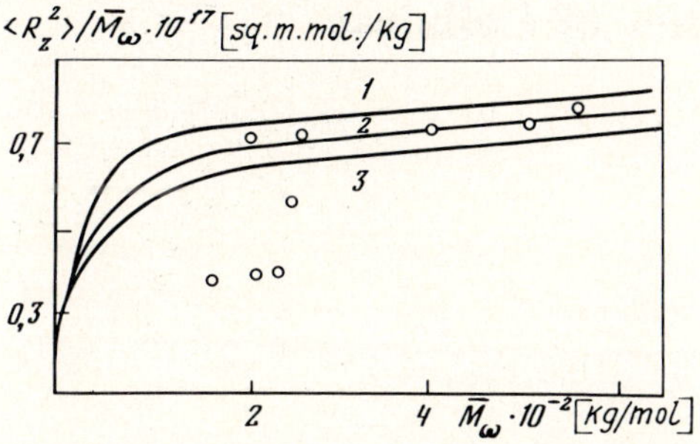

Fig. II.57. Plots of $\langle R_z^2 \rangle / M_w$ as a function of M for various nitrocellulose fractions in ethyl acetate. *1–3*—theoretical dependences according to Eq. (5.38) related to segment length, A = 270 Å (*2*) and 230 Å (*3*)

by the Gaussian coil. It can be seen from Fig. II.57 that at large $\bar{M}_w$, experimental points show the best fit to curve 2 which gives the value of $\lim \{\langle R_z^2 \rangle / \bar{M}_w\}_{\bar{M}_w \to \infty} = 0.76 \times 10^{17}$ m^2 × mole/kg and the Kuhn segment $A = 2.5 \times 10^{-8}$ m. This magnitude correlates with the rigidity determined by hydrodynamic and dynamooptic methods [75].

It is known that the cellulose chain and its esters are flexible due to internal rotation around two C–O bonds of the oxygen bridge binding two adjacent glucose moieties. It can be shown that the length of the Kuhn segment of the polyglucoside chain should be $A = 12 \times 10^{-10}$ m upon non-hindered rotation [70]. The experimental value for nitrocellulose $A = 2.5 \times 10^{-8}$ m corresponds to the degree of restrain $\sigma = (A/A_f)^{1/2} = 4.6 \pm 0.7$. This value is much higher than that for the flexible chain [103] and rigid chain aromatic polymers [70, 95, 110] and it is close to σ-values found earlier for other cellulose esters [70, 84]. Thus light scattering results together with hydrodynamic, dynamo- and electrooptical properties of solutions of cellulose derivatives demonstrate the high equilibrium and kinetic rigidity of cellulose esters. They are the quantitative measure of the high skeleton rigidity of their molecular chains as well as the higher degree of restraint.

For cellulose molecules containing ionogenic groups the skeleton rigidity can be increased by electrostatic interaction of the groups. It is possible to propose that intramolecular hydrogen bonds play an important role in restraining intramolecular rotation. Due to the complex composition of cellulose esters, their optical and hydrodynamic properties appear to be more influenced by the solvent nature. At the same time it is evident that the main factor defining all molecular properties of cellulose and its derivatives is the enhanced rigidity of the main chain.

6 Thermodynamic Properties of Cellulose and its Satellites

The results of thermochemical studies (Table II.14) clearly demonstrate the effect of the chemical and physical nonuniformity of various cellulose samples on the thermodynamic properties. The presence of certain structural changes in cellulose samples according to their prehistory is indicated in particular by the broad dispersion of experimental values of the combustion enthalpy of cellulose. According to data obtained by various authors the highest (in modulus) combustion enthalpy (4193 Kcal/mole) was found for very pure cellulose and wood. The difference between these values ($\sim$ 4–5%) over two order of magnitude exceeds the accuracy of calorimetry measurements (0.05–0.1%).

This fact demonstrates the applicability of thermochemical data, enthalpy of combustion in particular, as a source of information about chemical and physical nonuniformity (defectiveness) of cellulose. However in order to realise this possibility one should calculate the thermodynamic characteristics of cellulose itself and associated satellites—mannan, galactan, xylan, polyglucuronic acid, etc.

6.1 Modelling of Thermodynamic Properties of Cellulose and its Satellites

Modelling is the only reliable calculation method which allows us to determine the basic set of thermodynamic characteristics ΔH^{form}, ΔH^{comb}, S^0, ΔG^{form} for cellulose and its natural satellites—mannan, galactan, xylan. The application of any other method is hardly possible due to the complex chemical structure of these compounds. For determination of the thermodynamic parameters of cellulose, its hydrate, mannan, galactan and xylan it is convenient to use methods of finite differences [111, 112] which correspond to conventional transitions.

$$PI_{j^2}(\text{cond}) \xrightarrow{\Delta\varphi_1} D - PI_j(\text{cond}) + (1/n)[PI_j]_n(\text{cond}) \tag{6.1}$$

$$D - PI_j(\text{cond}) \xrightarrow{\Delta\varphi_2} \text{H}_2\text{O}(\text{cr.}) + (1/n)[PI_j]_n(\text{cond}) \tag{6.2}$$

Table II.14. Thermochemical characteristics of cellulose of different origin

Cellulose sample	Crystallinity (%)	$-\Delta H_{298}^{comb}$ (kcal/kg)	Reference	Cellulose sample	Crystallinity (%)	$-\Delta H_{298}^{comb}$ (kcal/kg)	Reference
Fibers	69–71	4190	[115]	Cotton			
Rami		4193	ibid	cellulose	67–71	4165	[231]
		4174	[230]	Wood			
		4185	[116]	cellulose	70	4173	[231]
Cotton	70	4110	[230]				
		4110	[115]				

Table II.15. Thermodynamic and thermochemical properties of cellulose and its main components calculated by modelling method

Polymer	Chain fragment of the polymer	Phase	Modeling: $-\Delta H_{298}^{form}$ (kcal/unit)	S_{298}^{0} (kcal/grad)	$-G_{298}^{form}$ (kcal/unit)	$-\Delta H_{298}^{kg}$ (kcal/unit)	$-\Delta H_{298}^{kg}$ (kcal/kg)	Experimental $-\Delta H^{comb}$ (kcal/kg)
1	2	3	4	5	6	7	8	9
Cellulose	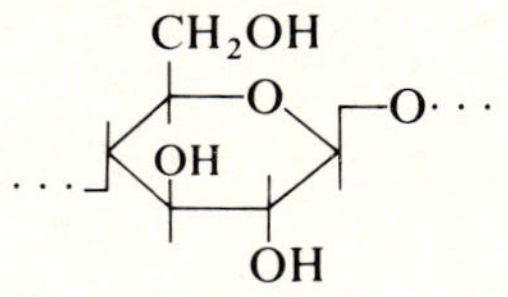	(crys) (am)	227.4 220.4	43.2 —	154.8 —	678.5 685.5	4185 4228	4185[a] 4193[b]
Cellulose hydrate	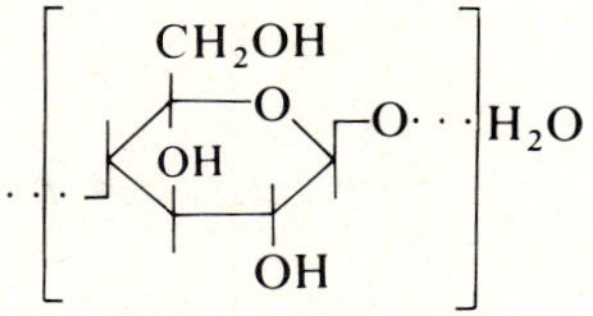	(crys)	298.6	—	—	675.6	3749	—
Mannan	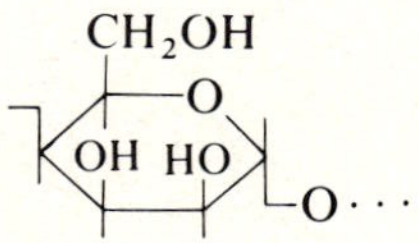	(crys)	224.7	—	—	681.2	4201	—
Galactan	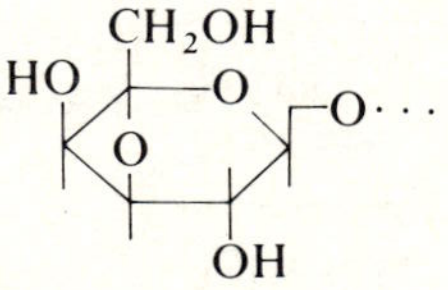	(crys)	230.0	41.6	156.9	675.9	4168	—

Xylan	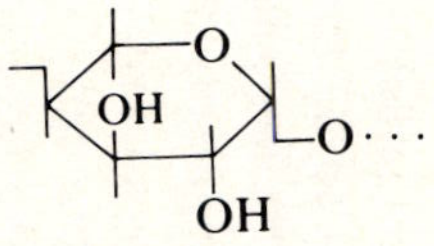	(crys)	173.6	—	—	569.5	4314	4260[c]
Polyglucuronic acid	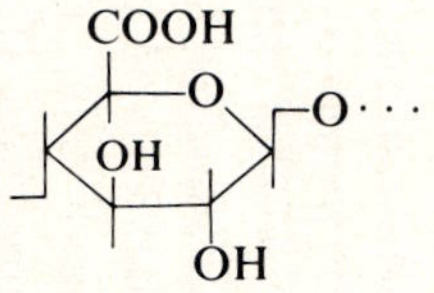	(crys)	276.7	44.4	205.5	560.9	3185	—
Polymer with dialdehyde unit	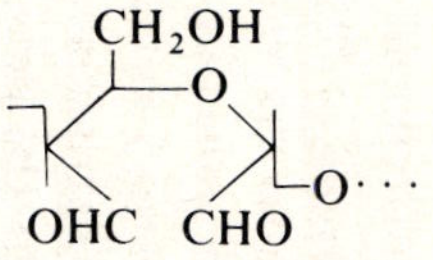	(crys)	190.0	—	—	647.6	4044	—
Cross-linked polymer	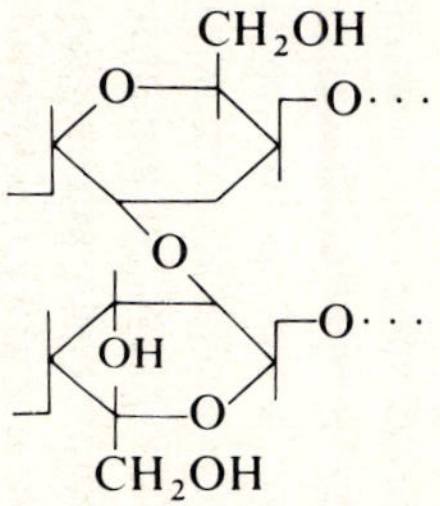	(crys)	377.6	78.9	246.7	1366	4459	—

Heats of combustion were obtained without amendments to Washbern and of process isothermic character.

[a] Data were taken from [115]
[b] Data were taken from [116]
[c] Data were taken from [117]

where $D-PI_j$, $[PI_j]_2$, are model mono- and disaccharides (pyranoses); $[PI_j]$–pyranose units of cellulose (Cell.), cellulose hydrate (Cell · H_2O), mannan (Mann), galactan (Gal) and xylan (Xyl).

On the basis of conventional transitions (6.1) and (6.2), thermodynamic characteristics of the above polysaccharides ($\varphi[PJ_j]$) and model compounds ($\varphi[PJ_j]_2$), (φ_{D-PI_j}) can be combined with the help of the following equations:

$$\varphi_{PI_j} = \Delta\varphi_1 + \varphi_{[PI_j]_2} - \varphi_{D-PI_j}; \tag{6.3}$$

$$\varphi_{PI_j} = \Delta\varphi_2 + \varphi_{D-PI_j} - \varphi_{H_2O} \tag{6.4}$$

where $\varphi = \Delta H^{\text{form}}, \; -\Delta H^{\text{comb}}, S^0, \Delta G^{\text{form}}$.

Using tabulated thermodynamic data for model sugars, (D-glucose, D-glucose monohydrate, D-mannose; D-galactose, D-xylose, cellulose, lactose and lactose monohydrate) one can determine, with the help of Eq. (6.3), thermodynamic parameters of cellulose, cellulose hydrate and galactan when $\Delta\varphi_1 = 0$, and by means of Eq. (6.4), the corresponding parameters of mannan and xylan when

$$\Delta\varphi_2(\text{D-Glc} \rightarrow [\text{Cell}]) = \Delta\varphi_2(\text{D-Mann} \rightarrow \text{Mann}) = \Delta\varphi_2(\text{D-Xyl} \rightarrow \text{Xyl}) \tag{6.5}$$

Table II.15 demonstrates the results of these calculations related to a possible error 0.1% and a confidence range of 99%.

The comparison of the calculated values of cellulose enthalpy of combustion at various crystallinities with corresponding experimental data (see Table II.14) shows the best correlation in case of the purest cellulose samples (trade mark "Rami"). The other experimental results obtained for cotton and wood cellulose samples are markedly lower than the calculated data.

6.2 The Influence of Adsorbed Water on Thermochemical Properties of Cellulose

The above mentioned underestimation of experimental heats of combustion is clearly related to the physical and chemical nonuniformity of the structure of actual cellulose samples. The chemical nonuniformity is mainly caused by the presence of "foreign" units in the cellulose macromolecules which are the residues of other polysaccharides (mannan, galactan) or oxidized fragments of the polymer chain. The physical nonuniformity is mainly caused by adsorbed water which may markedly affect the thermochemical data. This can be illustrated by the data [113] dealing with the determination of the heat of combustion for two samples of cotton cellulose, for which temperature of vacuum drying are changed but the period of drying (3 hours) and vacuum (10^{-3} bar) remain constant. As a result of different chemical treatment the samples had various degree of polymerization and crystallinity. The results of these experiments are represented as diagram $\Delta U^{\text{comb}} = f(T_{\text{drying}})$ in Fig. II.58.

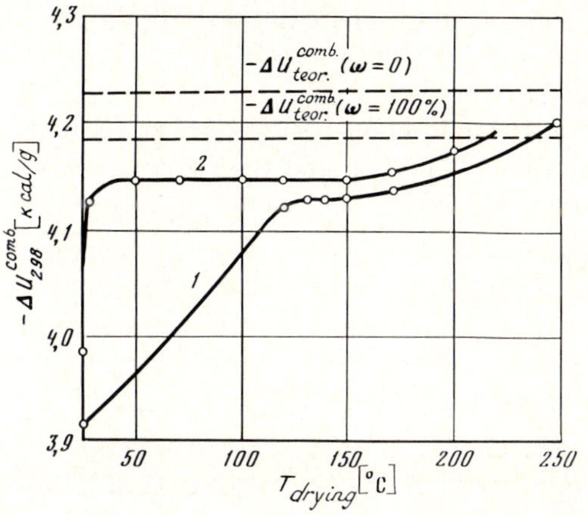

Fig. II.58. Plots of the heat of combustion of cotton cellulose as a function of preparation conditions and properties of the sample. Drying period 3 h, pressure 10^{-3} bar

1—$\omega = 6.9\%$, DP = 1.5×10^{-3}; *2*—$\omega = 90\%$; DP = 150; ($-\Delta U^{comb}_{298}$, Kcal/g; $-\Delta U^{comb}_{calc}$, T_{drying}, C°)

Two parallel lines drawn in this figure indicate the interval between limiting heats of combustion related to ideal crystalline and the amorphous state of pure cellulose. It can be seen from the diagrams that an increase of the drying temperature of cellulose samples results in significant growth of heat of combustion governed by the loss of physically bound water. However this growth has a certain limit (plateau on diagrams), which is lower than the calculated range for both cellulose samples. The observed difference between calculation and experiment in the temperature area of the plateau is not associated with adsorbed water, but with the chemical nonuniformity (defectiveness) of cellulose macromolecules.

Different dimensions of the plateau on diagrams of two samples can be explained by the diffusion accessibility of amorphous cellulose regions for water molecules. In the case of a cellulose sample with low molecular weight and high crystallinity ($DP = 150$, $W = 90\%$) it is naturally higher (the plateau is also higher) than for the sample with higher molecular weight and medium crystallinity ($DP = 1.5 \times 10^3$, $W = 67$ to 71%).

The second growth of the heat of combustion is observed in the diagrams $\Delta U^{comb.} = f(T)$ upon increase of the drying temperature from 150 to 160 °C. As was shown [114], at a temperature above 150 °C, the reduction of the ability of cellulose to absorb or retain water occurred due to intramolecular dehydration of the following type (Fig. II.59).

As well as for other rigid chain polymers, the start of dehydration of cellulose corresponds to a temperature close to the melting point of the monomer (crystalline D-glucose melts at 146 °C) [118].

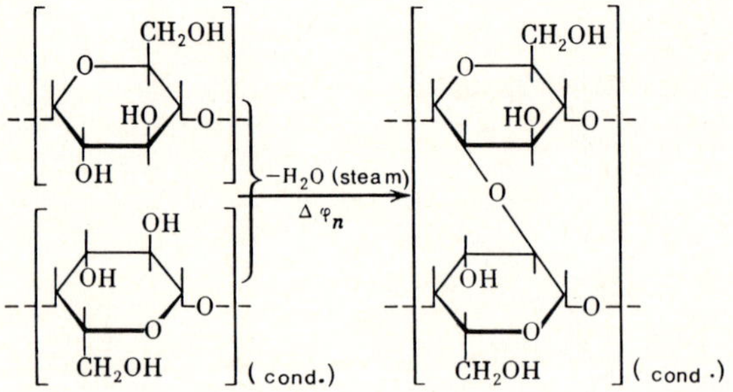

Fig. II.59. Typical scheme of the intermolecular dehydration reaction

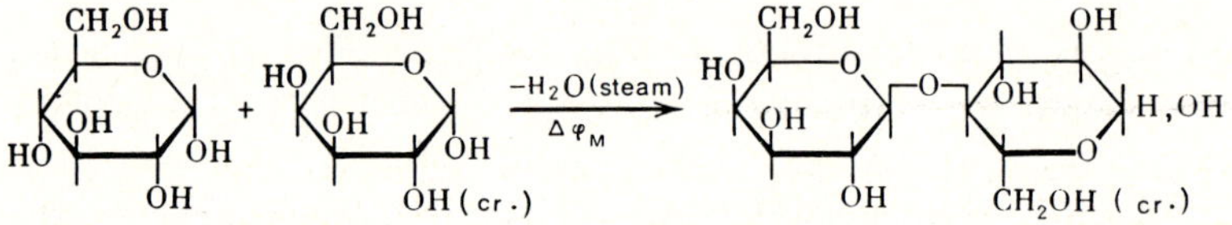

Fig. II.60. Model scheme of cellulose dehydration for estimation of thermodynamic parameters

Modeling of this process with the help of reaction (with thermodynamic parameters available for calculations [119, 120]) enables us to estimate on the basis of approximation $\Delta\varphi_n = \Delta\varphi_m$ the thermodynamic parameters and the specific heat of combustion of the product with cross-linkages, formed according to the scheme in Fig. II.59 and Fig. II.60 (see Table II.15). The specific heat of combustion calculated according to the schemes in Fig. II.59 and II.60 is 274 Kcal/kg higher than that of initial crystalline cellulose, that correlates with the increase of heat of combustion on diagrams $\Delta U^{\text{comb.}} = f(T)$ in the range 150–200 °C. In addition, the temperature dependence for Gibbs energy was evaluated for the cross-linking process on the basis of calculated data and specific heat results [119].

$$\Delta G = 21300 - 57.5 \times T + 0.0460 \times T^2 - 16.49 \times \lg(T/298) + 4.575 \times T \times \lg P \tag{6.6}$$

Provided partial water vapour pressure in Eq. (6.6) is equal to $P = 10^{-3}$ bar which is maintained in the course of the drying of the cellulose, the free energy of the process (Fig. II.61) upon attainment of the temperature range 130–140 °C is transferred from positive to negative values. This means that the formation of such cross-linkages is thermodynamically possible in the temperature range related to the second rise of the heat of combustion on diagrams $\Delta U^{\text{comb.}} = f(T)$.

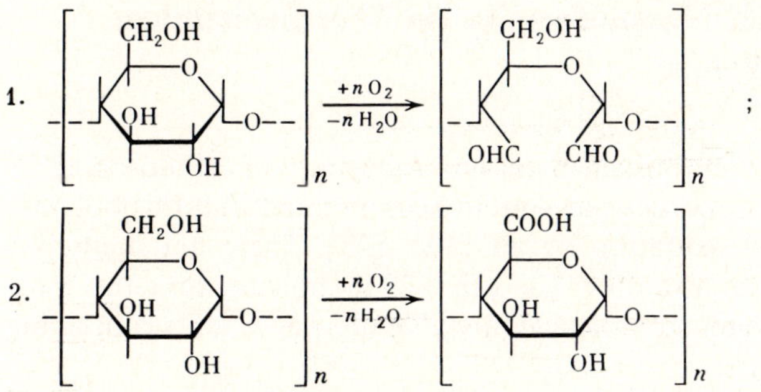

Fig. II.61. Configurational scheme of cellulose oxidation

It is also possible to show that the above diagrams are useful for the estimation of the equilibrium content of adsorbed water in initial non-dried cellulose provided the heat of combustion of the cellulose-water system is additive. This additivity was mentioned by Brauning [121] who proved that the change in humidity of the wood under investigation has no influence on its heating capacity determined in a calorimetric bomb with an accuracy of 0.5%, provided the heat values are calculated in relation to absolutely dry substance. Such a result is caused by the fact that water evaporating upon sample combustion is then condensed in the bomb and its latent heat of evaporation is restored.

Additivity of the heat of combustion in the cellulose-water system for the points on diagram $\Delta U^{\text{comb.}} = f(T)$ (below plateau) should be valid with an accuracy up to the energy of interaction of the cellulose and water which has maximum value 2.9 Kcal at an equimolar ratio of components. This value was obtained as the difference of molar heat of combustion of cellulose hydrate ($\text{Cell} \cdot H_2O$) and anhydrous cellulose (see Table II.15). Disregarding this energy, one can only make an approximate estimation of the equilibrium content of adsorbed water in non-dried cellulose samples with an accuracy of 0.1–0.4% according to additive relationship:

$$X_{H_2O/\text{weight}\%} = 10^2(\Delta U^{\text{comb.}}_{\text{Dr. Cell}} - \Delta U^{\text{comb.}}_{\text{Cell} \cdot XH_2O})/\Delta U^{\text{comb.}}_{\text{Dr. Cell}} \tag{6.7}$$

where $\Delta U^{\text{comb.}}_{\text{Cell} \cdot XH_2O}$, $\Delta U^{\text{comb.}}_{\text{Dr. Cell}}$—specific heats of combustion of initial and dried cellulose without the corrections from Washbern and for the isothermic character of the combustion process (the value $U^{\text{comb.}}_{\text{Dr. Cell}}$ corresponds to plateau level on diagram $\Delta U^{\text{comb.}} = f(T_{\text{drying}})$).

The application of Eq. (6.7) to the first high molecular weight sample of cotton cellulose gives a water content of about 5% and to the second low molecular weight sample—about 3.8%. With regard to the relatively low crystallinity index of the first sample ($W = 69\%$) and the high value of the second one ($W = 90\%$) the excess of water in the first sample in comparison with the second one can be related to excessive water in its amorphous phase.

6.3 Effect of Chemical Nonuniformity on Thermochemical Properties of Cellulose

When considering the character of chemical nonuniformity of cellulose and its effect on thermochemical properties one should keep in mind that this polymer exhibits high sensitivity to oxidation by air [121–123]. There are numerous experimental results (see for example: [3, 122, 123]) showing that the most abundant forms in cellulose macromolecules are glucopyranose units containing carboxyl or aldehyde groups.

The acid character of crude cellulose preparations is caused by the presence of the moieties of glucuronic acid. It is accepted [123] that oxidation of cellulose to polyglucuronic acid can not only proceed in the course of plant activity but during the preparation and storage of commercial cellulose. The following decarboxylation of polyglucuronic acid leads to the formation of xylan.

Cellulose preparations of wood origin may also contain mannan and galactan in small quantities.

The crude samples of cotton and wood cellulose together with isomeric structures (mannan, galactan) and oxidation products may contain lipids and gums. Wood cellulose also comprises lignin which is an oxygen-containing phenylpropane polymer [124].

Thermodynamic characteristics of the polymer structure were obtained by the method of finite differences (see Table II.15). In order to define mutual transformations of the polymers the models shown in Fig. II.61 were applied:

$$1.\ R_1\begin{cases}\text{CH—OH}\\ \text{CH—OH}\end{cases} + O_2 \longrightarrow R_1\begin{cases}\text{CHO}\\ \text{CHO}\end{cases} + H_2O$$

$$2.\ R_2CH_2OH + O_2 \longrightarrow R_2COOH + H_2$$

where $R_1 = (CH_2)_4$; $R_2 = \text{—}CH(CH_2)_5$, and $\Delta\varphi_{M,1}$, $\Delta\varphi_{M,2}$ are thermodynamic characteristics of transitions 1 and 2.

The thermodynamic parameters of pure cellulose obtained earlier were used as reference points for calculation. The corresponding tabulated parameters of the type of model compounds shown in Fig. II.61 were taken from review [119]. The thermodynamic parameters of transitions $\Delta\varphi_{M,1}$ and $\Delta\varphi_{M,2}$ estimated by this method were then equated to the characteristics of transitions (Fig. II.61).

In order to understand the effect of cellulose chemical non-uniformity on its thermochemical properties, it is important to know how the cellulose chemical composition changes at various stages of raw material purification. One can use, for this purpose, the typical data [125] on changes of composition of cotton fibers taking place during two steps of purification—alkaline pulping and bleaching (Table II.16). On the basis of data given in Table II.16 one can follow the alteration of the specific heat of combustion of cotton cellulose after each purification step.

The values of specific heat of combustion can be estimated provided they are proportional to the weight average portion of basic components of the fiber.

Table II.16. The alteration of composition of cotton cellulose after two steps of purification (pulping and bleaching) [125]

	Components composition (% from cellulose weight)		
Components	Initial raw material (1)	After alkaline pulping (2)	After pulping and bleaching by hypochlorite
Cellulose (w = 70%)	91.0	98.0–98.5	99.0–99.5
Pectins	3.0	1.0–1.2	0.8–1.0
Waxes and lipids	0.8	0.3	0.1–0.2
Proteins	1.2	0.1	0.08
Ash	4.0	0.4	0.2

Then the subsequent values of specific heat of combustion at each purification step can be obtained as:

$$-{}^{\Delta}H^{\text{COMB.}} = 0.910 \times {}_{\Delta}H_1^{\text{COMB.}} + 0.030 \times {}_{\Delta}H_2^{\text{COMB.}} + 0.008 \times {}_{\Delta}H_3^{\text{COMB.}} + 0.12 \times {}_{\Delta}H_4^{\text{COMB.}} + 0.040 \times {}_{\Delta}H_5^{\text{COMB.}} = 4051 \text{ Kcal/kg}$$

$$-{}^{\Delta}H^{\text{COMB.}} = 0.980 \times {}_{\Delta}H_1^{\text{COMB.}} + 0.012 \times {}_{\Delta}H_2^{\text{COMB.}} + 0.003 \times {}_{\Delta}H_3^{\text{COMB.}} + 0.001 \times {}_{\Delta}H_4^{\text{COMB.}} + 0.004 \times {}_{\Delta}H_5^{\text{COMB.}} = 4187 \text{ Kcal/kg}$$

$$-{}^{\Delta}H^{\text{COMB.}} = 0.988 \times {}_{\Delta}H_1^{\text{COMB.}} + 0.008 \times {}_{\Delta}H_2^{\text{COMB.}} + 0.0012 \times {}_{\Delta}H_3^{\text{COMB.}} + 0.0008 \times {}_{\Delta}H_4^{\text{COMB.}} + 0.002 \times {}_{\Delta}H_5^{\text{COMB.}} = 4189 \text{ Kcal/kg}$$

where ${}^{\Delta}H^{\text{COMB.}}$ is the enthalpy of combustion of the *i*-component of cotton cellulose, given in Table II.15.

The results of these calculations demonstrate that only after the second purification step does the heat of combustion of cotton fiber become close, with acceptable accuracy (0.2%), to that of pure cellulose—4198 Kcal/kg (at $W = 70\%$). The comparison of experimental data for cotton cellulose (see Table II.14) with results from calculations shows that all known experimental values of heat of combustion are within the interval of that for crude cotton (I) and for the cotton after alkaline pulping (II). The latter results mean that experimental samples of cotton cellulose contain basic components in quantities intermediate between state I and II. According to analysis the basic role in decrease of experimental heat of combustion in comparison with the calculated value is played the presence of incombustible admixtures e.g. salts and hydrates of Ca, Fe and other metals. These salts seem to appear in cellulose due to exchange of hydrogen of glucuronic acid moieties to cations of some metals. Part of the salts remains in cellulose after mercerization and other part—after washing of cellulose with water [125]. The presence of such admixtures (0.2–1%) in cotton cellulose samples may lead to decrease of specific heat of combustion on 8–42 Kcal/kg.

Table II.17. The change of components composition on various stages of preparation of sulphite cellulose from conifer wood (%) [125]

Purification step	Cellulose ($w = 70\%$)	Xylan	Mannan	Lignin	Lipids and gums	Ash
Initial (raw) wood	50–58	10–12	6–10	26–28	1–2	0.2–0.5
Sulphite pulping	88–89	4–5	6–8	1.5–2	1–2	0.3–0.8
Bleaching by chlorine and hypochlorine	87–89	2–4	5–6	0.3–0.7	0.3–0.5	0.1–0.2
Finishing	94–95	2–3	2–3	0.2–0.3	0.2–0.3	0.1

Using data presented in Table II.16 like in case of cotton cellulose one can follow the alteration of the heat of combustion of conifer cellulose in the course of four subsequent steps of sulphite method of wood purification.

$$-{}^{\Delta}H_{\mathrm{I}}^{\mathrm{COMB.}} = 0.537 \times {}_{\Delta}H_1^{\mathrm{COMB.}} + 0.11 \times {}_{\Delta}H_2^{\mathrm{COMB.}} + 0.08 \times {}_{\Delta}H_3^{\mathrm{COMB.}} + 0.26 \times {}_{\Delta}H_4^{\mathrm{COMB.}} + 0.01 \times {}_{\Delta}H_5^{\mathrm{COMB.}} + 0.003 \times {}_{\Delta}H_6^{\mathrm{COMB.}} = 4725\ \mathrm{Kcal/kg}$$

$$-{}^{\Delta}H_{\mathrm{II}}^{\mathrm{COMB.}} = 0.375 \times {}_{\Delta}H_1^{\mathrm{COMB.}} + 0.04 \times {}_{\Delta}H_2^{\mathrm{COMB.}} + 0.06 \times {}_{\Delta}H_3^{\mathrm{COMB.}} + 0.01 \times {}_{\Delta}H_4^{\mathrm{COMB.}} + 0.01 \times {}_{\Delta}H_5^{\mathrm{COMB.}} + 0.005 \times {}_{\Delta}H_6^{\mathrm{COMB.}} = 4258\ \mathrm{Kcal/kg}$$

$$-{}^{\Delta}H_{\mathrm{III}}^{\mathrm{COMB.}} = 0.889 \times {}_{\Delta}H_1^{\mathrm{COMB.}} + 0.04 \times {}_{\Delta}H_2^{\mathrm{COMB.}} + 0.06 \times {}_{\Delta}H_3^{\mathrm{COMB.}} + 0.006 \times {}_{\Delta}H_4^{\mathrm{COMB.}} + 0.004 \times {}_{\Delta}H_5^{\mathrm{COMB.}} + 0.001 \times {}_{\Delta}H_6^{\mathrm{COMB.}} = 4233\ \mathrm{Kcal/kg}$$

$$-{}^{\Delta}H_{\mathrm{IV}}^{\mathrm{COMB.}} = 0.944 \times {}_{\Delta}H_1^{\mathrm{COMB.}} + 0.025 \times {}_{\Delta}H_2^{\mathrm{COMB.}} + 0.025 \times {}_{\Delta}H_3^{\mathrm{COMB.}} + 0.003 \times {}_{\Delta}H_4^{\mathrm{COMB.}} + 0.002 \times {}_{\Delta}H_5^{\mathrm{COMB.}} + 0.001 \times {}_{\Delta}H_6^{\mathrm{COMB.}} = 4214\ \mathrm{Kcal/kg}$$

where ${}^{\Delta}H_i^{\mathrm{comb}}$ is the enthalpy of combustion of wood cellulose components given in Table II.17.

It is generally accepted in calculations that: ${}_{\Delta}H_1^{\mathrm{COMB.}} = -4198$ Kcal/kg; ${}_{\Delta}H_2^{\mathrm{COMB.}} = -4314$ Kcal/kg; ${}_{\Delta}H_3^{\mathrm{COMB.}} = -4201$ Kcal/kg; ${}_{\Delta}H_4^{\mathrm{COMB.}} = -6000$ Kcal/kg; ${}_{\Delta}H_5^{\mathrm{COMB.}} = -10000$ Kcal/kg; ${}_{\Delta}H_6^{\mathrm{COMB.}} = 0$.

It follows from these data that the heats of combustion of conifer wood decrease after every step of sulfite purification gradually approaching the calculated value of the heat of combustion of pure cellulose.

Thus the calorimetry of combustion together with modeling in the case of such a complex natural product as cellulose allows us not only to evaluate the stability of thermodynamic parameters but also to make some conclusions about the degree of chemical and physical non-uniformity of its macromolecule.

7 Theoretical Methods of Modeling of Conformational Properties of Cellulose and its Derivatives

One of the basic trends of modern studies of the physical chemistry of polymers is the wider application of modeling because the choice of corresponding models, their analysis, and optimization is performed on the basis of extensive use of computers [5, 126, 127]. This is specially important since cellulose does not form true solutions and all experimental data dealing with its conformation in solution have been obtained for solutions of cellulose derivatives, esters in particular (see Chap. 5). It is of great interest in this connection to develop the computing for reliable determination of corresponding parameters. Although the practical realization of such methods is difficult due to the very complex nature of the object of investigation, only these methods enable us to produce information about the conformation of cellulose itself, since there are no other ways of obtaining these data. The validity of this conclusion is confirmed so far by a few data related to some linear and cyclic macromolecules [128–134]. This also follows from the results related to the problem of conformation of the isolated cellulose coil (see Chap. 4).

7.1 Concentration Dependence of Medium Size Semirigid Macromolecules in Solutions

It follows from data given in Chap. 5 that cellulose and its esters are related to typical semirigid macromolecules which are characterized by a markedly higher length of statistical Kuhn segment (A) than its lateral size (d) which gives the parameter of local asymmetry $P = A/d$ which substantially exceeds unity ($P \gg 1$).

The specific feature of semirigid chains is their inability to form well-defined folds. Therefore the change of their conformation is performed by means of correlated turns of the units. It allows us to consider that in some cases the oriented order in solutions at the polymer concentrations exceeding a certain critical value is accompanied by a transition into an ordered liquid-crystalline state (see e.g. [135]). It is natural that the equilibrium rigidity of the chains expressed in terms of the Flory criterion f, should fit condition $f < 0.63$. Parameter f according to Flory is defined as the portion flexibly folded with respect to the rigid collinear linkages

$$f = ((\tilde{z} - 2)\exp(-\Delta F/KT))/(1 + (\tilde{z} - 2)\exp(-\Delta F/KT)) \quad (7.1)$$

where $\tilde{z}$ is the coordination number in the Flory model of the pseudolattice, F is the difference in free energies of the flexible and rigid conformations of a macromolecule. It should be noted that $f = 0$ for ultimately rigid rod-like molecules, and $f = 1$ for ultimately flexible ones [136]. Unfortunately the experimental determination of the average size of the chains in concentrated systems is faced with serious difficulties. Whereas the conclusions of modern analytical theories of the liquid crystalline state are valid only [137] for infinitely long chains and they cannot be directly transferred to real systems of semirigid chains, where the contour length L is often close to value A. The above difficulty was overcome in our paper [138] by means of computer modeling of semirigid fibers of nitrocellulose (NC) where $f = 0.074$, using the Monte–Carlo approach [136]. The dependence of root-mean-square sizes (R^2) of NC chains on the volume portion of the polymer (v) in the system was considered in this paper. The calculations were carried out on the basis of the model of the NC chain proposed by Kirste [139]. This model was successfully used earlier for calculations of mean sizes and some other characteristics of the isolated NC chains (at $v \to 0$). In accordance with this, the NC macromolecule was simulated by the chain of n vectors $\vec{i}$ of l length which were connected in series under angle γ. The mutual spatial positions of vectors (and thus conformation of the molecule) were defined by $(n - 1)$ angles γ as well as by $(n - 2)$ azimuth angles of internal rotation, which we supposed to be ranged between 0 and 2π. For each pair of joint points i and j of vectors $\vec{i}$ at $|i - j| > 1$ the potential of interaction was introduced

$$U(rij) = \begin{cases} \infty & \text{at } rij < d \\ -\varepsilon/KT & \text{at } d \leq rij \leq 1.5d \\ 0 & \text{at } rij > 1.5d \end{cases} \quad (7.2)$$

where rij is the distance between points i and j, ε is the energetic parameter similar to parameter χ in the Flory–Huggins theory of polymer solutions. The intermolecular interaction was described by the same way. The equations $d = 0$ and $\varepsilon = 0$ in Eq. (7.2) correspond to conditions of θ-solvent. In this case the root-mean-square chain lengths are defined by the equation

$$R_\theta^2/l^2 = n(1 + \alpha)/(1 - \alpha) - 2(1 - \alpha^2)/(1 - \alpha)^2 \quad (7.3)$$

where $\alpha = \cos(\pi - \gamma)$. The athermanous system was simulated at $\varepsilon = 0$ and $d = e$. The increase of ε (at $d = e$) was equivalent to the deterioration of the thermodynamic quality of the solvent.

The system from $m = 2$ to 6 interacting chains present within cubic cell with volume V was considered. The volume portion of the polymer was defined as $\upsilon = m(n + 1)d^3/V$. The periodic boundary conditions were applied to cell bounds so that the macroscopic translation periodic system was simulated. The required characteristics were averaged along the ensemble from $\Omega \sim 10^3 \div 10^4$ of independent system configurations, i.e. along Ω various ways of chain positions in space. Various chain conformations were generated using the Monte–Carlo method on a BESM-6 computer according to algorithm [140]. According to Flory the f-value may be estimated from relationship [136].

$$R_\theta^2/L = (2 - f)/f \tag{7.4}$$

Thus the value $f = 0.074$ for real NC chain [136, 141] is reproduced for our model, for example at $\alpha = 0.996$ and $n = 39$. In this case $R_\theta^2/l^2 = 1020$, $(L/l)^2 = 1050$ and $A/l = 57.7$. Provided l is identical to the length of the NC repeating unit equal to $D/2$ ($D = 1.02$ nm—length of lattice spacing [142]) then $A = 29$ nm which is close to the experimental value $A = 25$ nm for nitrocellulose with degree of substitution ~ 2.7 dissolved in ethyl acetate [106]. Taking into account the high rigidity of such a chain one can expect a weak effect on its conformation of distant intramolecular interactions. Actually as was shown by the "computer experiment", the R^2-value practically does not depend on ε and it is close to R^2 in diluted solutions at low v. Figure II.62 shows the dependence of R^2 on v obtained for athermanous system ($\varepsilon = 0$), i.e. for chains in a thermodynamically good solvent. As can be seen, the mean dimensions at low v remain close to the θ-dimensions. R^2 remains constant even at v-values markedly higher than the critical concentration of chains overlapping $\sim n(l/R)^3 \sim 10^{-3}$. However in the course of a further increase of v at $v \geq 0.1$, the mean dimensions increase rapidly (in the range $v = 0.05$ to 0.26 on about 12%) and $v = 0.3$ R^2 is only 30% less than L. It is of interest that $v \approx 0.1$ appears to be close to the critical value V^*, which in classic Flory theory [136] is responsible for the transition of stiff chain molecules from isotropic phase to nematic one. According to the same theory $V^* = (8/P)(1 - 2/P)$, where P is the degree of asymmetry. At $P = A/l$ for our model $V^* = 0.13$.

Thus, the well-defined extension of molecular chains is observed for the NC model system under study in the region of transition into the ordered state. This effect differentiates the behaviour of semi-flexible and flexible chains. The latter [143] shows a decrease rather than increase of R^2 with an increase of V in thermodynamically good solvents (in this case at $V \to 1$, $R^2 \to R_\theta^2$).

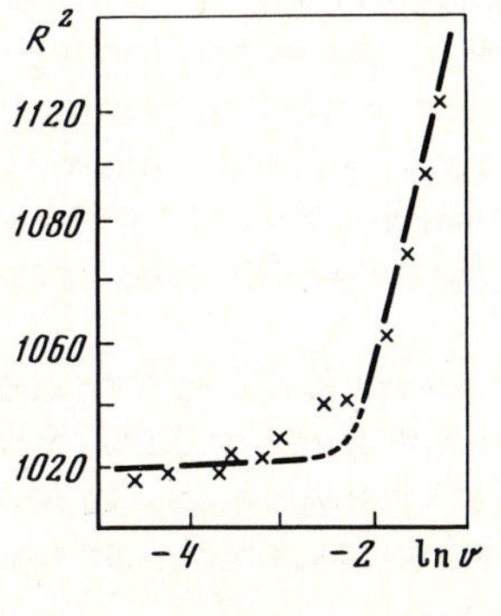

Fig. II.62. Plot of R^2 as a function of volume portion of the polymer at $\xi = 0$

Investigation of the effect of solvent thermodynamic quality on the dependence of R^2 on V, it was found that the increase of ε (in the considered range $\varepsilon/KT = 0$ to 0.3) was the same as that dependence shown in Fig. II.62. The slight increase of R^2 (not more than 2%) and some decrease of the V range (where $R^2 = R_\theta^2$) takes place. Thus the effects observed for semirigid macromolecules are weakly dependent on the intermolecular attraction forces, as in the case of low molecular weight liquid crystalline systems [144] the basic role is played by repulsive potential.

7.2 Asymmetry of Nitrocellulose Molecular Coils in Solutions

It has been shown previously (see Chap. 5) that macromolecular chains of nitrocellulose are rather stiff, and both equilibrium and kinetic rigidity of the chain markedly increase upon enhancement of the degree of substitution. The information about mean asymmetry of polymer coil is necessary [95] in the case of interpretation of experimental data on the hydrodynamic properties of polymer chains in solution in particular. It is not hard to obtain these data if the molecule can be represented by a Gauss chain. Provided orthogonal components (R_1^2, R_2^2, R_3^2) of the square of the radius of gyration R_z^2 (determined by molecular fixed coordinates) are used as characteristics of shape then the calculation shows [145] that average shape of Gauss coil can be approximated to a three-axis ellipsoid with a ratio of the axes squared equal to

$$\langle R_1^2\rangle/\langle R_2^2\rangle/\langle R_3^2\rangle = 1/2.69/11.7 \tag{7.5}$$

It is possible to propose, however, that the asymptotic ratio (7.5) would not be valid for stiff chains of finite length. Hence the coefficient would be different from the asymptotic value

$$b = (P^2 - 1)/(P^2 + 1) \tag{7.6}$$

where $P = [2\langle R_3^2\rangle/(\langle R_2^2\rangle + \langle R_1^2\rangle)]^{1/2}$ substituted into Eq. (5.21), which connects conformation parameters of the chain with experimentally measured characteristics (for example with the specific value of flow birefringence) [95].

The NC preparations usually studied have molecular weights $M \approx 10^4$ to 10^6 which related to $x \approx 10^2$ to 10^3 repeating units. With respect to the high rigidity of the NC chain, the determination of the average value of the shape of the macromolecular coil of this polymer is of great interest. The determination of the degree of shape asymmetry of macromolecular coils of cellulose trinitrate (CTN) in diluted solution as a function of molecular weight was performed using computerized Monte–Carlo calculations [126].

As a model of the equivalent chain, the model was accepted [139] with the following crystallographic parameters of the trinitrate cellulose monomer unit [146, 147]: $b_1 = b_3 = 0.139$ nm, $b_2 = 0.277$ nm, $\alpha_1 = 150.04°$, $\alpha_2 = 152.76°$, $\alpha_3 = 114°$, the distance $O_1 \ldots O_2$ i.e. the value $L_0 = 0.501$ nm, identity period

$D = 1.02$ nm. When all angles of internal rotation are equal to zero (ψ) such an equivalent chain has a flat conformation, which corresponds to the maximum extended real chain. The rotation is possible around bonds C_1–O–C_4 (angles ψ_1 and ψ_2); the turns around virtual bonds $O_1 \ldots O_4$ are prohibited ($\psi = 0$). It is known [148] that cellulose and its derivatives are related to so-called chains with localized correlation of internal rotation.

The rotations around pairs of C–O bonds coupling pyranose cycles are interrelated. The presence of long and rigid virtual bonds $O_1 \ldots O_4$ in the chain results in the fact than groups of atoms adjacent to rotated C–O bonds are separated by rather large distances and practically do not interact. In this case conformation of the chain can be characterized by only potential map $E(\psi_1, \psi_2)$ related to the chain fragment from two pyranose rings coupled via a "bridge" oxygen atom (Fig. II.63). These data demonstrate 11 stable conformations with energy $E < 4.6$ KJ/mole. These results were used for the generation of the CTN chain by Monte–Carlo calculations. For this purpose the values of potential energies were chosen without taking into account the electrostatic interactions, since the electrostatic component of potential energy has no practical effect on the conformational parameters of the chain [148].

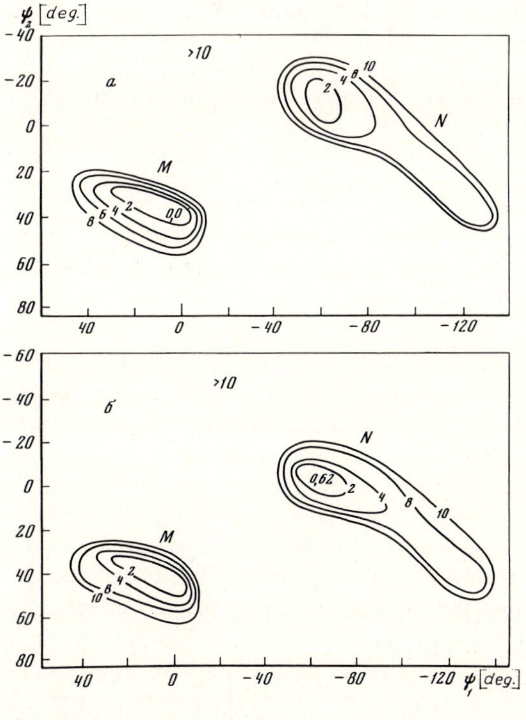

Fig. II.63. Conformation maps of the fragment of cellulose trinitrate chain without (**a**) and with (**b**) consideration of electrostatic interactions [233]

Thus the conformation of the CTN chain is defined by pair of angles $\psi_1^{(1)}$, $\psi_2^{(1)} \ldots \psi_2^{(x)}, \psi_2^{(x)}$ (the number of pairs is equal to that of repeating units in the chain x), and the values of angles in each pair are chosen with a probability governed by Boltzmann factor $\exp[-E(\psi_1, \psi_2)/RT]$. The required mean characteristics were obtained as a result of averaging a rather large number (10^3 to 10^4) of conformations generated independently by computer. The volume effects were neglected since their role, even for the highest molecular weight fractions of nitrocellulose (see Chapter 5), is negligible (the coefficient of osmotic swelling α in thermodynamic good solvent is close to unity). The algorithm of calculation in this case is the following. Let us numerate the coupling points of the bonds of X-unit chain ($b_1, b_2, b_3 \ldots$): $1, 2, \ldots, N$ ($N = 3x$). We determine the symmetric tensor of the second range with components for each conformation in coordinates $OX_1X_2X_3$ whose directions of axis are governed by the two first linkages b_1 and b_2 [148].

$$NX_{ij} = \sum_{m=1}^{N} (X_i^{(m)} - X_i^*)(X_j^{(m)} - X_j^*) \tag{7.7}$$

where $X_i^* = N^{-1} \sum_{m=1}^{N} X_i^{(m)}$ is the projection of the center of masses of the polymer coil on the axis OX_i, $X_1^{(m)}, X_2^{(m)}, X_3^{(m)}$—the coordinates in the system $OX_1X_2X_3$. Then tensor of inertia can be written as follows [149]:

$$T = NSpXE - NX \tag{7.8}$$

where E denotes the single matrix and SpX—trace of tensor X. After reduction of $\tilde{T}$ to diagonal form we obtain for the components of the square of the radius of gyration

$$R_i^2 = (tj + tk + ti)/2N \tag{7.9}$$

$$(i, j, k = 1, 2, 3 \quad \text{and} \quad i \neq j \neq k)$$

Here t is the main central moments of tensor of inertia. Now we make the following transformation of coordinates X_1^m, X_2^m, X_3^m:

$$\begin{pmatrix} X_1^{(m)} \\ X_2^{(m)} \\ X_3^{(m)} \end{pmatrix} = \Gamma^{-1} \begin{pmatrix} X_1^{(m)} \\ X_2^{(m)} \\ X_3^{(m)} \end{pmatrix} \tag{7.10}$$

where Γ denotes the modal matrix defined by the transformation of similarity $\tilde{T} = \Gamma^{-1} T \Gamma$. Transformation (7.10) transfers coordinates from system $OX_1X_2X_3$ into $O^*X_1X_2X_3$, the directions of the axes which coincide with those of vectors of tensor T_1 and the beginning is superposed with the coil center of masses. The moment of inertia of the axis O^*X_3 is assumed to be maximal in each conformation and that of the axis O^*X_1—minimal. Thus "the molecular-immobile" system of coordinates is taken in which $t_3 \geq t_2 \geq t_1 \geq 0$ or $R_1^2 \geq R_2^2 \geq R_3^2$. Note that shape asymmetry of individual conformations of the chain can be revealed only by such (or similar) "internal" coordinates [145, 149].

After averaging, the components R^2 and t-values characterize directly both mean shape asymmetry of the polymer coil and its mean size (since $\langle R^2 \rangle = \langle R_1^2 \rangle + \langle R_2^2 \rangle + \langle R_3^2 \rangle = \langle 2N \rangle^{-1} SpT$). Table II.18 presents the results of ratio calculations of the mean ordered components of the square of the radius of gyration, of the ratio of ordered moments of tensor of inertia, as well as the values $\langle K_1 \rangle = \langle t_1/t_3 \rangle$ and $\langle K_2 \rangle = \langle t_2/t_3 \rangle$. Figure II.64 demonstrates the double distribution system $\rho(K_1, K_3)$ which is the measure of spacial asymmetry of the individual conformations of the chain, it also defines their portion from the general number of generated conformations with a given ratio $K_1 = t_1/t_3$ and $K_2 = t_2/t_3$. The function $\rho(K_1, K_2)$ is presented in Fig. II.64 as projection on the plane $K_1 K_2$ (as isolines).

The data obtained shows that the shape asymmetry of the polymer coil is the most pronounced for the short chains. However the asymmetry remains very high even for the longest considered chains ($X = 400$). One can deduce first of all, that the coil shape resembles strongly the body extended along the axis of inertia. Actually, the inertial features of the rigid rod, oriented along the axis O^*X_1 are defined as $t_1 = 0$, $t_2 = t_3$ or $K_1 = 0$ and $K_2 = 1$. As it can be seen from Table II.18 the equations $\langle t_2 \rangle \langle t_3 \rangle$ and $\langle K_1 \rangle$ 1 are valid with rather good precision (for small x in particular). Figure II.64 also demonstrates that most of the chains possess conformations which are strongly extended along the main axis of inertia O^*X_1. The value of K_1 does not exceed 0.6 for any of the generated conformations at $x = 200$, in the case of shorter chains it takes even lower values (it can be confirmed by comparison of β-function (K_1, K_2) for $x = 200$ and $x = 100$ (see Fig. II.64)). There is, on the contrary, a high probability that $K_2 = 1$. Note that the equations $K_2 = 1$, $K_1 = K_2$ mean that the mass distribution in the system may be considered as a solid body with axial symmetry (i.e. as the ellipsoid of rotation) while equality to unity of K_1 and K_2 values (or the same $t_1 = t_2 = t_3$) denotes a spherical–symmetrical distribution of masses with regard to the center of the coil. For the chains of TNC, probability of appearance of spherical–symmetrical conformation is close to zero (see Fig. II.64). The coils with the shape close to a flattened ellipsoid (where K_1, K_2)

Table II.18. Characteristics of the shape asymmetry of trinitrate cellulose chain in solution

The number of repeating units in the chain (x)	$\langle R_3^2 \rangle \langle R_2^2 \rangle \langle R_1^2 \rangle$	$\langle t_2 \rangle \langle t_2 \rangle \langle t_3 \rangle$	$\langle K_1 \rangle$	$\langle K_2 \rangle$
25	1:11.2:703	1:57.6:58.4	0.0176	0.985
50	1:11.4:366	1:29.5:30.3	0.0346	0.971
100	1:10.2:184	1:16.4:17.3	0.0630	0.948
150	1:8.86:119	1:12.2:13.0	0.0863	0.931
200	1:8.16:87.4	1:9.66:10.4	0.0111	0.913
400	1:6.22:37.1	—	—	—
∞	1:2.69:11.7	—	—	—

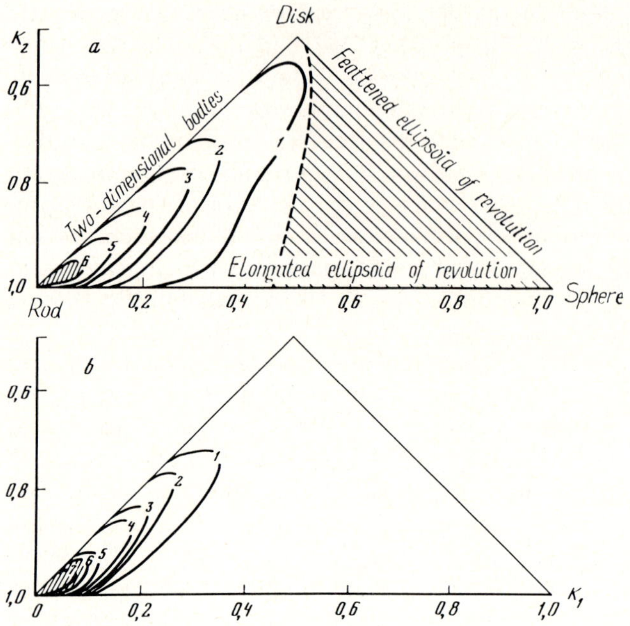

Fig. II.64. The function $\rho(K_1, K_2)$

a: $x = 200$, isolines: *1*—0.002; *2*—0.02, *3*—0.06, *4*—0.2, *5*—0.4 and *6*—0.6; **b**: $x = 100$, isolines: *1*—0.002, *2*—0.01, *3*—0.06; *4*—0.1, *5*—0.2, *6*—0.6 and *7*—0.8 (rod, two-dimensional bodies, disk, flattened ellipsoid of revolution, sphere, extended ellipsoid of rotation)

are also highly unlikely. However, the extended ellipsoid conformations with axial symmetry ($K_2 = 1$) are rather probable.

Besides, the average shape of the coils appears to be like the extended ellipsoid of rotation (for small x we have $\langle K_2 \rangle$ 1 or $\langle t_2 \rangle$ $\langle t_3 \rangle$ (see Table II.18). The short chains ($X \leq 100$) more often than the long ones possess the conformation which corresponds in shape to the two-dimensional (flat) body; actually, the sum $\langle K_1 \rangle + \langle K_2 \rangle$ for them is close to unity (see Table II.18) and the probability that $K_1 + K_2 = 1$ is rather high (see Fig. II.64). Nevertheless it can be shown that such "two-dimensional" conformations have nothing in common with the disk shape (for a proper disk $t_1 = t_2 = t_3/2$ or $K_1 = K_2 = 1/2$). Although there is a high probability that the shape of CTN macromolecules in solution corresponds to the extended ellipsoid of rotation, the average shape appears to be more similar to a three axis ellipsoid (especially for relatively large x). This is confirmed by the character of changes of the values presented in Table II.18 as a function of the chain length. It can be seen in particular that with increase of x the sum $\langle K_1 \rangle + \langle K_2 \rangle$ exceeds 1 more and more. It should be stressed, in general, that for the coils corresponding to long chains, a significantly wider variety of shapes is possible; the scope of the realized values of function

$\rho(K_1, K_2)$ is much broader for them (see Fig. II.64). The long chains also possess higher conformational instabilities in comparison with the short ones.

The inertial characteristics of the polymer coil depending on the spacial position of all chain units has been previously discussed. However the shape of the coil can also be characterized by the mutual position of the most remote units, i.e. by the overall dimensions of the coil. It is possible to introduce for this purpose the maximum spans of the coil $\xi_1 \geq \xi_2 \geq \xi_3$ in three orthogonal directions for each chain conformation. Let us define the values $\xi_i\,(i = 1, 2, 3)$ according to [131] as follows

$$\xi_i = \max(X^m_{(i)}) - \min(X^m_{(i)}) \qquad (7.11)$$
$$(i = 1, 2, 3; \quad m = 1, 2, \ldots, N),$$

where $\max(X^m_{(i)})$ and $\min(X^m_{(i)})$ are the maximum and minimum projection of the units coordinates on the basic axis of inertia $O^*X_i\,(i = 1, 2, 3)$. The conformational averages $\omega_1 = \xi_3/\xi_1$ and $\omega_2 = \xi_2/\xi_1$ were also calculated in Khalatur's paper. The results of calculation as a function of the reciprocate number of repeating units in the chain are shown in Fig. II.65.

Note that the asymptotic values of ω_1 and ω_2 at $1/x = 0$ for a Gauss coil are $\omega_1 = 0.382$ and $\omega_2 = 0.596$ (for ball-shape body $\omega_1 = \omega_2 = 1$) [132]. As can be seen from Fig. II.65, the average shape of the polymer coil characterized by its linear dimensions is also specified by high asymmetry. So, one can say for a 400-unit chain that the maximum length of the macromolecular coil is twice its width on an average and 5 times its thickness. Nevertheless asymmetry defined by coil "spans" appears to be slightly lower than in the case of inertial characteristics. The reason for such a difference as shown by Khalatur [149] is based on the fact that type t or R values are dependent on the mutual position of all chain units, while ξ-values are dependent only on the positions of the most remote parts of

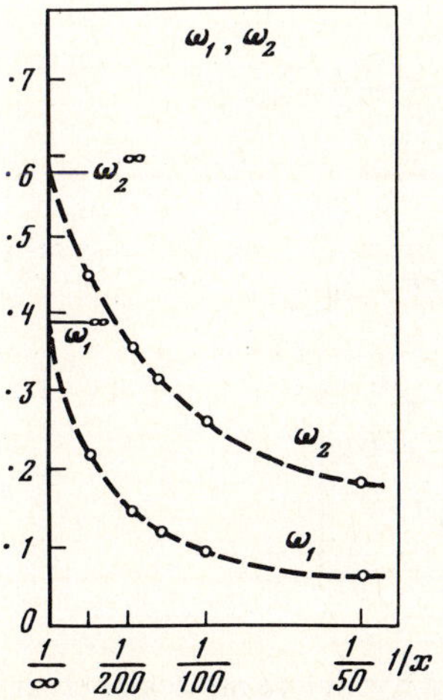

Fig. II.65. Dependence of ω_1 and ω_2 on $1/x_0^{\infty}$—asymptotic values

the chain. Short fragments possessing the highest asymmetry make their contribution to t or R and thus the resulting asymmetry of the coil shape defined by its inertial characteristics is always expressed in higher terms.

The results obtained show that the shape of the coil changes rather slowly on increase of the CTN chain length. Asymptotic values of parameters corresponding to the Gauss coil fail to be reached even for rather long CTN chains with a molecular weight exceeding 10^5. These data require conservative estimates of experimental results on the basis of particular models of the shape of the CTN coil. This conclusion is not unexpected since it is in good correlation with the geometrical peculiarity of the cellulose macromolecule. One can illustrate this with a projection of a cellobiose fragment and the scheme of the macromolecular chain of cellulose constructed with its application (Fig. II.66a, b). The hydrogen bonds 05 . . . HO3 (Fig. II.66a) and 01 . . . HO6 (Fig. II.66b) as well as coordination linkage 6–1 (Fig. II.66b) shown in this figure provide the stabilizing effect of macromolecule shape and thus the high degree of its asymmetry. This

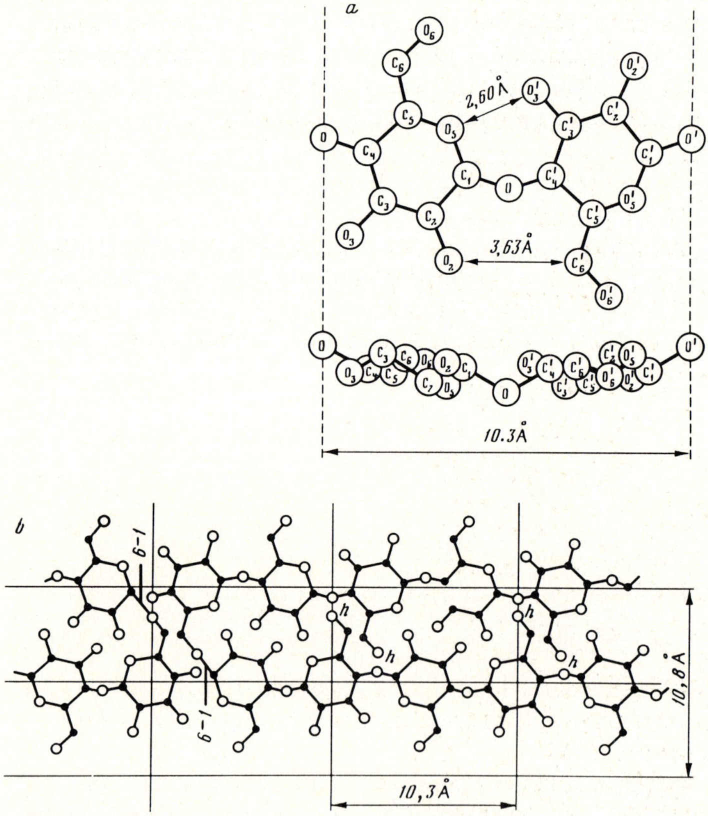

Fig. II.66. The projections of **a** cellobiose fragment [234] and **b** cellulose molecular chain

also follows from above data about the equilibrium and kinetic rigidity of some cellulose derivatives (see Chap. 5).

7.3 Dynamics and the Mechanism of Conformational Rearrangements of the Cellulose Macromolecule

Substantial progress was made recently in the understanding of the mechanism of conformational rearrangements and the dynamic properties of the complex biopolymers: globular proteins, solutions of polypeptides and polynucleotides. The progress in this field is mainly due to the development and application of the methods of direct computer simulation—Monte–Carlo calculations and Brownian dynamics for investigation of the above topics. For example, dynamic numerical experiments were successfully used for the detailed analysis of various types of local motions in proteins. The first results of computer modeling for dynamic studies of minor conformational rearrangements of cellulose were published by us [127]. The calculations were carried out for the macromolecule comprising eight repeating units i.e. 168 atoms shown in Fig. II.66. The lengths of chemical bonds and valent angles were proposed as being similar to cellobiose in the crystalline state [147]. The chemical bonds were assumed to be absolutely rigid. The valent angles were fixed around the equilibrium position by the harmonic potential

$$U_\theta(r) = \mathscr{E}_\theta(r_{ij}^2 - r_\theta^2)^2 \tag{7.12}$$

where r_{ij} is the distance between terminal atoms i and j within three atoms $i - k - j$ forming angle θ; r_θ is the equilibrium state corresponding to a given θ-value; $\mathscr{E}_\theta = 83680$ Kj/mol nm^4. The interaction of any atom pair (i and j) having no mutual connections and free from valent angle formation has been described by the potential [150].

$$U(r) = \begin{cases} \mathscr{E}[(\sigma/r)^{12} - (\sigma/r)^6] \\ \mathscr{E}[3(r-\sigma)^2/(r_c-\sigma)^2 - 2(r-\sigma)^3/(r_c-\sigma)^3] \\ 0 \end{cases} \tag{7.13}$$

The calculation using the method of molecular dynamics (MD) involves numerical integration of motion equations written for all ($N = 168$) atoms

$$m_i \mathrm{d}^2 r_i/\mathrm{d}t^2 = F_i + G_i; \quad i = 1, 2, \ldots, N \tag{7.14}$$

where r_i is the radius-vector defining position of i-atom; m_i is its mass; F_i is the force corresponding to potential of intermolecular interaction; G is the force of reactions of rigid chemical bonds.

The Brownian dynamics (BD) method allows us to study the effect of medium viscosity. In this case the right side of Eq. (7.14) involves two additional terms: 1) random force Ri with Gaussian distribution of components, which simulates the chaotic impacts of medium atoms and 2) dissipative term $m_i \gamma_1 \, dr_i/dt$ governing friction effects. Assuming that Stokes' law is valid, the friction coefficient can be written approximately as $\gamma_i = 3\pi\sigma\eta^2/m_i$, where the medium viscosity is proposed as being $\eta = 2 \times 10^{-4}$ N $\times$ s/m^2.

The motion equations were integrated in Cartesian coordinates with intervals $\Delta t = 0.002$ ps. The total energy fluctuations in the MD method do not exceed 0.2%. The bond lengths remain constant with relative accuracy 10^{-6}. After rather long relaxation period the tracks from 10^4 steps were plotted, which corresponds to a time interval of about 20 ps. The following values of temperatures 347 $\pm$ 15 and 342 $\pm$ 17 K were obtained after calculations on the basis of MD and BD methods using a BESM-6/7 computer for selected parameters.

The visual analysis of "instant shots" plotted by the printer shows that cellulose macromolecule retains a strongly extended conformation during all computer experiments. Such behavior is defined by the geometrical peculiarities of the polymer chain and the intermolecular hydrogen bondings. Nevertheless, atoms undergo rather intensive thermal motions. In order to characterize the collective motions of the atoms it is convenient to use dihedral angles of internal rotation φ which are shown in Fig. II.66. Note that twisted position of the corresponding bonds (*trans*-conformation) is related to $\varphi = 0$, whereas $\varphi = \pm 180°$ corresponds to *gauche* (*cis*) conformation. Fig. II.67 and II.68 demonstrate the alteration in time of the angles $\varphi_1 - \varphi_6$ which determine the conformation of pyranose ring and the angle φ_7 which describes the rotation of the hydroxymethyl group CH_2OH. The results presented are related to the repeating fragment of the studied macromolecule. The short fluctuations of the angles φ with a typical duration of 0.01 ps can be easily seen. These fast vibrations define the saw-type profile of the curves presented in Fig. II.67 and II.68. The other slower type of motions with a typical duration of 0.1 ps refers to the twisted vibrations φ around positions of local equilibrium. Finally one can stress the drastic changes of φ resulted from serious conformational rearrangements of some atomic groups. The viscous solvent (in the BD method) has an influence mainly on these collective motions, increasing the lifetime of stable conformations. At the same time, the medium viscosity has no practical effect on the average conformations. Thus the most stable conformation of pyranose is the chair conformation C1 which has minimal potential energy and it is experimentally observed in the crystalline state [56, 147]. However the C1 conformation is not the only possible one and in the course of the dynamic computer experiment it is possible to follow its inversion into another chair conformation 1C (for symbols see Chap. 4). Transition C1$\rightleftarrows$1C is accompanied by the appearance of the intermediate twist-forms of the ring. Although such transitions are rather rare they proceed very fast (a duration of around 0.1 ps). The lifetime of the intermediate twist-forms usually does not exceed several picoseconds. The calculations show that the most flexible angles are φ_5 and φ_6 which are rotated around C—O bonds in the ring. As a rule, it is their fluctuations which induce rearrangements of the heterocycle. In this sense, these angles are the so-called "active center". The significant mobility of the angles φ_5

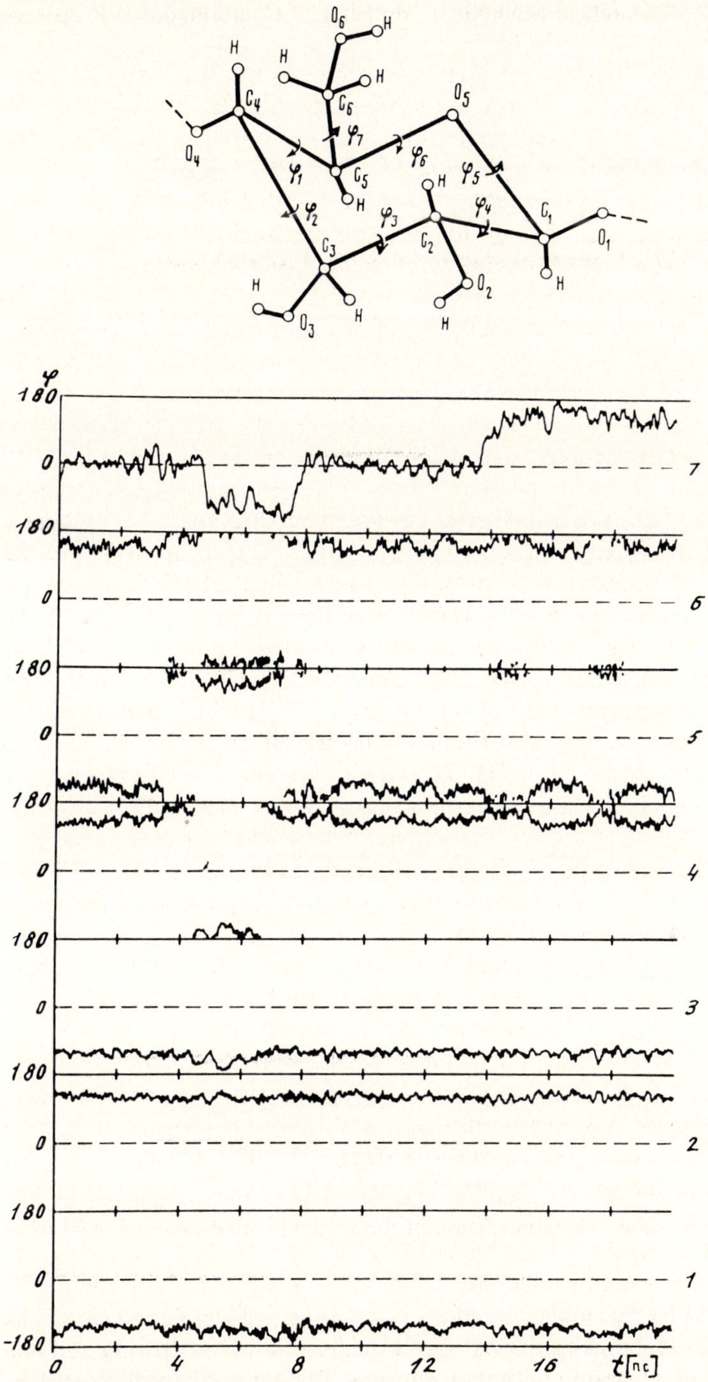

Fig. II.67. Repeating unit of the cellulose macromolecule and angles of internal rotation as a function of time according to molecular dynamic calculations. The figures on the curves correspond to the numbers of the angles of internal rotation (φ_i)

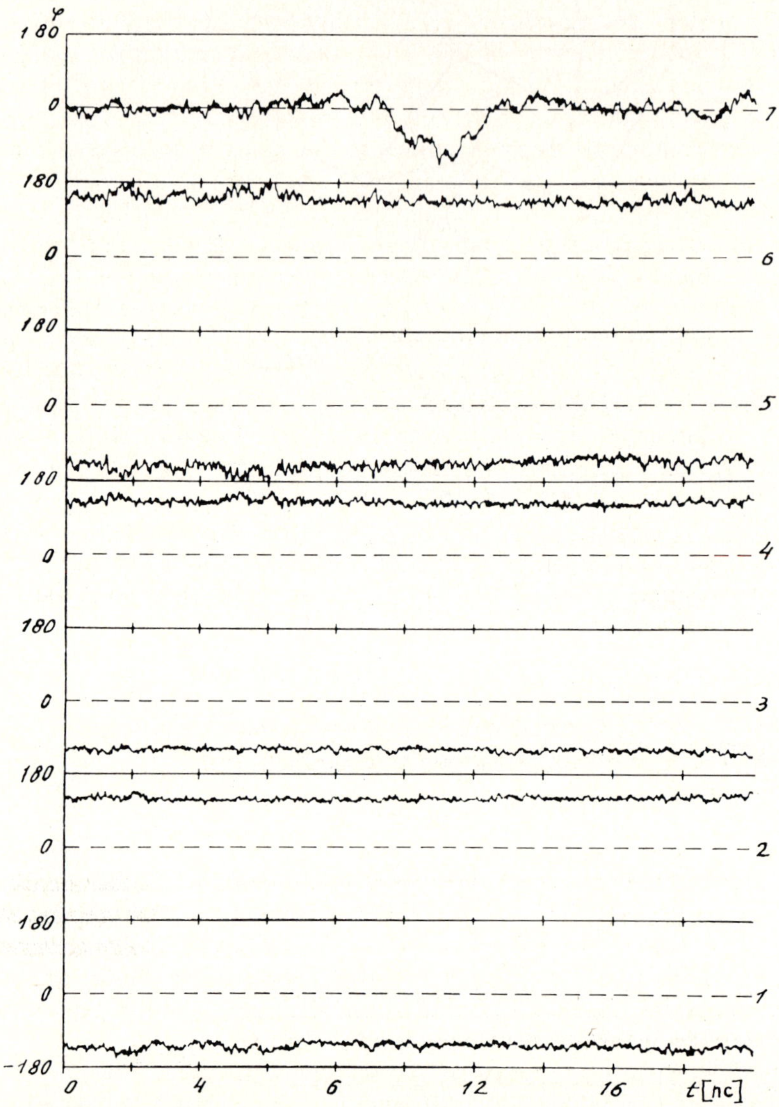

Fig. II.68. The angles of internal rotation as a function of time calculated by the Brownian dynamics method. The numeration of the angles is presented in Fig. II.67

and φ_6 is explained by the higher freedom of rotation around C—O bonds in comparison with C—C bonds, since oxygen is the freest atom in the ring and has the smallest number of bonds with other atoms. Dynamic of conformational rearrangements of the ring is characterized by a significant correlation in changes of torsion angles. For example, it can be easily seen from Fig. II.67 that variations of pairs of φ_4, φ_5 and φ_5, φ_6 are almost similar, equal in magnitude but opposite in sign. The viscous media efficiently decreases the frequency of conformational rearrangements of the ring, thus in the above fragment (Fig. II.68) the ring retains the C1 shape during the whole 20 psec of the computer experiment undergoing relatively small fluctuations. The effect of the medium on the rotation of the side groups is the same. Figure II.68 clearly demonstrates the sequence of *trans*(*t*)–*gauche*(*g*) conformational transitions of hydroxymethyl group.

The duration of $t \rightleftarrows g$ transitions for cellulose in vacuum is less than 1 ps. This time is slightly changed when taking into account the viscosity of the medium. In both computer experiments, the CH_2OH group retains the *gauche*-conformation for about 75% of the calculation period. Note, that preference for this conformation is caused by the favorable conditions for hydrogen bond formation between the H atom of the CH_2OH hydroxyl group and the O_1 and O_3 atoms. The transitions $g^+ \rightleftarrows g^-$ proceed, as a rule, via formation of the *trans*-conformer. However the transition via the potential barrier in the *cis*-conformation ($\varphi_7 = \pm 180°$) is sometimes possible.

Considering the mutual rotation of pyranose cycles around C—O bonds, one does not observe any significant changes of the related torsion angles. The conformational changes of the macromolecule as a whole are rather extended in time due to the required migration of many atoms.

7.4 Molecular and Crystalline Structure of Cellulose. The Most Probable Models

In spite of a great number of crystallographic studies of cellulose the detailed crystallographic structure of this polymer remains still unclear. This refers to all known polymorphous modifications of cellulose including the most examined modifications—cellulose I (natural) and cellulose II (mercerized or regenerated). This fact is connected primarily with non-informative nature of the diffraction pictures of oriented cellulose samples.

Typical X-ray patterns of cellulose I and II exhibit only a few tens of reflections and that is totally insufficient for determining all the atomic parameters by standard crystallographic methods. In order to increase the ratio of the number of observed values (reflections) to that of the model parameters, one uses various stereochemical limitations. Thus the bonds lengths and valent angles are fixed in the course of structure verification and the shortest non-valent contacts are limited by the sum of Van-der-Waals radii. However, even in this case, the diffraction data available are very ambiguous for making an unequivocal choice between several competing models. Such difficulties in the interpretation of diffraction data force us to employ some non-crystallographic criteria. One of them is the most evident one—the total potential energy of the crystal should be minimal in the case of a correct structural model.

It is rather evident that the energy of such a complex system as cellulose cannot be estimated in the scope of simple empirical approximations like those used by Sarko [50, 52]. This approximation is the sum of fine functions as

$$f(X_i) = W_i(X_i - X_i^0)^2 \qquad (7.15)$$

where X_i is the conformational parameter of the macromolecule or the distance between valent–not bonded atoms; X_i^0—the related standard or equilibrium value; W_i is the weight multiplier. This model of the potential of the system is very simple and it is hardly useful for quantitative estimations of potential energy. The more realistic model can be created on the basis of the method of atom–atomic potentials [55], which has proved to be quite reliable for the calculation of the structure and properties of organic crystals. Its efficiency was also demonstrated (see Chap. 4) in the case of analysis of the conformation of the isolated coil of the cellulose macromolecule. This is the reason why the method of atom–atomic potentials was applied for the search of the most probable structural models of cellulose I and II [64]. In this case, the geometrical model of cellulose is the set of deformable cellulose chains packed according to the observed symmetry of the crystal.

As the initial approximation for the search of the structure of the crystalline area of cellulose I and II, the geometrical parameters of six stable conformations of the isolated cellulose coil were accepted (see Fig. II.27, Tables II.3 and II.4) their search algorithm is presented in Chap. 4.

Note that all six conformers of the isolated chain (see Fig. II.27) possess equilibrium spacing similar to the observed identity period. Besides, all conformers exhibit free vacancies for the formation of intermolecular hydrogen bonds. Therefore each of the conformers may exist principally in the crystal. Their positioning in the lattice spacing was performed according to the observed symmetry $P2_1$. One chain was positioned at the beginning of crystallographic coordinate system ("angular" chain) parallel to the *c* axis, the other one—in the center of the *ab* facet ("central" chain). The lattice spacings were fixed in this case as the corresponding "mean" values. Another five parameters φ^1, φ^2, P^1, P^2 and S were introduced at this stage, which described orientation, direction and relative shift of the chains. The angles φ_1 and φ_2 defined the rotation of the angular and central chains around their axes and they were determined as the angles formed by vectors ρ with the crystallographic axis $a^* = a \cdot \cos(\gamma - \pi/2)$. The position of atoms 04 with the relative coordinates $(0, -y, 0)$ for the angular chain and $(1/2, 1/2 - y, S)$ for the central chain corresponded to the equation $\varphi_1 = \varphi_2 = -\pi/2$. The rotation from *a* to *b* was taken as a positive one. The direction of the chain was proposed as positive ($P = +1$) for $Z_{C1} > Z_{C4}$ and as negative ($P = -1$) in the opposite case. The parameter S was defined as the translation of the beginning of the central chain along the *c* axis (in positions of period *c*). Since angular and central chains were crystallographically non-equivalent, 15 varied parameters: τ_i^k, δ^k, φ^k, S and P^k ($i = 1, \ldots, 4$; $k = 1, 2$) were taken into account in the calculation of the cellulose model. The parameter h required for chain construction was fixed at $c/2$. Together with this, one another non-geometrical parameter was introduced—that is the mean isotropic temperature multiplier B, which was involved in the equation for structural amplitudes (Eq. 7.20). The quality of any tentative set of parameters was estimated on the basis of the magnitude of target function

$$\Phi = U + WR'' \tag{7.16}$$

where U is the potential energy of the system, R'' is the crystallographic R''-factor, W is the weight multiplier. The potential energy comprised intramolecular (intrachain) and intermolecular (interchain) terms

$$U = U_{\text{intra}} + U_{\text{inter}} \tag{7.17}$$

The intermolecular energy U_{inter} has been calculated by the method of atom–atomic potentials as the sum of the nonvalent interactions of the atoms belonging to various chains; the intramolecular energy was determined as

$$U_{\text{intra}} = U_{\text{mon}} + U_{\text{junct}} \tag{7.18}$$

In this expression U_{mon} is the conformation energy of the monomer units, including torsion and nonvalent contributions from all the dihedral angles and nonvalent atom–atomic contacts which are changed upon variation of parameter τ_i^k; U_{junct} involves the deformation energy of the valent angle β, the torsion contributions of the dihedral angles in coupling of monomer units as well as all atom–atomic interactions between the two sequential monomer units along every chain. The potential functions used in energy calculations are given in Table II.19.

The crystallographic R''-factor in Eq. (7.16) is defined by the expression

$$R'' = \left(\sum_{m=1}^{M} W_m |F_m^{\text{obs}} - F_m^{\text{cal}}|^2 \Bigg/ \sum_{m=1}^{M} W_m |F_m^{\text{obs}}|^2 \right)^{1/2} \tag{7.19}$$

Table II.19. Parameters of the potential functions used in calculations of the most probable cellulose models

Type of interaction	Intramolecular			Intermolecular		
	A × 10¹	B × 10⁴	α	A × 10²	B × 10⁴	α
H . . . H	4.01	2.86	5.200	2.90	0.49	4.290
H . . . C	12.1	3.28	4.130	11.80	1.86	3.940
H . . . O	12.2	5.75	4.727	8.67	1.95	4.235
C . . . C	47.6	3·77	3.513	42.10	7.16	3.680
C . . . O	44.1	6.37	3.881	33.10	7.46	3.465
O . . . O	34.6	9.65	4.333	25.19	7.77	4.180

Measuring units: parameters A—kcal/mole
B—kcal/mole Å⁶
α—dimensionless value

where F^{obs} and F^{cal} are the observed and calculated structural amplitudes, W_m are the weight multipliers, M is the number of the observed reflections. F_m^{cal} is the function of varied parameters of the model and it is calculated according to the formula:

$$F_m^{calc} = K\left\{\sum_{hkl}\left[F_{hkl}^{calc}\exp\left(-B\rho_m^2/4\right)\right]^2\right\}^{1/2} \tag{7.20}$$

Summation in this equation is performed along all the planes hkl contributing to m-reflection; ρ_m is the inverse interplane distance, K is the scale multiplier. The search for optimal structural model is this case is formally reduced to the finding of a global minimum of the target function (7.16), in 16 dimension space, of the variable parameters. Such a search, in general, is an extremely complicated problem. However, it can be significantly simplified by the application of simple packing and symmetry approaches. Thus, based on the shape of the chain in the projection along the c axis, it can be easily shown that the reasonable (non-overlapping) disposition of the chains can be obtained only when the planes of the glucopyranose cycles are positioned almost parallel to the ac facet of the lattice spacing. This corresponds to the alteration of the φ_K angles in a relatively narrow range around $\varphi^k \approx \pm\pi/2$ at $P^k = 1$, or $\varphi^k = 0, \pi$ at $P^k = -1$ in the case of cellulose II and $\varphi^k = \pm\pi/4$ at $P^k = 1$ or $\varphi^k = 0, \pi$ at $P^k = -1$ for cellulose I.

The volume of the configuration space to be scanned can be markedly decreased using the following relationships of invariance taking place in $P2_1$ symmetry group:

$$\begin{aligned}\Phi(\eta', \varphi', P'; \eta^2, P^2, S) &= \Phi(\eta^2, \varphi^2, P^2; \eta^1, \varphi', P', -S)\\ &= \Phi(\eta', \varphi' + \pi, P'; \eta^2, \varphi^2 + \pi, P^2, S) = \Phi(\eta', \varphi', P'; \eta^2, \varphi^2\\ &\quad + \pi, P^2, S \pm 1/2) = \Phi(\eta^1, \varphi^1 + \pi, P^1; \eta^2, \varphi^2, P^2, S \pm 1/2);\end{aligned} \tag{7.21}$$

where η^k is the vector whose components are the conformation parameters of K-chain.

$$\eta^k = \{\tau_i^k, \delta^k\}$$

On the basis of relationship (7.21) one can select three symmetric independent fields in the subspace of parameters, φ^K, P^K and S:

a) $P^1 = 1;\quad P^2 = -1;\quad \varphi_1 = -\pi/2;\quad \varphi_2 = 0;\quad 0 \le S \le 1/2$

b) $P^1 = P^2 = 1;\quad 0 < S \le 1/2;\quad \varphi^1 \approx -\pi/2;\quad \varphi^2 \approx \pm\pi/2$

c) $P^1 = P^2 = 1;\quad 0 < S \le 1/2;\quad \varphi^1 \approx 0;\quad \varphi^2 \approx 0, \pi$

Each of the areas (a, b and c) involves practically two independent subareas distinguished by φ^2 values. The two parameters of direction P^1 and P^2 can be formally reduced to one parameter P having values: $0(P^1 = 1; P^2 = -1)$, $1(P^1 = P^2 = 1)$ and $-1(P^1 = P^2 = -1)$. The value $P = 0$ corresponds to an antiparallel chain stacking whereas $P = \pm 1$—to a parallel one.

The further decrease of scanning area is possible due to aforementioned fact (see Chap. 4) that reasonable values of the angles δ^k, governing rotation of the monomer units around the virtual bond, are limited to rather narrow range $\delta^k \approx 45 \pm 10°$. The variation of δ^k beyond this range results in either shortened contacts 02 . . . 06′, or unreal values of the valent angle at the glucoside oxygen atom. Finally the optimal value of the mean isotrope temperature multiplier B appears to be close to 23 Å for cellulose and to 32 Å for cellulose and it is slightly dependent on the magnitudes of other model parameters. Thus the accepted cellulose model involves only nine "significant" parameters τ_i^k and S which should be scanned along the whole range of their variation. Selection of these parameters was performed using a uniformly random pattern. Every tentative set of parameters of the model was checked by local minimization of the target function. The detailed scanning of the area of parameters in the case of low values of the target function was carried out by means of alternate variation of the "significant" parameters τ_i^k and S with a simultaneous local optimization of all the other parameters of the model.

For the consideration of cellulose crystalline structure it is expedient to begin from the analysis of the crystal potential energy of the most stable polymorph-cellulose II.

7.4.1 Single-Chain Models of Cellulose II

Using the method of the variable virtual bond and the stereochemical limitations considered in Chap. 4 the search for an optimal single chain model [151] consists of the retrieval of a global minimum of the target function Φ (Eq. 7.13) in the space of eight variable parameters: $\tau_i (i = 1, \ldots, 4)$, δ, φ, P and B.

The results of the retrieval are presented in Tables II.20 and II.21. The characteristics of the deepest local minima Φ found in the course of global search and subsequent local optimization are given in Table II.20. Minima denoted as P_j correspond to polarity $P = 1$, and $\bar{P}_j$—to polarity $P = -1$ [151].

The parameters of hydrogen bonds in structures related to the deepest minima are shown in Table II.21. The angle λ in Table II.21 defines the dichroism of OH bond valent vibration and it is determined as the angle between the direction of the bond and the crystallographic axis c. Only crystallographic non-equivalent bonds are given. The lower index in notations of hydrogen bonds denotes the number of the basis monomer unit with respect to the unit numbers in Fig. II.69. The upper index denotes the translation by which this monomer unit was obtained from the basis one.

It is interesting that the conformation of the cellulose macromolecule in the majority of the most probable single-chain models is similar to that typical for the isolated state. Thus in the case of models P_2 and $\bar{P}_4$, we deal with conformer A_1, for P_1 and $\bar{P}_1$—with B_1, for P_3 and $\bar{P}_5$—with B_2. Rather significant differences from the conformation of the isolated state are observed in the cases of models P_4 and $\bar{P}_3$ (conformer C1) and model $\bar{P}_2$ (conformer C2). However the scheme of intramolecular bonds even in this case is identical to that in isolated chains.

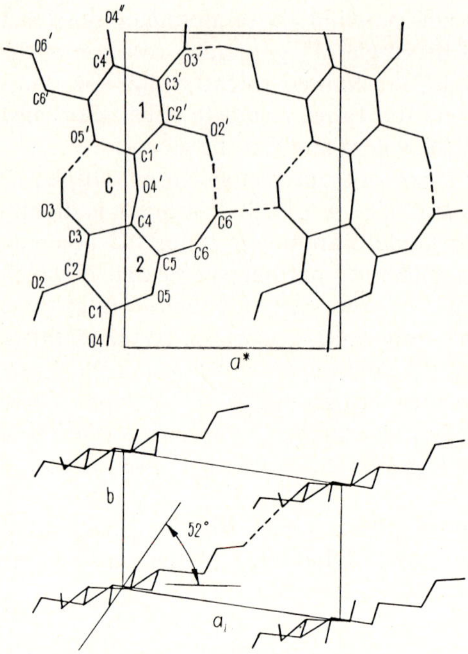

Fig. II.69. Single-chain structural model of cellulose II related to the deepest minimum of the target function (here and further atoms adjacent to cycle are omitted)

Among P_j type models (with positive polarity) the P_1 model is the most preferable one according to Φ. This model is characterized by *gt* orientation of the hydroxymethyl group. Conformation of the chain is approximately related to the B_1 conformer. It is this model which according to its parameters and system of hydrogen linkages is similar to the model proposed by Hunter and Dweltz [54] ($\delta = 40°$, $\varphi^1 = -41°$). However, as can be seen from Table II.20, the model P_1 is not the best according to the R''-factor and it is inferior according to the energy of the best models of $\bar{P}_j$ type.

In spite of Hunter and Dweltz's opinion [54] that it is impossible within the scope of a single-chain model to obtain the stable structure with *gt*-orientation of the hydroxymethyl group and reasonable values of R''-factor. We found such a model. This model is shown in Table II.20 as $\bar{P}_2$. It is energetically close to P_1 and it exhibits the same correlation with diffraction data.

As can be seen from Table II.20, the model P_1 is optimal with respect to Φ. This model is similar to P_1 according to the conformational parameters and the system of hydrogen bonds, but it is distinguished from it by the direction of chains. The model P_1 comprises two intramolecular hydrogen bonds $03_2H \ldots 05_1$ and $02_1H \ldots 06_2$ with parallel dichroism and intermolecular bond $06_1H \ldots 03_1^{a+b}$ with perpendicular dichroism. The projection of this model on the plane a^*c and ab of elementary unit is given in Fig. II.69 [151].

Table II.20. Some structural and energetic parameters of the most probable models of cellulose II. Single-chain variant of the structure

Model	τ_1	τ_2	τ_3	τ_4	δ	φ^1	B	β	φ	ψ	R''	U	ϕ
P_1	63	148	−74	175	41	−29	14.1	117.4	26.3	−27.6	0.37	−17.1	−11.6
P_2	63	−39	−156	180	37	−33	13.9	116.3	23.7	−24.9	0.37	−15.1	−9.5
P_3	−13	−171	−45	182	36	−37	13.0	115.7	23.5	−24.6	0.37	−14.0	−8.4
P_4	−157	77	−105	178	39	−35	9.1	116.9	25.2	−26.4	0.36	−11.7	−6.3
$\bar{P}_1$	69	168	−58	175	41	52	15.6	117.3	26.0	−27.3	0.40	−19.2	−13.1
$\bar{P}_2$	−163	72	−175	185	34	62	14.4	115.7	22.5	−23.7	0.39	−17.4	−11.5
$\bar{P}_3$	−158	74	−94	180	39	59	14.4	116.8	24.8	−26.1	0.40	−17.0	−11.1
$\bar{P}_4$	70	−52	181	180	35	60	15.0	116.0	23.0	−24.2	0.40	−16.7	−10.8
$\bar{P}_5$	−15	−171	−44	190	34	64	14.3	115.8	22.4	−23.6	0.40	−15.9	−8.9

Table II.21. Parameters of hydrogen bonds in structures of cellulose II related to the deepest minima Φ (single-chain variant of the structure)

Model	Bond type	$R_{O \ldots O}$ (Å)	$R_{O \ldots H}$ (Å)	$U_{O \ldots H}$ (kcal/mole)	λ (grad)	Expected dichroism
	03_2H . . . 05_1	2.66	1.84	− 3.95	20	II
P_1	02_1H . . . 06_2	2.63	1.77	− 3.97	40	II
	06_1H . . . 03_1^a	2.46	1.85	− 3.91	94	I
	03_2H . . . 05_1	2.66	1.84	− 3.93	20	II
$\bar{P}_1$	02_1H . . . 06_2	2.73	1.78	− 4.00	26	II
	06_1H . . . 03^{a+b}	2.60	1.79	− 4.00	78	I
	03_2H . . . 05_1	2.74	1.87	− 3.84	15	II
	03_2H . . . 06_1	2.65	2.42	− 1.20	15	
$\bar{P}_2$	06_1H . . . 02_1^a	2.53	1.86	− 3.90	44	II
	06_1H . . . H02^{a+b}	2.50	1.82	− 3.98	61	I

I and II—perpendicular and parallel dichroism respectively

In general the results of the search of optimal structure of cellulose II within the scope of a single chain model can be hardly considered as satisfactory. Table II.20 shows that the best single-chain models possess R''-factor values significantly higher than those typical for the structural analysis of polymers (15–20%). Besides, as will be seen from further data, the single-chain packing version is "inconvenient" from the energetic point.

7.4.2 Double-Chain Models of Cellulose II. Mercerized Cellulose

The characteristics of the deepest minima found in the course of the search for the optimal structure of mercerized cellulose (MC) are given in Table II.22 [64, 152]. Table II.23 represents parameters of hydrogen linkages which occur in the best models. The nomenclature is the same as in Tables II.20 and II.21. The parallel models P_j and $\bar{P}_j$ correspond to the average polarity $P = (P^1 + P^2)/2 = \pm 1$ and the antiparallel models a_j—to polarity $P = 0$.

It follows from Table II.22 that the best parallel model P_1 optimized within double-chain unit although significantly better than the best single-chain model is markedly inferior with respect to (Eq. 7.16) to the best antiparallel models. The differences in target function between parallel and antiparallel variants are rather too high to reject the hypothesis of parallel stacking of MC chains.

It is noteworthy that the antiparallel version of CII structure has been supported [153] by analysis of the corresponding high resolution ^{13}C NMR spectra in the solid phase.

As it can be seen from Table II.22, the values of the R''-factor in the above MC models vary within a relatively narrow range. The ratio of R''-factors does not exceed 1.6 for all models apart from a_7. This means that none of the models (apart from a_7) can be rejected at the significance level of 99% and higher. Otherwise, it can be deduced that the applied diffraction data and stereochemical limitations are "insufficient" for elucidation of the MC structure.

The fact that low values of the R''-factor cannot be of great importance is illustrated by the data for models a_{14} and a_{15}. As can be seen from Table II.22, the models a_{14} and a_{15} possess very low R''-factors but they are absolutely unacceptable from the energetic point of view.

Analysis of the data from Table II.23 allows us to conclude that the MC structure cannot be determined unambiguously on the basis of spectral data. All the structures given in Table II.23 correlate with the data on IR dichroism of valence vibrations, which require that MC should exhibit not less than two bands

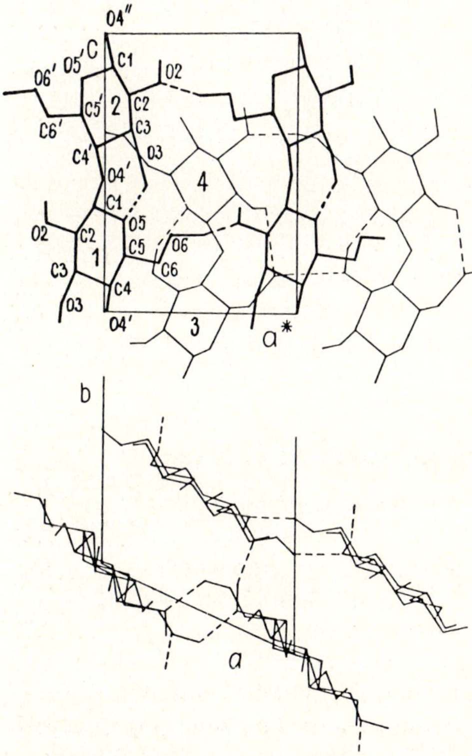

Fig. II.70. Structural model a_6 related to one of the local minima on the energetic surface of the MC crystal

Table II.22. Parameters of the most favorable with respect to ϕ structural MC models. (Double chain version of CII structure)

Model	k	τ_1^k	τ_2^k	τ_3^k	τ_4^k	δ^k	φ^k	S	R''	U	ϕ
1	2	3	4	5	6	7	8	9	10	11	12
a_1	1	66	− 51	− 151	174	43	− 82	0.106	0.179	− 21.4	− 18.7
	2	69	162	− 53	163	46	− 3				
a_2	1	65	57	− 79	177	42	− 83	0.122	0.178	− 21.2	− 18.5
	2	71	166	− 52	175	46	− 2				
a_3	1	68	− 52	− 152	176	42	− 83	0.102	0.183	− 21.2	− 18.5
	2	71	103	− 57	163	45	− 2				
a_4	1	− 63	173	− 161	182	37	− 90	0.125	0.164	− 20.1	− 17.6
	2	69	163	− 52	174	45	− 2				
a_5	1	67	− 50	− 152	− 55	42	− 83	0.100	0.186	− 20.4	− 17.6
	2	68	102	− 56	161	47	− 6				
a_6	1	178	164	− 97	174	48	− 81	0.226	0.124	− 19.4	− 17.5
	2	72	161	− 56	164	42	1				
a_7	1	67	− 52	− 153	177	42	− 86	0.116	0.208	− 20.5	− 17.4
	2	70	34	− 54	72	45	1				
a_8	1	177	163	− 97	175	43	− 82	0.228	0.122	− 19.1	− 17.3
	2	71	105	− 58	168	42	1				
a_9	1	67	− 51	− 156	176	43	− 82	0.110	0.163	− 19.6	− 17.2
	2	77	176	175	160	51	− 8				
a_{10}	1	174	166	− 123	178	42	− 82	0.217	0.126	− 18.3	− 16.4
	2	67	− 55	177	166	43	− 1				

a_{11}	1	− 183	162	− 109	176	43	− 82	0.226	0.121	− 17.6	− 15.8
	2	76	172	− 183	167	43	0				
a_{12}	1	175	− 85	− 98	177	42	− 83	0.231	0.132	− 17.6	− 15.6
	2	70	163	56	− 150	43	0				
a_{13}	1	182	168	− 88	174	41	− 83	0.245	0.133	− 17.1	− 15.1
	2	− 164	− 62	− 35	167	43	0				
a_{14}	1	77	− 48	− 156	173	57	− 61	0.107	0.110	− 9.7	− 8.0
	2	102	162	− 52	160	44	− 7				
a_{15}	1	155	164	− 81	165	43	− 80	0.235	0.108	− 9.5	− 7.9
	2	157	− 80	− 35	167	43	0				
P_1	1	64	64	− 70	172	45	− 74	0.136	0.167	− 19.4	− 16.9
	2	− 170	173	− 79	169	45	− 80				
P_2	1	64	62	− 70	172	45	− 74	0.138	0.168	− 18.9	− 16.4
	2	− 168	95	− 85	169	45	− 83				
$\bar{P}_1$	1	65	158	− 42	169	48	175	0.288	0.164	− 17.5	− 15.1
	2	− 166	84	− 161	174	46	177				
$\bar{P}_2$	1	67	113	− 54	173	46	178	0.261	0.267	− 18.6	− 14.5
	2	67	170	− 64	172	46	180				
$\bar{P}_3$	1	65	160	− 51	170	49	− 52	0.270	0.239	− 16.9	− 13.3
	2	68	− 56	161	173	45	0				

Table II.23. Parameters of hydrogen bonds present in MC models corresponding to the deepest ϕ minima

Model	H-bond type	$R_{O\ldots O}$	$R_{O\ldots H}$	$U_{O\ldots H}$	λ	Dichroism	Model	H-bond type	$R_{O\ldots O}$	$R_{O\ldots H}$	$U_{O\ldots H}$	λ	Expected dichroism
1	2	3	4	5	6	7	8	9	10	11	12	13	14
	$05_1 \ldots H03_2$	2.66	1.84	– 3.94	19.4	II		$05_1 \ldots H03_2$	2.72	1.87	– 3.87	15.7	II
	$06_2 \cdots 04_1$	2.73	2.09	– 2.65	30.6	II }		$05_4 \ldots H03_3$	2.65	1.82	– 3.98	19.0	II
	$06_2 \cdots 02_1$	2.72	1.86	– 3.89	30.6	II }		$02H_4 \ldots 06_3$	2.77	1.82	– 3.99	21.7	II
a_1	$03H_3 \cdots 05_4$	2.64	1.92	– 3.65	26.5	II }		$06H_1 \cdots 02_1^a$	2.84	2.42	– 1.15	71.2	I }
	$03H_3 \cdots 06_1^b$	2.67	2.30	– 1.57	26.5	II }	a_4	$06H_1 \cdots 03_1^a$	3.14	2.18	– 2.17	71.2	I }
	$02H_4 \ldots 06_3$	2.78	1.84	– 3.96	22.7	II		$02H_1 \cdots 06_1^{-a}$	2.84	2.23	– 1.88	69.7	I }
	$02H_1 \ldots 02_4^{-a}$	2.80	1.98	– 3.29	76.0	I		$02H_1 \cdots 02_4^{-a}$	2.66	1.87	– 3.87	69.7	I }
	$06H_3 \ldots 03_3^a$	2.67	1.88	– 3.81	83.0	I		$06H_3 \ldots 03_3^a$	2.72	1.88	– 3.81	81.7	I
	$05_1 \ldots H03_2$	2.67	1.84	– 3.94	18.3	II		$06H_2 \cdots 04_1$	2.73	2.09	– 2.63	32.0	II }
	$02H_1 \cdots 06_2$	2.68	1.84	– 3.95	44.4	II }		$06H_2 \cdots 02_1$	2.69	1.87	– 3.84	32.0	II }
	$02H_1 \cdots 02_4^{-a}$	2.80	2.38	– 1.20	44.4	II }		$03H_3 \cdots 05_4$	2.63	1.91	– 3.67	27.7	II }
a_2	$H0_3 \ldots 05_4$	2.65	1.82	– 3.99	18.6	II	a_5	$03H_3 \cdots 06_1$	2.64	2.25	– 1.80	27.7	II }
	$02H_4 \ldots 06_3$	2.80	1.85	– 3.91	21.9	II		$02H_4 \ldots 06_3$	2.79	1.83	– 3.97	24.8	II
	$03_3 \cdots H06_1^b$	2.69	1.97	– 3.37	92.1	I }		$06_3 \ldots H03_1$	3.00	2.13	– 2.39	84.5	I
	$03_3 \cdots H06_3^{-a}$	2.69	1.86	– 3.90	80.8	I }		$02H_1 \ldots 02_4^{-a}$	2.77	1.95	– 3.49	75.3	I
								$06H_3 \ldots 03_3^a$	2.78	1.93	– 3.57	82.0	I

	$05_1 \dots H03_2$	2.67	1.84	− 3.93	18.8	II
	$06H_2 \dots 04_1$	2.74	2.11	− 2.53	30.8	II }
	$06H_2 \dots 02_1$	2.73	1.87	− 3.86	30.8	II }
	$05_4 \dots H03_3$	2.66	1.92	− 3.63	26.8	II }
a_3	$06_1^b \dots H03_3$	2.64	2.26	− 1.78	26.8	II }
	$02H_1 \dots 02_4^{-a}$	2.81	1.99	− 3.26	75.7	I
	$02H_4 \dots 06_3$	2.72	1.85	− 3.93	25.3	II
	$06H_3 \dots 03_3^a$	2.71	1.89	− 3.77	79.3	I
	$03_2H \dots 05_1$	2.65	1.84	− 3.96	20	II
	$02_1H \dots 06_2$	2.69	1.80	− 4.00	36	II }
	$02_1H \dots 06_3^{-a}$	3.06	2.36	− 1.37	36	II }
	$03_4H \dots 05_3$	2.63	1.84	− 3.96	22	II }
P_1	$03_4H \dots 06_3$	3.08	2.30	− 1.59	22	II }
	$06_1H \dots 03_3^{-b}$	2.74	2.05	− 2.86	89	I }
	$06_1H \dots 06_4$	2.58	1.87	− 3.87	89	I }
	$06_3H \dots 02_3^a$	2.53	1.84	− 3.96	88	I
	05_1 $H03_2$	2.66	1.84	− 3.96	19.4	II
	05_4	2.68	1.92	− 3.66	24.7	II
	$03H_3$ 06_1^b	2.94	2.29	− 1.65	24.7	II
	$02H_4$ 06_3	2.78	1.82	− 3.98	24.7	II
a_6	$06H_1$ 02_1	2.63	1.88	− 3.82	86.7	I
	$02H_1$ 02_4	2.87	1.99	− 3.26	59.7	I
	$06H_3$ 03_3^a	2.68	1.82	− 3.98	83.0	I

with parallel dichroism and not less than three bands with perpendicular dichroism.

Considering the structural models given in Table II.22, we note first of all the model a_6. According to the nomenclature of Stipanovic and Sarko, this model is similar to that of the (*gt*, *tg*) type where *gt* and *tg* denote orientation of hydroxymethyl groups in angular and central chains respectively. According to parameters τ_1^k, φ^k and S this model is close to that of Kolpak et al. [53] ($\tau_1^1 = -175°, \tau_1^2 = 70°, \varphi^1 = -91°, \varphi^2 = 4°, S = 0.227$ and $R'' = 0.168$), which is considered now as the most probable structural MC model. The projection of the model on the planes a^*c and ab is depicted in Fig. II.70. The structure is a stack of alternating layers which are parallel to the plane ac and are formed by translation of identical chains. The central chains forming layer (020) are in a conformation similar to B_1 of the isolated coil. As can be seen from Table II.23, the layer comprises the strong intramolecular hydrogen bonds $05_4 \ldots H03_3$, $02_4 \ldots 06_3$ and intermolecular bond $06_3H \ldots 03_3^a$. Conformation of the chains in the layer (010) is not typical for the isolated macromolecule: there is only one strong intramolecular bond $05_1 \ldots H03_2$, whereas the hydroxymethyl group is involved in intermolecular intralayer bond $06_1H \ldots 02_1^a$. The adjacent layers in the model a_6 are combined by means of the strong intermolecular bond $02_4 \ldots H02_1^a$ and the weak one $03_3H \ldots 06_1^6$. In this case the proton at 03_3 is involved in biacceptor bond

$$03_3H \begin{array}{l} \cdot\cdot\, 06_1^b \\ \cdot\cdot\, 05_4 \end{array}$$

In comparison with the model of Kolpak et al. [53] the model a_6 has the additional bond $03_3H \ldots 06_1^b$. This bond also exists in the model of Stipanovic and Sarko [154] which is optimized on the basis of diffraction data on regenerated cellulose fibers. This model also has the interlayer bond $06_3H \ldots 03_1^a$ involved in the biacceptor bond

$$06_3H \begin{array}{l} \cdot\cdot\, 03_1^a \\ \cdot\cdot\, 03_3^a \end{array}$$

We failed to register this bond together with all the others—all the structures with the system of hydrogen bonds according to Stipanovič and Sarko were unstable and they transferred into a_6 upon Φ minimization.

The system of hydrogen bonds similar to that in model a_6 is observed in model a_{10}. The only difference lies in the formation of the bond $03_3H \ldots 06_1^b$ via the proton at 06_1^b which also participates in the formation of the weak bond $06_1^b \ldots 05_4$. The following interesting chain of the hydrogen bonds appears as a result

$$03_3H \cdot\cdot\, 05_4 \cdot\cdot\, H06_1^b \cdot\cdot\, 03_3 \cdot\cdot\, H06_3$$

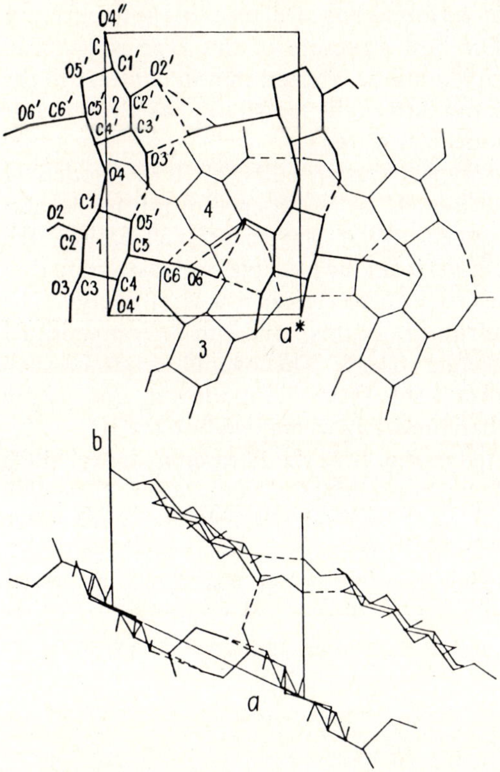

Fig. II.71. Structural model a_4 related to a local minimum on the energetic surface of the MC crystal

In this chain, atoms 05_1 and 03_3 are involved in bidonor linkages while hydroxyl $HO6_1$ participates in the formation of biacceptor bond.

In spite of Stipanovič and Sarko's claim that models (*gt*, *tg*) were markedly preferable in energy than other models of "mixed" type (that is with various orientations of hydroxymethyl groups in angular and central chains) we found the model of (*gg*, *tg*) type has a low value of potential energy and a quite reasonable magnitude of R''-factor. This model is shown in Table II.22 as a_4. The model a_4 is equivalent to a_6 regarding the Φ-value. The projections of the structure a_4 on the plane a^*c and ab are given in Fig. II.71. As can be seen from Table II.23, the model a_4, also has three-centered hydrogen bonds. These are biacceptor bonds of the following type:

$$06_1H \vdots \begin{matrix} \cdot 02_1^a \\ \cdot 03_1^a \end{matrix}$$

and

$$02_1H \vdots \begin{matrix} \cdot 06_1 \\ \cdot 02_4 \end{matrix}$$

Neither Stipanovič and Sarko nor we found any structures of the (*gt*, *gt*) type with rather low stacking energy. The best structure of this type is given in Table II.22 as model a_{13}. However, it should be stressed that in contrast to the corresponding models of Stipanovič and Sarko, the model a_{13} has a rather low R''-factor, exceeding only that of model a_6 on 0.011.

Considering models of the (*tg*, *tg*) type, let us note first of all the models a_5 and a_7. The specific feature of these models is the absence of intramolecular bonds 05 . . . H03′ in one of the layers. As it was mentioned earlier, the structures of this type are usually excluded from the possible ones, assuming that the rupture of 03H . . . 05′ bond results in a too great a loss of energy. The calculation shows, however, that the rupture of this bond can be compensated by the formation of new intermolecular bonds. The overall result of such compensation can be easily seen from Table II.23; the models a_5 and a_7 are about 1 Kcal/mole more favorable than model a_6 where two bonds 03H . . . 05′ are present. In the structure a_5 depicted in Fig. II.72 the released hydroxyl at C3 is involved in intermolecular interlayer bond 03_1^aH . . . 06_3. Apart from this

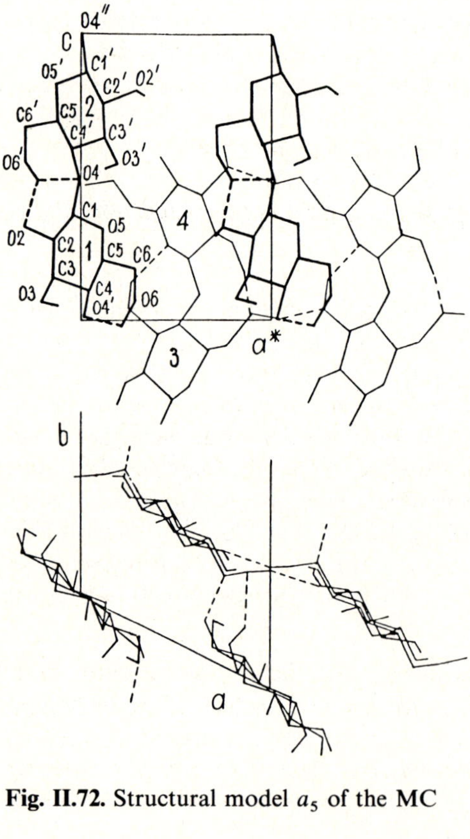

Fig. II.72. Structural model a_5 of the MC

bond the model a_5 has two biacceptor bonds:

$$06_2H \begin{array}{l} \cdot\cdot\cdot 04_2 \\ \cdot\cdot\cdot 02_1 \end{array}$$

and

$$03_3H \begin{array}{l} \cdot\cdot\cdot 05_4 \\ \cdot\cdot\cdot 06_1 \end{array}$$

as well as series of hydrogen bonds, which are given in Table II.23.

In the structure a_7 the cleavage of intramolecular bond 03H . . . 05′ is compensated by the formation of an intermolecular bond which is the binding element of the following type:

$$06_3H \cdots 06_1^{b+a} \cdots H03_3^a \cdots 06_3 \cdots H02_4$$

The model a_1 corresponds to the global minimum of target function. The projection of the model a_1 on the planes a^*b and ab is depicted in Fig. II.73. As can be seen from the figure the layer (020) of the model a_1 is similar to that of (020) of the model a_6. The conformation of the molecules in the layer is similar to the *B*1 conformation of the isolated coil. The layer has intramolecular hydrogen bondings $03_3H \ldots 05_4$, $02_4H \ldots 06_3$ and the intermolecular bond $06_3H \ldots 03_3^a$, which are almost equivalent in energy to the bonds in layer (020) of the model a_6. The molecules in the layer (010) exist in the conformation close to *A*1; there is an intramolecular bond $05_1 \ldots H03_2$ and typical for *A*1, the conformer biacceptor bond

$$06_2H \begin{array}{l} \cdot\cdot\cdot 04_2 \\ \cdot\cdot\cdot 02_1 \end{array}$$

Intermolecular bonds in the layer (010) are absent. The layers (010) and (020) are bonded by intermolecular bonds $03_3H \ldots 06_1$ and $02_1H \ldots 02_4$. Thus hydroxyl 03_3H is involved in three-centered bond

$$03_3H \begin{array}{l} \cdot\cdot\cdot 05_4 \\ \cdot\cdot\cdot 06_1 \end{array}$$

As can be seen from Table II.23, the model a_1 can be transformed into model a_2 by variation of the orientation of the hydroxyls H06 and H02 of the angular chain (that is of parameters τ_2^1 and τ_3^1) with practically unchanged positions of all the other atoms. The reconstruction of the system of hydrogen bondings occurring in the case of transition a_1 to a_2 can be illustrated by the following

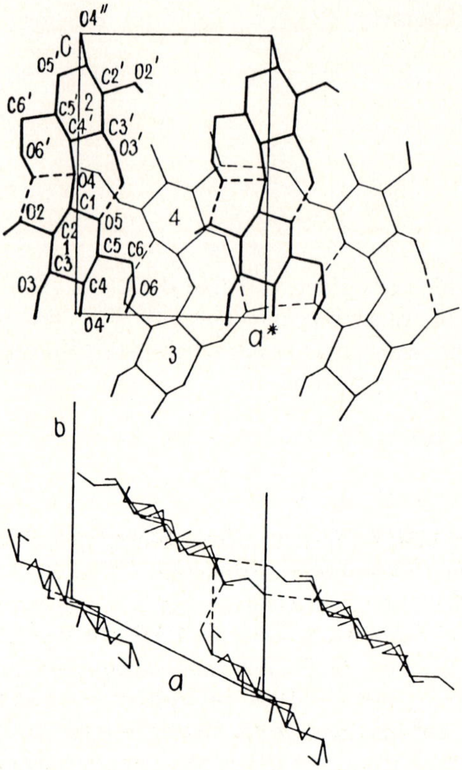

Fig. II.73. Structural model a_1 corresponding to global minimum of the target function Φ

scheme [64]:

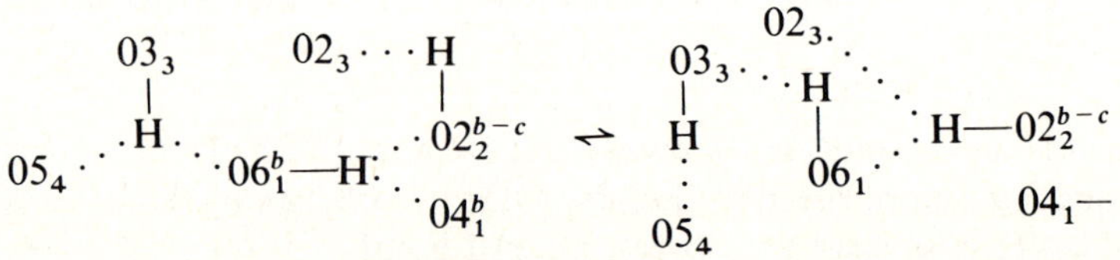

As can be seen from this scheme, the three-centered hydrogen bondings play an important role in the reconstruction of the system of hydrogen bondings.

Similarly the model a_1 can be transferred into model a_3 by rotation of the hydroxyl of the central chain (parameter τ_2^2). In this case the new system of hydrogen bonds does not appear but the position of the proton is only changed in relation to the line connecting donor 06_3 with acceptor 03_3^4 (a similar situation occurs with the transition $a_6 \rightarrow a_8$).

It is important that the structures a_1, a_2, and a_3 are practically equivalent with regard to energy. The calculation of the transition $a_1 \rightarrow a_2$ and $a_1 \rightarrow a_3$ barriers performed by alternating variation of one of transition parameters with

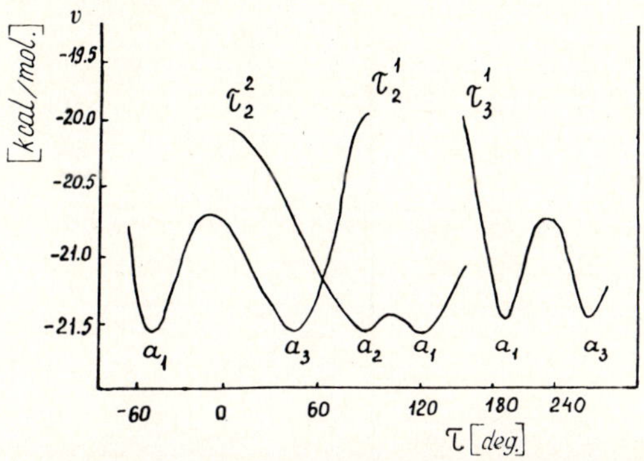

Fig. II.74. Energetic barriers of transitions $a_1 \leftrightarrow a_2$ and $a_1 \leftrightarrow a_3$

simultaneous optimization of other parameters gives values less than 0.9 and 0.2 Kcal/mole respectively (see Fig. II.74). This fact allows us to propose that the structures a_1, a_2, and a_3 can coexist simultaneously and each can change to the other by means of the thermal migration of protons. Otherwise, the models a_1, a_2, and a_3 may be considered as a common model with a labile system of hydrogen bondings (and furthermore also the model a_0).

The presence of "soft" degree of freedom makes the model a_0 also favorable from entropy point of view.

After comparison of the model a_0 with the model a_6-analogue of the model of Kolpak et al., it can be seen that at a quite reasonable value of the R''-factor, the model a_0 has almost 2 Kcal/mole lower energy than a_6. In general, energy of about 1–2 Kcal/mole is in the accuracy limit of the calculations of the absolute values by the atom–atomic potential method. Nevertheless the difference in 2 Kcal/mole between models a_0 and a_6 seems to be quite significant. The point is that the errors introduced into the potential energy of the system by assumptions and approximations by the atom–atomic potential method are usually systematic ones and they are compensated to a great extent by the calculation of the difference in energies between alternative structural models of the same substance. This fact was demonstrated by Bernstein and Hagler [155] for three modifications of *N*-(*p*-chloro-benzylidene)-*p*-chloroaniline.

The model a_0 is strongly supported by the calculation of structural amplitudes of meridian reflexes (002) and (004) which are observed in experiment as moderately weak and strong respectively [64]. The ratio ($|F002|/|F004|$) for the model a_0 is about 1.4/27.8 whereas for the model a_6 the calculation gives 26.3/21.8. It is clear that if the precise determination of the intensities of the above reflexes and their optimization were possible, the advantage of model a_0 over a_6 with respect to the Φ-value would be even greater.

7.4.3 Regenerated Cellulose

The results of the search for the optimal RC structure are shown in Tables II.24, 25 [156]. The comparison of Tables II.22 and II.24 shows that the substitution

Table II.24. Geometric and energetic parameters of the most probable structural models of RC

Model	K	τ_1^k	τ_2^k	τ_3^k	τ_4^k	δ^k	φ^k	S	R''	U	ϕ
a_1	1	66	−50	−151	176	42	−84	0.117	0.216	−21.5	−18.3
	2	69	162	−53	163	47	−5				
a_2	1	64	60	−86	177	40	−83	0.144	0.212	−21.8	−18.6
	2	70	170	−51	170	47	−3				
a_3	1	67	−51	−152	176	41	−86	0.116	0.216	−21.5	−18.3
	2	69	101	−55	162	45	−2				
a_4	1	−63	171	−158	178	39	−89	0.125	0.191	−20.3	−17.4
	2	67	162	−52	172	46	−3				
a_5	1	67	−52	−154	−54	41	−85	0.111	0.219	−20.5	−17.3
	2	66	101	−55	160	47	−5				
a_6	1	−181	162	−97	176	42	−83	0.216	0.161	−19.7	−17.3
	2	70	162	−56	168	43	−1				
a_7	1	67	−52	−153	176	41	−88	0.128	0.232	−20.9	−17.3
	2	69	37	−53	69	46	0				
P_1	1	62	68	−70	171	45	−75	0.133	0.183	−19.6	−16.9
	2	−168	171	−75	165	48	79				

Table II.25. Parameters of hydrogen bonds present in the most probable models of RC

Model	Bond type	$R_{O\ldots O}$	$R_{O\ldots H}$	$U_{O\ldots H}$	λ	Expected dichroism	Model	Bond type	$R_{O\ldots O}$	$R_{O\ldots H}$	$U_{O\ldots H}$	λ	Expected dichroism
1	2	3	4	5	6	7	8	9	10	11	12	13	14
	$03H_2\ldots 06_1$	2.67	1.84	− 3.95	18	II		$03H\ldots 05_1$	2.66	1.84	− 3.96	19	II
	$\cdot 04$	2.72	2.08	− 2.71	32	II		$\cdot 04$	2.73	2.10	− 2.57	32	II
	$06_2H\colon$							$06_2H\colon$					
	$\cdot 02_1$	2.68	1.87	− 3.87	32	II		$\cdot 02_1$	2.69	1.87	− 3.84	32	II
	$\cdot 05_4$	2.63	1.89	− 3.75	26	II	a_3	$\cdot 05_4$	2.63	1.91	− 3.65	28	II
a_1	$03_3H\colon$							$03_3H\colon$					
	$\cdot 06_1^b$	2.65	2.35	− 1.40	26	II		$\cdot 06_1^b$	2.61	2.27	− 1.73	28	II
	$02_1H\ldots 02_4^{-a}$	2.75	1.92	− 3.65	76	I		$02_1H\ldots 02_4^{-a}$	2.77	1.94	− 3.53	75	I
	$02_4H\ldots 06_3$	2.80	1.84	− 3.95	22	II		$02_4H\ldots 06_3$	2.80	1.84	− 3.95	24	II
	$06_3H\ldots 03_3^a$	2.73	1.90	− 3.73	82	I		$06_3H\ldots 03_3^a$	2.71	1.92	− 3.62	80	I
	$03_2H\ldots 05_1$	2.67	1.84	− 3.94	19	II		$03_2H\ldots 05_1$	2.69	1.85	− 3.91	19	II
a_2	$\cdot 06_2$	2.61	1.88	− 3.81	51	II		$03_3H\ldots 05_4$	2.62	1.81	− 4.00	21	II
	$02_1H\colon$						a_4						
	$\cdot 02_4$	2.70	2.09	− 2.64	51	II		$02_4H\ldots 06_3$	2.77	1.81	− 4.00	22	II
	$03_3H\ldots 05_4$	2.61	1.82	− 3.99	22	II		$\cdot 02_1^a$	2.79	2.38	− 1.27	70	I
								$06_1H\colon$					
	$02_4H\ldots 06_3$	2.87	1.88	− 3.82	22	II		$\cdot 03_1^a$	3.12	2.19	− 2.10	70	I
	$\cdot H06_3^{-a}$	2.68	1.85	− 3.91	79	I	a_4	$\cdot 06_1^{-a}$	2.86	2.19	− 2.11	81	I
a_2	$03_3\colon$							$02_1H\colon$					
	$\colon H06_1^b$	2.64	1.98	− 3.31	89	I		$\cdot 03_1^{-a}$	2.70	1.89	− 3.75	81	I
	$06_3^{-a}\cdot$	2.99	2.24	− 1.84	89	I		$06_3H\ldots 03_3^a$	2.73	1.88	− 3.82	82	I

Table II.25. (Cont.)

Model	Bond type	$R_{O\ldots O}$	$R_{O\ldots H}$	$U_{O\ldots H}$	λ	Expected dichroism	Model	Bond type	$R_{O\ldots O}$	$R_{O\ldots H}$	$U_{O\ldots H}$	λ	Expected dichroism
1	2	3	4	5	6	7	8	9	10	11	12	13	14
a_5	$06_2H \cdots 04$	2.73	2.11	− 2.55	31	II	P_1	$03_2H \ldots 05_1$	2.63	1.83	− 3.47	22	II
	$06_2H \cdots 02_1$	2.70	1.87	− 3.86	31	II		$02_1H \cdots 06_2$	2.66	1.79	− 4.00	37	II
	$03_3H \cdots 05_4$	2.60	1.90	− 3.74	29	II		$02_1H \cdots 06_3^{-a}$	2.98	2.34	− 1.44	37	II
	$03_3H \cdots 06_1^b$	2.62	2.26	− 1.78	29	II		$03_1H \cdots 05_3$	2.60	1.85	− 3.93	25	II
	$02_4H \ldots 06_3$	2.79	1.84	− 3.96	24	II		$03_1H \cdots 06_3$	3.04	2.75	− 1.82	25	II
	$06_3 \ldots H03_1^a$	3.00	2.14	− 2.34	85	I		$06_2^aH \ldots 06_3$	2.58	1.85	− 3.93	88	I
	$02_1H \ldots 02_4^{-a}$	2.73	1.91	− 3.69	77	I		$03_3 \ldots H06_1^b$	2.76	2.07	− 2.76	88	I
	$06_3H \ldots 03_3^a$	2.77	1.94	− 3.51	83	I		$06_3H \ldots 03_1^a$	2.53	1.80	− 4.00	90	I
a_6	$03_2H \ldots 05_1$	2.65	1.82	− 4.00	19	II							
	$03_3H \cdots 05_4$	2.64	1.87	− 3.84	24	II							
	$03_3H \cdots 06_1^b$	3.01	2.35	− 1.41	24	II							
	$02_3H \ldots 06_4$	2.79	1.83	− 3.97	25	II							
	$06_1H \ldots 02_1^a$	2.64	1.87	− 3.86	86	I							
	$02_1H \ldots 02_4^{-a}$	2.84	1.96	− 3.40	60	I							
	$06_3H \ldots 03_3^a$	2.70	1.85	− 3.93	82	I							

of the diffraction data for MC for the data for RC upon correction of the CII structure has no practical effect—neither on the position of local minima nor their relative depth. The difference in geometric parameters between corresponding structures of MC and RC is about 2° for angles and $0.02C$ for relative shift of the chains. The serious deviation is that for average isotrope temperature multiplier (B), its optimal value is $32A^2$ for MC and $20A^2$ for RC. This fact has been already mentioned in the literature and it has been interpreted as a result of the differences in degree of ordering between these types of cellulose.

Analysis of Table II.25 where the parameters of hydrogen bonding are given for the best RC models does not show any significant differences in comparison with the results for MC. The only difference, which should be emphasized, occurs in model a_2. Here the three-centered bond $06_1^b\mathrm{H} \genfrac{}{}{0pt}{}{\cdot\cdot\, 03_3}{\cdot\cdot\, 06_3^a}$ is formed instead of the linear bond $06_1^b\mathrm{H} \ldots 03_3$ present in MC.

7.4.4 The Most Probable Models of Cellulose I

The analysis of the surface by the target function (7.13) in cellulose I as well as in case of cellulose II exhibits a great number of local minima with rather low Φ-values. The characteristics of the deepest minima are given in Table II.26, whereas Table II.27 represents the parameters of hydrogen bonds present in the most favorable models [151, 152, 156].

Analysis of Tables II.26 and II.27 allows us to make some important conclusions. Firstly it can be seen from Table II.26 that, as well as in case of CII, the R''-factor is an insufficient criterion for the unequivocal selection of the optimal structural model. The application of statistical Hamilton tests shows that the majority of structural models in Table II.26 are statistically indistinguishable at the significance level of 99%. The models P_5, $\bar{P}_5$, $\bar{P}_7$, a_3, a_4, a_7 and a_{13} are the exceptions. As well as in the case of CII, the calculation demonstrates that several structural models ($\bar{P}_7$, P_5 and a_{13}) which give very low values of the R''-factor are evidently unsuitable from the energetic point of view.

Dichroism of the valent vibrations of the OH-groups is also an insufficient criterion. As can be seen from Table II.27 all models presented in this Table correlate with the IR-dichroism experimental data.

Comparison of energy values from Table II.22 and II.26 shows that the best structural models of CI are markedly inferior in energy to the best models of CII (by about 1.5 Kcal/mole).

Considering the system of hydrogen bonds present in the most favorable structural models of CI, one should mention the following: firstly the change of parameters of the unit cell upon transition from CI to CII does not affect the geometry and energetics of intramolecular hydrogen bonds. Secondly, as can be seen from Table II.27, CI is characterized by a less widespread system of interlayer hydrogen bonds. The majority of intermolecular bonds in CI are localized in the planes (010) and (020). Thirdly, the relative proportion of the three-centered hydrogen bonds in CI is markedly less than in CII.

Table II.26. Geometric and energetic characteristics of the most probable structural models for cellulose I

Model	K	τ_1^k	τ_2^k	τ_3^k	τ_4^k	δ^k	φ^k	S	R''	U	Φ
1	2	3	4	5	6	7	8	9	10	11	12
$\bar{P}_1$	1	70	164	− 54	176	45	27	0.218	0.165	− 19.9	− 17.4
	2	71	164	− 54	177	44	− 150				
$\bar{P}_2$	1	70	163	− 56	175	43	27	0.202	0.187	− 19.7	− 16.9
	2	70	76	− 56	174	42	− 143				
$\bar{P}_3$	1	83	182	173	171	46	21	0.199	0.182	− 17.9	− 15.2
	2	70	77	− 56	174	42	− 142				
$\bar{P}_4$	1	69	172	66	175	43	27	0.181	0.200	− 17.5	− 14.5
	2	71	69	− 55	174	42	− 143				
$\bar{P}_5$	1	68	175	− 57	172	44	− 6	0.126	0.274	− 17.5	− 13.4
	2	61	76	− 58	167	48	138				
$\bar{P}_6$	1	76	164	− 57	185	35	− 41	0.303	0.188	− 15.3	− 12.5
	2	71	105	− 56	177	41	− 56				
$\bar{P}_7$	1	74	136	− 131	108	60	10	0.209	0.113	4.7	6.4
	2	45	172	− 191	88	65	163				
P_1	1	70	164	− 54	176	43	− 43	0.302	0.167	− 17.1	− 14.6
	2	67	167	− 52	172	46	− 43				
P_2	1	67	167	− 58	173	44	63	0.270	0.196	− 17.5	− 14.6
	2	77	162	59	187	33	39				
P_3	1	75	164	− 56	181	38	− 48	0.291	0.240	− 17.5	− 13.9
	2	73	164	− 56	180	38	134				
P_4	1	70	163	− 54	176	53	− 48	0.294	0.152	− 14.1	− 11.8
	2	65	164	− 52	172	52	− 34				
P_5	1	51	148	− 170	119	59	− 33	0.283	0.127	11.8	13.7
	2	65	147	− 152	102	65	− 23				

a_1	1	68	163	− 52	174	43	− 39	0.155	0.213	− 19.7	− 16.5
	2	67	163	− 53	173	46	25				
a_2	1	67	166	− 51	174	44	− 41	− 0.355	0.155	− 18.6	− 16.1
	2	68	165	− 53	169	48	23				
a_3	1	75	163	− 54	180	38	− 49	0.084	0.240	− 19.7	− 16.0
	2	73	68	− 56	177	39	35				
a_4	1	74	163	− 55	180	38	− 48	0.083	0.240	− 19.7	− 16.0
	2	75	61	− 56	62	38	37				
a_5	1	61	− 52	92	163	48	− 34	0.210	0.218	− 18.9	− 15.6
	2	70	162	− 56	176	43	26				
a_6	1	69	161	− 53	175	43	− 38	0.173	0.208	− 18.5	− 15.4
	2	65	− 56	− 151	169	47	21				
a_7	1	67	163	− 52	174	52	− 34	− 0.360	0.135	− 17.1	− 15.1
	2	66	163	− 53	171	54	19				
a_8	1	77	69	− 54	177	38	− 54	0.036	0.260	− 19.1	− 15.2
	2	78	161	− 56	182	37	36				
a_9	1	− 149	143	− 80	165	49	− 35	− 0.292	0.187	− 17.7	− 14.9
	2	69	161	− 55	170	46	22				
a_{10}	1	69	162	− 55	174	44	− 38	0.173	0.210	− 18.0	− 14.8
	2	78	180	162	170	47	21				
a_{11}	1	85	191	169	179	44	− 38	0.098	0.222	− 18.2	− 14.8
	2	70	162	− 54	175	42	27				
a_{12}	1	69	165	− 53	172	45	− 31	0.154	0.150	− 16.7	− 14.4
	2	81	186	162	163	54	195				
a_{13}	1	50	151	− 191	120	65	− 26	0.143	0.109	14.9	16.5
	2	63	155	− 158	88	60	196				

Table II.27. Characteristics of hydrogen bonds of structural models of cellulose I corresponded to the deepest minima of target function ϕ

Model	Bond type	$R_{O\ldots O}$	$R_{O\ldots H}$	$U_{O\ldots H}$	λ	Expected dichroism	Model	Bond type	$R_{O\ldots O}$	$R_{O\ldots H}$	$U_{O\ldots H}$	λ	Expected dichroism
1	2	3	4	5	6	7	8	9	10	11	12	13	14
	$03_2H\ldots05_1$	2.65	1.82	−4.00	18	II		$03_2\ldots05_1$	2.62	1.82	−4.00	21	II
	$02_1\ldots06_2$	2.78	1.82	−4.00	23	II		$03_4\ldots05_3$	2.65	1.83	−3.96	20	II
$\bar{P}_1$	$03_4H\ldots05_3$	2.64	1.82	−4.00	18	II	$\bar{P}_3$	$02_3H\ldots06_4$	2.77	1.81	−4.00	25	II
	$02_3H\ldots06_4$	2.77	1.83	−3.97	23	II		$06_3H\ldots04_1$	3.01	2.20	−2.03	81	I
	$06_1H\ldots03_1^a$	2.81	1.84	−3.95	81	I		$06_3H\ldots05_2$	2.98	2.10	−2.60	81	I
	$06_3H\ldots03_3^{-a}$	2.86	1.90	−3.75	84	I		$06_1H\ldots03_1^a$	2.70	1.83	−3.97	73	I
								$02_1H\ldots06_1^a$	2.91	2.19	−2.10	53	II
	$03_2H\ldots05_1$	2.65	1.82	−4.00	19	II		$03_2H\ldots05_1$	2.64	1.81	−4.00	19	II
	$02_1H\ldots06_2$	2.78	1.83	−3.97	25	II		$02_1H\ldots06_2$	2.78	1.82	−4.00	24	II
$\bar{P}_2$	$03_4H\ldots05_3$	2.64	1.84	−3.95	20	II	$\bar{P}_1$	$03_3H\ldots05_4$	2.62	1.80	−4.00	21	II
	$02_3H\ldots06_4$	2.78	1.82	−4.00	25	II		$02_4H\ldots06_3$	2.78	1.82	−4.00	22	II
	$06_1H\ldots03_1^a$	2.78	1.83	−3.97	82	I		$06_1H\ldots03_1^a$	2.83	1.86	−3.90	81	I
	$06_3H\ldots01_1$	2.99	2.20	−2.04	81	I		$06_3H\ldots03_3^a$	2.93	2.02	−3.10	80	I
	$06_3H\ldots05_2$	2.98	2.14	−2.36	81	I							
	$03_2H\ldots05_1$	2.64	1.82	−4.00	20	II		$03_2H\ldots05_1$	2.69	1.85	−3.93	17	II
	$02_1H\ldots06_2$	2.76	1.80	−4.00	22	II		$02_1H\ldots06_2$	2.78	1.82	−4.00	24	II
	$03_3H\ldots05_4$	2.62	1.81	−4.00	21	II		$02_3H\ldots06_4$	2.79	1.82	−4.00	25	II
a_1	$02_4H\ldots06_3$	2.76	1.80	−4.00	23	II	a_4	$06_1H\ldots03_1$	2.83	1.87	−3.86	82	I
	$06_1H\ldots03_1^a$	2.81	1.84	−3.95	81	I		$03_3H\ldots06_3^{-a}$	2.84	1.87	−3.85	82	I
	$06_3H\ldots03_3^a$	2.83	1.86	−3.90	82	I		$01_1\ldots H06_3^{-a-b}$	3.01	2.05	−2.90	80	I

	$03_2H \ldots 05_1$	2.63	1.81	− 4.00	20	II
	$02_1H \ldots 06_2$	2.81	1.84	− 3.95	22	II
a_2	$03_3H \ldots 05_4$	2.60	1.81	− 4.00	23	II
	$02_4H \ldots 06_3$	2.82	1.86	− 3.90	22	II
	$06_1H \ldots 03_1^a$	2.84	1.87	− 3.86	80	I
	$06_3H \ldots 03_3^a$	2.83	1.86	− 3.90	81	I
	$03_2H \ldots 05_1$	2.70	1.85	− 3.92	17	II
	$02_1H \ldots 06_2$	2.79	1.82	− 4.00	24	II
	$03_4H \ldots 05_3$	2.69	1.86	− 3.90	19	II
a_3	$02_3H \ldots 06_4$	2.78	1.82	− 4.00	25	II
	$06_4H \ldots 04_2$	3.01	2.07	− 2.80	82	I
	$06_1H \ldots 03_1^a$	2.83	1.87	− 3.86	80	I
	$03_2H \ldots 05_1$	2.64	1.82	− 4.00	19	II
	$02_1H \ldots 06_2$	2.77	1.81	− 4.00	22	II
a_7	$03_3H \ldots 05_4$	2.61	1.82	− 4.00	23	II
	$06_4H \ldots 01_1$	2.70	2.11	− 2.60	29	II
	$06_4H \ldots 02_3$	2.76	1.87	− 3.90	29	II
	$06_1H \ldots 03_1^a$	2.79	1.83	− 3.97	83	I

	$03_2H \ldots 05_1$	2.60	1.81	− 4.00	22	II
	$06_2H \ldots 04_1$	2.92	2.04	− 3.93	32	II
a_5	$06_2H \ldots 02_1$	2.72	1.84	− 3.95	32	II
	$03_3H \ldots 05_4$	2.64	1.82	− 4.00	19	II
	$02_4H \ldots 06_3$	2.77	1.82	− 4.00	25	II
	$04_3 \ldots H02_1^{a+b}$	2.99	2.21	− 2.00	71	I
	$06_3H \ldots 03_3^a$	2.79	1.83	− 3.98	82	I
	$03_2H \ldots 05_1$	2.57	1.73	− 3.80	19	II
	$02_1H \ldots 06_2$	2.88	1.92	− 3.65	20	II
	$03_3H \ldots 05_4$	2.56	1.73	− 3.80	20	II
a_6	$02_3H \ldots 06_4$	2.87	1.92	− 3.64	22	II
	$06_1H \ldots 03_1^a$	2.88	1.91	− 3.67	71	I
	$06_3H \ldots 03_3^a$	2.88	1.90	− 3.70	82	I
	$03_2H \ldots 05_1$	2.59	1.83	− 3.97	25	II
a_8	$03_2H \ldots 06_1$	2.97	2.13	− 2.40	25	II
	$05_4 \ldots H03_3$	2.61	1.82	− 4.00	23	II
	$02_4H \ldots 06_3$	2.81	1.85	− 3.90	25	II
	$02_1H \ldots 06_1^{-a}$	2.44	1.94	− 3.55	45	II
	$06_1H \ldots 01_4$	3.01	2.44	− 1.10	88	I
	$06_1H \ldots 05_4$	2.98	2.01	− 3.10	88	I
	$06_3H \ldots 03_3^{-a}$	2.80	1.85	− 3.92	83	I

The conclusion that CI is characterized by weaker coupling of layers than CII is qualitatively correlated with *C* behaviour in such heterogeneous reactions as methylation, acetylation, deuteration etc. CI is known to be a more "mild" reagent than CII and it is possible that this property is caused by easier diffusion of liquid reagents within the interlayer space.

Starting with consideration of some structural models of *C*1, let us examine parallel model P_1. This model according to its parameters and system of hydrogen bonds is similar to the structural models of Gardner and Blackwell [157] ($\tau_1^{1,2} = 80°$, $\varphi^1 = -52°$, $\varphi^2 = -52°$, $S = 0.266$, $R'' = 0.210$) and Woodcock and Sarko [232] ($\tau_1^{1,2} \approx 60°$, $\varphi^1 \approx \varphi^2 \approx 130°$, $S = -0.251$, $R'' = 0.193$). Model P_1 is the best parallel model among the models with positive polarity, but it is markedly inferior to the models of type $\bar{P}_j$ and a_j with respect to the value of the target function.

Among models of the $\bar{P}_j$ type, we can mention model $\bar{P}_3$ which has an R''-factor statistically equivalent to that of model P_1 but it is markedly more favorable than this model according to its energy. Conformation of the central chain in model $\bar{P}_3$ is close to the *B*1 conformation of the isolated coil. Conformation of the angular chain is not typical for the isolated macromolecule; there is only one intramolecular bond $03_2H \ldots 05_1$ whereas the hydroxyl at *C*(2) is involved in the intermolecular bond $02_1H \ldots 06_{\bar{1}}^{-a}$. It is noteworthy that the intermolecular linkage $02_1H \ldots 06_{\bar{1}}^{-a}$ exhibits, in this case, parallel dichroism (see Table II.27). Angular and central chains in the model $\bar{P}_3$ are coupled by the biacceptor bond $06_3H \begin{matrix} \cdots O4_1 \\ \cdots O5_1 \end{matrix}$ whereas atom 05 is involved in the bidonor bond $O5_1 \begin{matrix} \cdots OH3_2 \\ \cdots OH6_3 \end{matrix}$

The parallel model $\bar{P}_1$ is related to the global minimum Φ (II.74). The projections of this model in the planes *a**c* and *ab* are depicted in Fig. II.75.

As it can be seen from this figure the model $\bar{P}_1$ has a clearly lamellar character. The layer (010) formed by the angular chains and the layer (020) formed by central chains are practically identical with respect to their parameters and system of hydrogen bonds, and they correspond to the *B*1 conformer. Separate macromolecules within the layers are linked by intermolecular bonds of the type $06_1H \ldots O3_1^a$ and $06_3H \ldots 03_{\bar{3}}^{-a}$. The interlayer hydrogen bonds are absent in the model $\bar{P}_1$.

It can be seen from Table II.26 that the model $\bar{P}_1$ can be transferred into model $\bar{P}_2$ by changing angle τ_2^2 at 90° while other parameters remain unchanged. In this case the hydrogen bond $06_3H \ldots 03_{\bar{3}}^{-a}$ in the layer (020) is cleaved and the new interlayer biacceptor bond $06_3H \begin{matrix} \cdot 04_2 \\ \cdot 05_2 \end{matrix}$ lying in the plane (200) is formed. (The same bond occurs in model $\bar{P}_3$ considered earlier.)

The projections of model $\bar{P}_2$ in the planes *a**c* and *ab* are depicted in Fig. II.76.

The models $\bar{P}_1$ and $\bar{P}_2$ as can be seen from Table II.26 are practically equivalent in energy. The calculation of the transition $\bar{P}_1 \to \bar{P}_2$ barrier per-

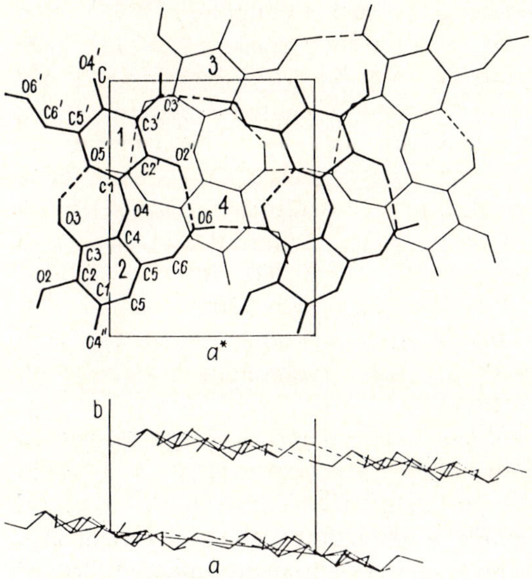

Fig. II.75. The projections of the model $\bar{P}_1$ corresponding to the global minimum of the target function Φ

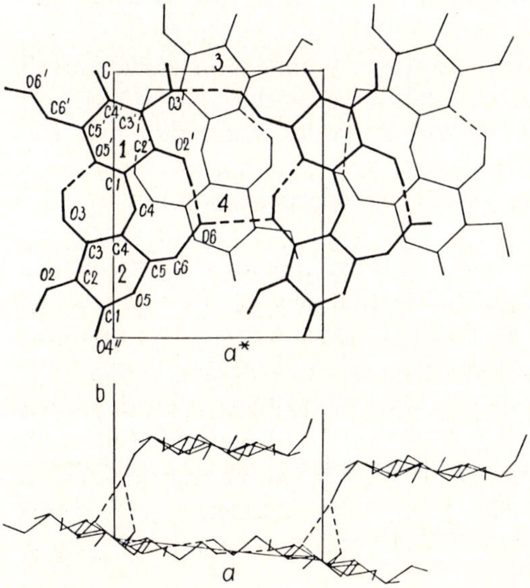

Fig. II.76. The projections of the model $\bar{P}_2$ equivalent in energy to model $\bar{P}_1$

formed by scanning of the parameters τ_2^2 with simultaneous optimization of the other parameters affords a value of less than 0.7 Kcal/mole (see Fig. II.77).

As in the case of CII, this fact allows us to consider models P_1 and P_2 as a unified structural model $\bar{P}_{12}$ with a labile system of hydrogen bonds.

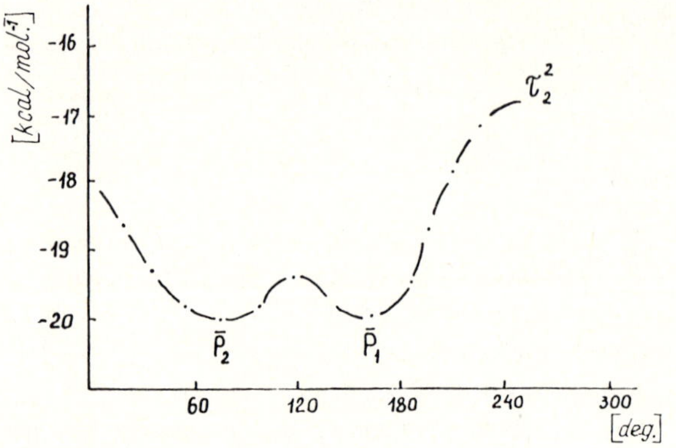

Fig. II.77. The barrier of transition $\bar{P}_1 \leftrightarrow \bar{P}_2$

Among antiparallel models, the model a_2 should be mentioned first of all, it is practically an analog of the structural model of French [158] ($\tau_1^{1,2} = 63°$, $\varphi^1 = -29°$, $S = -0.360$, $\varphi^2 = 29°$, $R'' = 0.158$). The conformation of the angular and central chains in the model a_2 is similar to $B1$, whereas the system of hydrogen bonds is practically identical to that of model $\bar{P}_1$. The model a_2 as well as similar models has rather low values of the R''-factor, however it is inferior in energy by about 1 Kcal/mole to the best antiparallel models.

The models a_5 and a_6 should also be mentioned among antiparallel models. The conformation of the chains of these models in one of the layers is similar to the $A1$ conformation of the isolated coil with its typical biacceptor hydrogen bond $06_2 \vdots\begin{smallmatrix}\cdot 04 \\ \cdot 02_1\end{smallmatrix}$. As well as in case of CII the incorporation of hydroxyl at $C6$ into system of intramolecular bonds results in the absence of intermolecular hydrogen bonds in the planes (010) (or (020) for model a_6). In this case the remaining "free" proton at 02 participates in formation of the interlayer hydrogen bond of the rather rare type $02_1\text{H} \ldots 01_3$.

An interesting chain of hydrogen bonds is observed in the model a_9. This model may be referred to as one of the models of "mixed" type ($gt + tg$). Conformation of the chain forming layer (010) is similar to the $C2$ conformation of the isolated macromolecule with its typical biacceptor bond $03_2\text{H} \vdots\begin{smallmatrix}\cdot 05_1 \\ \cdot 06_1\end{smallmatrix}$

In models a_5 and a_6 the formation of the biacceptor bonds of intramolecular type in one of macromolecule by donor (or acceptor) also results in cleavage of intermolecular bonds in the corresponding layer. However the break of the intermolecular intralayer bond is compensated, as a rule, by creation of an interlayer bond. The bonds in the layer (010) of the model a_9 are absent although the hydroxyls at $C(2)$ and $C(6)$ atoms form intermolecular three-centered hydrogen bonds of the following type:

$$\begin{array}{lll} & \cdot\cdot 05_1 & \\ 03_2\text{—H}\colon & & \cdot\cdot 04_4 \\ & \colon 06_1\text{—H}\colon & \\ 02_1^a\text{—H}\cdot\cdot & & \cdot 05_4 \end{array}$$

Thus the specific feature of the above structural models of *C*1 is the presence of intermolecular bonds with acceptors which fail to have their own protons (glucoside and semiacetal oxygen atoms i.e. 04 and 05).

The model a_1 is the most preferable according to Φ among the antiparallel models, it is mainly distinguished from a_2 by the magnitude of parameter *S*. The central chain in model a_1 is shifted "up" by about half-period in relation to its position in a_2. This fact results in more than 1 Kcal/mole gain in system energy. In case of above mentioned shift the net of hydrogen bonds remains practically untouched. The projections of model a_1 in the planes a^*c and ab are depicted in Fig. II.78.

As can be seen from Table II.26, the model a_1 can be transferred to model a_3 by variation of parameter τ_2^2. The scanning curve of τ_2^2 with simultaneous local optimization of other parameters is presented in Fig. II.79. The barrier of transition $a_1 \rightarrow a_3$ can be seen from the Figure, is rather low (0.4 Kcal/mole). The transition $a_1 \rightarrow a_3$ is combined with the rupture of intermolecular bond

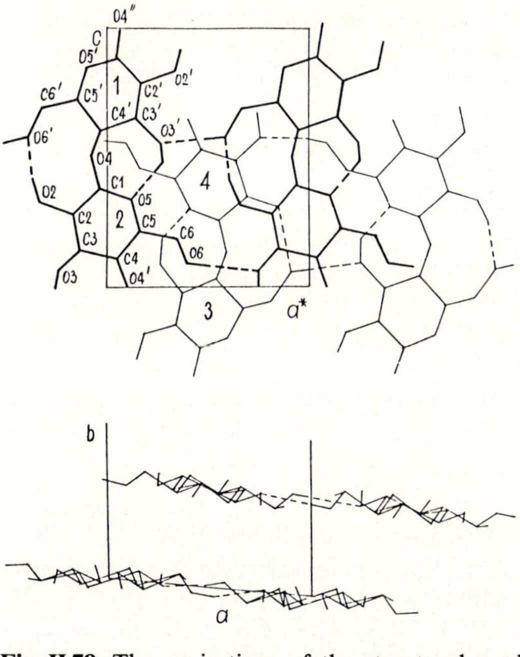

Fig. II.78. The projections of the structural model a_1—the most favorable one with respect to energy in the scope of antiparallel stacking of the chains

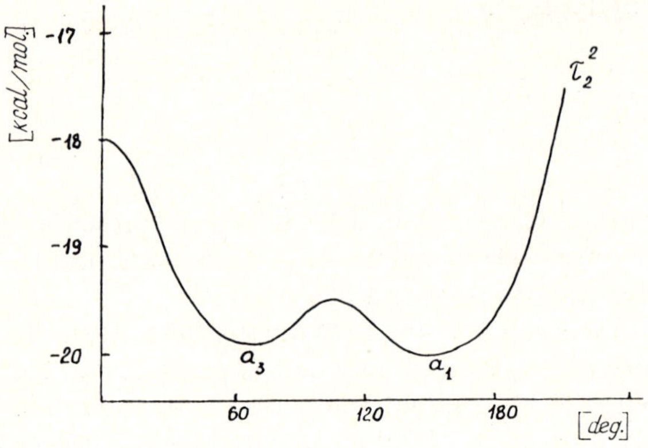

Fig. II.79. Energetic barrier of transition $a_1 \leftrightarrow a_3$

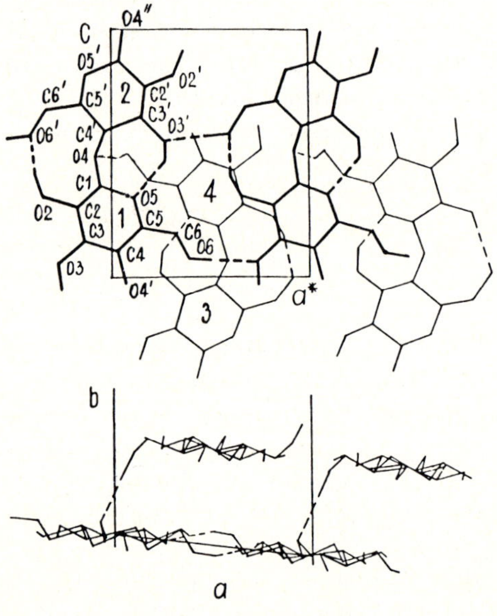

Fig. II.80. The projections of the model a_3 which is equivalent in energy to model a_1

$06_3H \ldots 03$ and formation of interlayer bond $06_3H \ldots 04_1$. The projections of the model a_3 on the planes a^*C and ab are depicted in Fig. II.80.

The model a_1 can be also transformed into a_4 by variation of parameters τ_2^2 and τ_4^2 (by about 100°) and by shifting central chain on about 0.7 Å. The scanning curves of parameters τ_4^2 and τ_2^2 are depicted in Fig. II.81. The barrier of transition $a_1 \rightarrow a_4$ is about 1.1 Kcal/mole. This transition is accompanied by the reconstruction of the system of hydrogen bonds, which can be illustrated by the

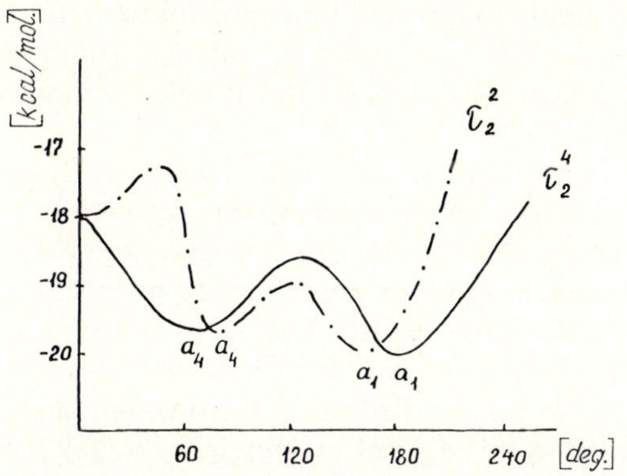

Fig. II.81. Scanning curves of the parameters τ_2^2 and τ_4^2. The barrier of transition $a_1 \leftrightarrow a_4$

following scheme [151]:

$$
\begin{array}{l} 03^{-a}\text{—H}\cdots 05_4^{-a} \\ \vdots \\ \text{H—}06_3\cdots \text{H}02_4 \end{array}
\quad \longleftrightarrow \quad
\begin{array}{l} 03_3^{-a} \qquad\qquad\qquad 05_4^{-a} \\ | \\ \text{H}\cdots 06_3\cdots \text{H}02_4 \\ \qquad | \\ \qquad \text{H}\cdots 04_1^{a+B} \end{array}
$$

As can be seen from this scheme the break of intramolecular bond 03 H . . . 05 is compensated by the formation of intermolecular bonds 03 H . . . 06 and 06_3H . . . 04 (a similar situation takes place in the model a_4 of CII).

Although models a_1, a_3 and a_4 are similar in energy and distinguished by small energetic barriers, the probability of transitions $a_1 \rightarrow a_3$ and $a_1 \rightarrow a_4$ seems to be small. It follows from the high value of the shift of chains (0.7 Å) required for transition of the model a_1 into a_3 or a_4. The realization of such a shift requires a collectively coordinated shift of the atoms within the whole layer which is a very improbable event. It formally means that the pre-exponential multiplier in the expression for transitions $a_1 \rightarrow a_3$ and $a_1 \rightarrow a_4$ is very small.

As was seen in the previous chapters, cellulose macromolecules possess a great number of variants in conformation both for the isolated state and for the crystals of its two basic polymorphs [159–161]. This fact makes it very difficult to determine the molecular-crystalline structure of this natural polymer. It follows from the above calculations that stereochemical and packing limitations, as well as the results of X-ray diffraction on highly oriented cellulose fibers, are evidently insufficient for an unambiguous solution to the structural problem—there is a series of structurally different models, which correlate to the same extent with diffraction data. The above models, as a rule, are in line with the results of IR-dichroism of the OH group valent vibrations. It is these facts which can be used for explanation of the variety of cellulose structural models proposed by different authors.

The energetic criterion which sets limits to the number of possible structural variants helps substantially in the search for the optimal cellulose structure. However, even in this case, the solution of the structural problem is often ambiguous.

The situation is more definite in the case of CII where there are three equivalent structural models a_1, a_2 and a_3, which are markedly distinguished from other variants in relation to energy. The models a_1, a_2 and a_3 are almost equivalent from the energetic point of view and they are separated by less than 1 Kcal/mole transition barriers. In addition, the transitions $a_1 \rightarrow a_2$ and $a_1 \rightarrow a_3$ are accompanied only by reorientation of one or two hydroxyls that affect the network of hydrogen bonds but have no practical effect on the position of the nonhydrogen atoms. That is, the models a_1, a_2 and a_3 comprise in fact one structural model with a labile system of hydrogen bonds.

Whether the model obtained adequately describes the real structure of crystalline cellulose II depends on the reliability of estimations of the system potential energy, that is on the correctness of the calculation that the structures a_1, a_2 and a_3 are similar with respect to energy. As has already been mentioned, the errors introduced into values of energy by the method of atom–atomic potentials are mainly systematic and they are compensated for, to a great extent, by comparison of different packing variants of the same molecules. One can expect that such compensation will occur mostly upon comparison of structurally similar packing variants as it is in the case of models a_1, a_2 and a_3.

Since various states of the hydrogen bonds system may exist in cellulose II, it is of interest to estimate the distribution of these states in the crystalline areas of the polymer. It is possible to propose that only one state is realized within one crystallite, covering its whole volume. Assume that E_1 and E_2 are total energies of the crystallite for the states 1 and 2 respectively. Provided $E_1 < E_2$ the state 2 can be realized only when $E_2 - E_1 \sim RT$. Converting to energies U_1 and U_2 expressed in calculation per mole of monomer units we obtain $(U_2 - U_1) \times n/N_A \sim KT$ or $(U_1 - U_2) \times n \sim RT$, where n is the number of monomer units in the crystallite and N_A is Avogadro number. Provided the average lateral size of the crystallite to be equal to ~ 40 Å and its length to ~ 100 Å, we obtain the magnitude of about 10^3 for the number of monomer units n. This means that the state 2 can be realized together with the state 1 in the whole crystallite volume only when energy U_1 and U_2 differs by not more than 10^{-3} to 10^{-4} Kcal/mole. Such coincidence seems to be, however, very doubtful and therefore it is reasonable to propose that the real number of monomer units involved into structural reconstructions is markedly less than (N).

The situation seems to be more realistic when the state with high energy either appears inside the crystallite in very limited regions (such as domains) or is realized as discrete conformation defects. Since the barriers of transitions between structural states a_1, a_2 and a_3 are small, the above areas or defects can be excitated spontaneously by thermal motion or they can migrate within the crystallite.

The selection of the optimal structural model in the case of cellulose I (CI) is rather difficult. In contrast to CII the most probable models of which (a_1, a_2 and a_3) were distinguished from the others by a rather significant energetic gap, the most favorable models of CI had more or less uniform energy. And the best parallel (P_1) and antiparallel (a_1) models of CI are practically equivalent with respect to energy and they possess statistically similar values of the R''-factor ($R''_{a_1}/R''_{\bar{p}_1} = 1.3$).

An important argument supporting the antiparallel variant of CI structure is the fact that only slight swelling of the fibers occurs upon mercerization, resulting in the polymorph transition of CI into CII. In this case, the significant reconstruction of the structure (restacking of the chains in particular) cannot proceed and thus the structures of CI and CII should have a definite similarity. Comparison of the structural parameters of the models a_1 of CI and CII demonstrates that these models are distinguished mainly by orientations of the protons at 06 and 02. As for other conformation parameters and the relative shift of the layers, they are minimal.

8 Morphological Structure of Cellulose

The specific nature of the morphological structure of cellulose is defined primarily by its function within the cell wall. Having no mineral frame the substance surrounding the plant cell provides it with resistance against exposure to the considerable mass forces. This ability is realized by secretion of cellulose by the membrane of the plant cell as tightly packed fibrils, which afford something like a crystalline lattice. At the same time the fibrils providing cell strength are present in a complex relationship with the amorphous matrix of other polysaccharides, which are a rather complicated group of substances having specific features. Among them the most widespread are hemicelluloses comprising xylans—linear polymers of D-xylose binded by β-1,4-linkages with side chains of 4-*O*-methylglucuronic acid and (or) arabinose residues. Together with hemicelluloses the matrix comprises unbound pectins—polymers of methyl esters of galacturonic acid. Matrix of some plant species contain D-galactan, L-arabinan and other heteroglycans. Cellulose fibrils of the matrix interact strongly with protein—extensin containing a great amount of hydroxyproline. It is important that arabinose or galactose residues of hemicelluloses are bound by the glycoside bond with some hydroxyl groups of hydroxyproline [162] (proline is hydroxylated after its intercalation into the peptide chain). The formation and orientation of extensin as a rigid left-twisted helix may define the growth of cellulose macromolecules and direction of their ordering upon fibril formation. It is noteworthy that the structural strength of such packing which is a rather unusual matrix in composition and additives is capable of withstanding a high osmotic pressure—20000 kg/cm^2.

8.1 Elementary Fibril, Microfibril

The model of the elementary fibril, according to Dennis and Preston, is shown in Fig. II.82 is now generally accepted. In accordance with this model, the molecular chains of cellulose form a crystalline nucleus with a high degree of ordering. The degree of ordering of macromolecular chains is markedly decreased in the area adjacent to nucleus. Here the macromolecules comprising other monosaccharides may be present together with polyglucoside chains. The elementary fibrils under actual conditions are always associated into aggregates of a great number of elementary fibrils forming microfibrils (Fig. II.83, 84).

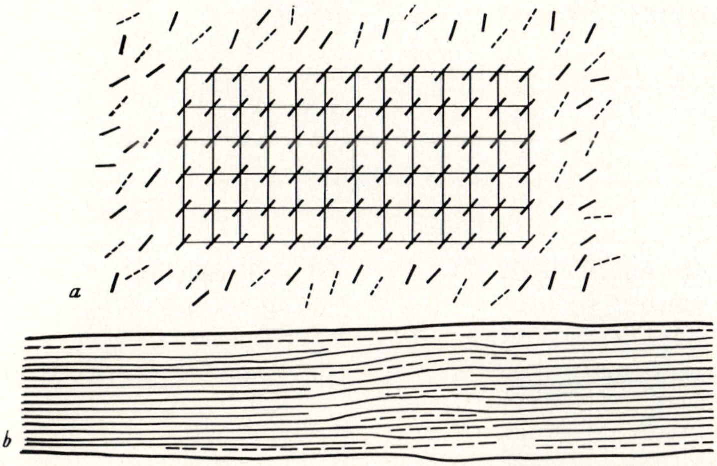

Fig. II.82. The model of elementary fibril of natural cellulose. **a**—cross-section; the inclines denote the position of macromolecules comprising other monosaccharides. The crystalline nucleus of the elementary fibril is depicted in the center. **b**—axial section of elementary fibril; *dotted line* denotes position of macromolecules consisting of non-glucose residue

A microphotograph of the primary cell wall of a cotton fiber is presented in Fig. II.83a, and those of two layers of the secondary cell wall of cotton fiber are shown in Fig. II.83b, c. These data demonstrate a rather random orientation of microfibrils in the primary cell wall and their strict orientation in the secondary cell wall. It is notable that the uniformity of distribution of the orientation degree of microfibrils in the secondary cell wall does not markedly depend on its thickness. Frey–Wyssling's model of a microfibril presented in Fig. II.84 [163] is based on the results of X-ray investigations [164] and comprises four elementary microfibrils with cross-section 70 × 30 Å combined in paracrystalline phase. Thus the microfibril is characterized by a 200 to 100 Å cross-section. According to this model, the cellulose microfibril is formed by rather extended, highly crystalline elementary fibrils present in a specific matrix of less ordered macromolecular formations.

8.2 Crystallinity of Cellulose

The aforementioned models of the elementary fibril and the microfibril (see Fig. II.82, 84) are based on the definite crystallinity of these structures. It follows from them also that rather less ordered macromolecular chains are positioned in the interfibrillar area. It is of interest to mention here the opinion of V. A. Kargin who supposed that cellulose be considered on the basis of the modern concepts

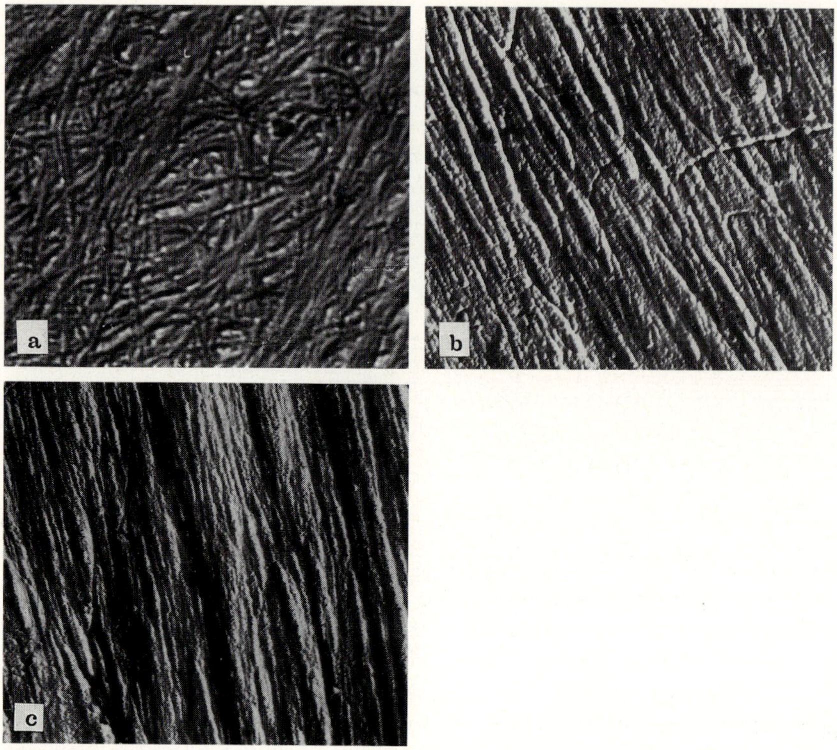

Fig. II.83. Cell wall of cotton fiber [235]

a—the primary cell wall; **b**—the first layer of the secondary cell wall; **c**—the second layer of the secondary cell wall

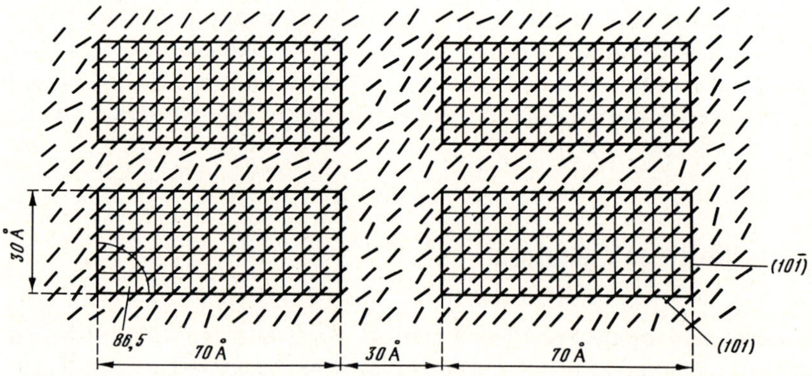

Fig. II.84. The model of the microfibril of natural cellulose. The crystalline nucleus formed by uniformly oriented macromolecules is surrounded by the paracrystalline phase. The interplane (*101* and *10$\bar{1}$*) distances in the lattice are shown in the cross-section

of macromolecule structure as typical amorphous polymer whose properties fail to conform to the hypothesis of its crystalline structure [165]. This point of view was not confirmed since it was in conflict with the experimental data available. These controversies were summarized by Rogovin [166]. He claimed that there were a series of indisputable aspects of cellulose structure which seems to agree with the opinions of all the scientists working in this field. These aspects are: a) the preparations of both natural and regenerated cellulose are structurally heterogeneous; b) structural heterogeneity of all cellulose preparations may be changed within wide limits depending on the conditions and methods of their treatment; c) the majority of spontaneous processes for preparations of cellulose and its esters (their rate increases upon enhancement of temperature or swelling degree) result in an increase of the orientation (or crystallization) degree of these preparations.

The validity of this conclusion is accepted without question now. It is sufficient to mention reviews [167–169] where the formation of the so-called microcrystals—the products with a high degree of crystallinity are recorded for various cellulose preparations. The ability of cellulose to form distinctly fixed crystallites (Fig. II.85) after a certain type of treatment is also well-known.

It is also worth mentioning the very interesting results of Honjo and Watanabe who obtained electron diffraction patterns of a natural specimen of Valonia cellulose containing 400 reflexes [170]—this clearly confirms the crystallinity of these preparations.

Fig. II.85. Crystallites obtained from cotton cellulose [236]

The results obtained up to now cause us to consider natural and highly oriented regenerated celluloses at least as being linear polymers having alternating zones with high and low degrees of orientation. In this case, the structure of the zone with high degree of orientation is similar to the crystalline structure of other linear polymers [171–173]. The crystallinity of zones in cellulose with a high degree of orientation was proved by the following methods: birefringence, magnetic anisotropy, fluorescence, the piezoelectric effect and infrared dichroism (for a review of these results see, for example, [11, 157, 171]). The analysis of X-ray patterns of true crystals of oligosaccharides obtained by heavy cleavage of cellulose lead to the same conclusion (Table II.28).

It is important that the X-ray patterns of powders of oligosaccharide homologous starting from cellotetraose and higher are close that of cellulose II [175, 176]. It is of great interest in this connection to mention a few papers on the preparation of crystals of linear polysaccharides (amylose [177–179], inulin and mannan [180, 181], xylan [181, 182, 188] and, especially important, cellulose itself [183–187]). Cellulose crystals as rhombic plates were obtained by mild saponification of aqueous solutions of low molecular weight (DP = 50 to 400) acetylcellulose. X-ray and electron diffraction study of these crystals shows that they are a system of parallel plates with a thickness of about 100 Å with, perpendicular to this plane position, cellulose macromolecules. Similar crystals with respect to shape and dimensions were obtained for amylose [178, 179] and acetylcellulose [185, 187].

Manley has studied in detail monocrystals of cellulose triacetate with degree of polymerization 300. Crystals were obtained from 0.002 to 0.05% solutions in nitromethane using *n*-butanol as precipitant. It is typical that without precipitant the crystallization results in the formation of spherulites or hedrites but with it—in the formation of almost square lamellar crystals (Fig. II.86a).

The dendritic structures (Fig. II.86b) were formed upon enhancement of the polymer concentration. The thickness of the lamellae was 180 to 200 Å regardless of molecular weight and temperature of crystallization. The conclusion about the spiral mechanism of growth was made on the basis of data obtained. These results demonstrate the normal position with respect to the lamella plane of the macromolecules of cellulose triacetate which with regard to the ratio of lamella size and mean length of the macromolecule indicates the folding of chains upon crystal formation. According to Manley [187], the fold may involve only four glucose residues inspite of the high rigidity of acetylcellulose chains. It is important that crystals can be saponified with the transformation to cellulose II without disturbance of their appearance. A 30% contraction of size in the lateral direction upon retention of the macromolecules orientation has been observed [8]. It is of interest in this connection to mention a paper [186] which up to now is unique and deals with the crystallization of cellulose itself from diluted solutions. The isothermal crystallization from diluted solutions of cellulose with various degrees of polymerization in sodium xanthate or in aqueous solutions of sodium hydroxide has been performed and are described in this paper. The authors found a pronounced increase in specific viscosity of the

Table II.28. Parameters of crystalline structure of some oligosaccharides [11]

	Parameters of elementary unit[a]								Parameters of crystalline structure		
	Size of axis in the unit				Angle value in the unit, grad						
Oligosaccharide	*a*	*b*	*c*	*d*	β	γ	Z_0[b]	ρ_0[c]	Possible spacial group	Number of reflexes	Factor *R*
β-glucose	9.29	12.65	6.70	90	90	90	4	1.53	$P2_12_12_1$	850	0.11
β-cellobiose	10.99	13.02	5.06	90	90.8	90	2	1.56	$P2_1$	1116	0.06
	10.94	13.05	5.11	90	90 ± 0.5	90	2	1.55	$P2_1$	986	0.17
	10.97	13.05	5.09	90	90.8 ± 0.05	90	2	1.56	$P2_1$	1239	0.07
β-cellotriose	29.6	15.3	4.91	90	90	90	4	1.56	$P2_12_12$	—	—
β-cellotetraose	15.9	44.9	18.1	96	116.6	90	16	1.50	—	—	—
	7.3	22.5	4.5	91	98	95	1	1.49	P_1	85	0.23
	(8.04)	—	(9.05)	—	(116.8)	—	—	—	—	—	—
β-cellopentaose	16.1	55.9	18.0	90	116.7	90	16	1.52	—	—	—
	(8.00)	—	(9.06)	—	(116.6)	—	—	—	—	—	—

[a] All dimensions given in parenthesis are determined from positions of reflexes corresponding to those (002), (101) and $(\overline{101})$ present on X-ray pattern of cellulose II

[b] The number of elementary units in the cell

[c] Measured density (g/cm^3)

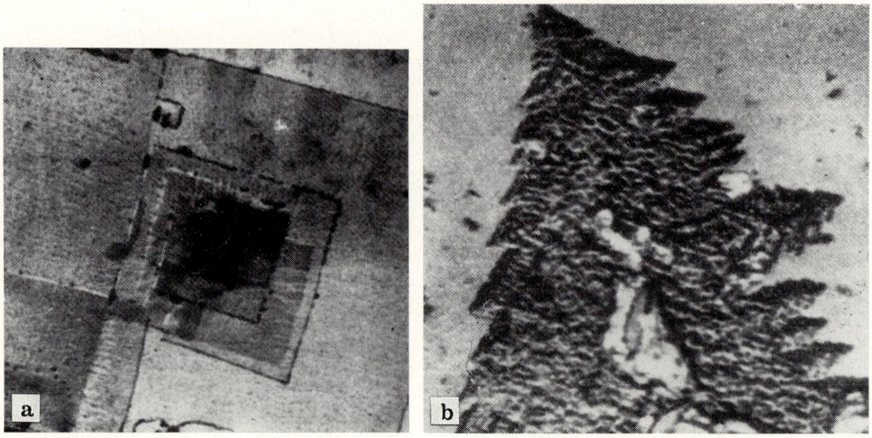

Fig. II.86. a single and **b** dendrite crystals of cellulose triacetate [185]

solutions just before the beginning of crystallization which, according to their opinion, is caused by the formation of intermolecular hydrogen bonds which increase the rigidity of the macromolecular chains. It is important that electron-microscopic data allow us to suggest that the crystalline structures formed are the crystals with straightened chains. The interesting attempt to perform crystallization from the solution in conditions which exclude the formation of intramolecular hydrogen bonds was described in this paper. This was achieved by using an equimolar mixture of aqueous sodium hydroxide and methanol as a solvent. It is typical that the increase of viscosity preceding crystallization was not observed in this case, but the crystals formed were lamellar ones with folded chains. The above examples of crystallization by two different mechanisms (straightened and folded chains) depending on the nature of solvent is accounted for by the specific effect of the hydrogen bonds.

At the same time, one can deduce from it that such level of backbone rigidity of the macromolecular chain may take place so that crystallization becomes possible only in the case of the aggregation of straightened chains. It is important that the X-ray study of cellulose crystals [188] determines them to be highly oriented cellulose hydrate fibers.

The crystallization from a diluted solution of acetylated and alkylated (methyl-, ethyl-) celluloses had been described earlier [10]. The crystallization was carried out in this case from a mixture of chloroform and alcohol in ratio 1:1 (trimethylcellulose); benzene and alcohol in the same ratio (acetylcellulose) and benzene (triethylcellulose). Analysis of these data demonstrates however that morphological forms of the corresponding cellulose derivatives observed in these papers can be correctly qualified as spherulites and dendrites (triethylcellulose), crystallites and hedrites (triacetylcellulose), spherulites and hedrites (trimethylcellulose) but not as polymer monocrystals.

Nevertheless all these results demonstrate that submolecular cellulose structure is characterized by the whole set of morphological forms inherent to a crystallized polymer. This set involves both forms characterized by a very high level of orientation ordering (monocrystals, hedrites) and those regions with high degree of ordering and with a rather low orientation order (crystallites, spherulites).

8.3 Character of Disposition of Cellulose Macromolecules in Polymer Monocrystal

The experimentally observed shape of polymeric monocrystals of cellulose (plate-like lamellaes with thickness of 180 to 200 Å), also inherent to other linear macromolecules [8, 189, 190], is possible, as has already been mentioned, provided folding of the chain macromolecules takes place upon their aggregation in the course of the formation of the lamellar crystal. The principal scheme of arrangement of macromolecule folds in polymer monocrystals is depicted in Fig. II.87.

The folding of macromolecules was confirmed by the detection of dislocation which arise due to interactions between the folds [191]. The confirmation of the tight folding model (see Fig. II.87) is considered [192] to be a), the often observed aggregation of lamellar crystals proceeding according to the mechanism of spiral dislocation and b), the epitaxial growth of polymeric monocrystals on the substrate of low molecular weight or polymer crystals. At the same time there is confirmation of significant non-uniformity of both separate monocrystals and the structure of folds in some regions of the same monocrystal [193]. It is of interest, in this connection, to consider the results of an investigation by NMR (broad lines) of the surface layer of a monocrystal containing folds. It has been shown that on the surface of the monocrystal, there is a layer of polymer whose properties are similar to those of an amorphous

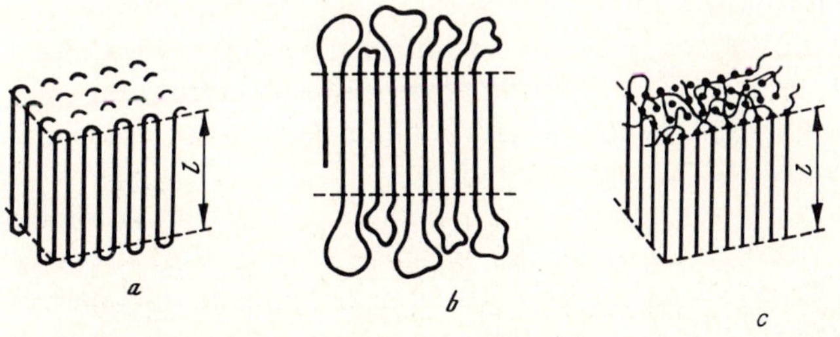

Fig. II.87. Models of polymeric monocrystal with folding chains

a—the model of tight folding [203]; **b**—Fisher model of long (light) loops [237]; **c**—Flory model of "disturbing panel" [238] sample [8, 194, 195]. It is important that realization of various crystalline forms in the course of crystallization can be due to a systematic change of the chemical structure of the macromolecule (the relative ratio of non-crystallizable component, the presence of branching etc.) [196, 197]. This conclusion was made on the basis of a detailed study of morphological forms obtained by crystallization from diluted solutions of polyethylene modified by insertion of side alkyl (methyl-, ethyl-, propyl) substituents [198].

The most essential result was the fact that these crystallization conditions may insert branching points into the crystalline lattice resulting in a logical increase in the size of the elementary unit in proportion to the increase of the number of branchings in the chain. According to Kawai [5] the roughness of the monocrystal surface observed in some cases during electronmicroscopic studies may be considered as confirmation of the existence of long loops (see Fig. II.87). One can make the conclusion, with regard to this, that the "rough" surface of monocrystals is formed upon crystallization of macromolecules at enhanced temperatures, whereas at reduced temperatures, the crystalline rim is produced. Its profile allows one to suggest a very low number of long loops and they are formed from lower molecular weight fractions of the crystallized polymer [199, 200]. In this case "intramolecular crystallization" proceeds primarily due to folding of the chains and formation of the ribbon with a thickness corresponding to monomolecular layer. Then "intermolecular crystallization" takes place because of the aggregation of ribbon resulting in the formation of crystal-lamella. An alternative mechanism of polymer crystal formation consists, according to Kawai [199, 201–203], of the intermolecular crystallization of straightened macromolecular chains with the production of so-called "packet" crystals. As an example of "packet" crystals one thinks of either crystals where the macromolecule is completely incorporated into the crystalline lattice or those comprising only separate parts of the chains and having unordered regions.

The effect of molecular weight, as well as the molecular weight distribution, on the crystallization process is quite substantial. The possibility of the formation of crystals with straightened chains in the case of low molecular weight macromolecules has been demonstrated by the investigation of some paraffins [204]. In this case, the formation of crystals with straightened chains with a length of up to 100 Å was observed, whereas for longer macromolecules, chain folding occurs. It was found upon investigation of crystallization from the melt [205] that low molecular weight fractions with a chain length not exceeding 1000 Å also formed crystals with straightened chains.

We agree with Kawai's proposition [5] that the influence of chain terminals grows according to the decrease of molecular weight (see Fig. II.87). It is of interest to mention data [206] about the discrete change of lamella thickness of low molecular weight fractions of polyoxyethylene having values related to 1, 1/2, 1/3 etc. of the length of the totally extended macromolecule with decreasing temperature of crystallization from the melt. The theory which explains this phenomenon has been proposed [207].

It is noteworthy that polyoxyethylene samples with a narrow molecular weight distribution ($\bar{M}_w/\bar{M}_n \sim 1.1$) were used for the above experiments. These results indicate the importance of both molecular weight values and the molecular weight distribution character in the course of crystallization favoring the formation of highly oriented ordering.

This conclusion seems to be natural since realization of the mechanism of intermolecular crystallization of straightened chains with the formation of packet crystals is certainly possible only in conditions of preliminary orientation ordering of macromolecules. This is equally valid for crystallization conditions from solution and from the melt. It should be stressed that folding of the macromolecule, irrespective of whether it is shaped like the statistical helix or if it is characterized by the high asymmetry of shape that is typical of rigid chain polymers of the cellulose type, is accompanied by serious conformation changes affecting the molecule as a whole. This seems to be the main reason leading to crystallization of straightened chains in the conditions when the macromolecules possess limited mobility, which naturally hinders their folding. It is possible that this is the reason for the formation of transfer chains which was demonstrated by Keller and his coauthors [208].

It should be noted that the discovery of polymeric folded crystals [209] and the following systematic study of polymer crystallization [202, 210, 211] posed the question about the reasons of regularity of folding of macromolecular chains. The papers dedicated to this problem have been analysed in detail by Jale [8]. These papers were summarized in the review in the following way: 1) the majority, possibly all, crystallizable polymers are crystallized from solution and from melts as lamellae or as their unfinished form-ribbons, where folding of macromolecules occurs; 2) the period of folding depends on the crystallization temperature, increasing with the enhancement of temperature. The value of the folding length is close to 100 Å for many polymers; 3) at a given crystallization temperature, the observed number of orders of low angle X-ray reflections shows that the folding length for crystals raised from solution varies from crystal to crystal within limits not exceeding 8%; 4) development of a crystalline structure in the shape of a hollow pyramid allows us to suggest that the precise folding length is typical for such a single crystal; 5) discrete changes of crystallization temperature both increase and decrease result in corresponding stepwise alterations of the growing lamella; 6) the folding length is increased if a lamella crystallized from solution or melt is heated at temperatures above the primary crystallization temperature. At temperature a little higher than glass temperature seems to be the limit temperature lower than which the repacking of molecules with formations of folds of new length cannot occur. In addition there is a temperature above which the folding length will be increased with time indefinitely, and lowering of this temperature it will attain a certain value; 7) extended (orientated) polymers often possess clearly expressed regularity (100 Å). The reasons leading to this regularity have no connections with those causing high regularity of folds in lamellae; 8) studies of oligomers show that polymer molecules from crystals are folded regularly provided the degree of polymerization just reaches the value of the folding length or exceeds it. The foldingback of the molecule in such crystals should take place close to the site of its outlet and the ends of the macromolecule should also be situated adjacent to each other. Provided several repeating units are in the disordered surface layer they will be crystallized and the final thickness of the crystal will correlate with the total length of the molecule.

The latter fact is of special interest with regard to the rigid-chain cellulose-type polymers, since their thermodynamic high backbone rigidity should hinder, to a great extent, the formation of folds resulting in increase of the number of molecules which fit in the lattice. In this connection, it is of interest to mention Flory's opinion [212] that the crystallization of rigid-chain polymers may be realized only because of reduction of free volume resulting from this increase in lengthwise order due to the rigidity of macromolecules in turn due to the effects of intramolecular interaction. The intermolecular interactions in this case enable only the tightest packing of chain macromolecules in the crystalline lattice to occur. On this basis, the crystals of polymers with absolutely rigid macromolecules, in the absence of free volume, cannot be melted without decomposition. In this case only the destructive breaking of the backbone can destroy the three-dimensional order [7].

Data dealing with this problem of macromolecular chains folding upon crystallization forces us to consider carefully the problem of cellulose polymorphism. Taking into account the results obtained from investigations of the polymers markedly different from cellulose, the consideration of the products of rather severe chemical treatment disturbing both morphological and molecular cellulose structure as separate cellulose polymorphs cannot be considered fully valid.

8.4 Parameters of the Unit Cell of the Natural Cellulose Polymeric Monocrystal

The model of the unit cell of the natural cellulose crystalline lattice proposed by Meyer and Misch is generally accepted [6]. The evaluation of the parameters of this model (see Fig. II.88) performed during investigation of natural celluloses from various sources (cotton, flax, ramie, bacterial cellulose) demonstrated [176, 213] that there are certain differences in lattice parameters. Thus the a-values were in the range 8.17–8.28 Å, C-values were in the range 7.85–7.96 Å, the values of β-angle were in the range 81.8–83.6°. The observed differences in parameters of the crystalline lattice are small, nevertheless they exist and they demonstrate the variability of the dimensions of the unit cell for various cellulose preparations.

It was postulated, according to this model, that the cellulose macromolecule has a two-fold screw axis and an identity length 10.3 Å along the chain axis. It was suggested by the authors of this model that the sequential glucopyranose

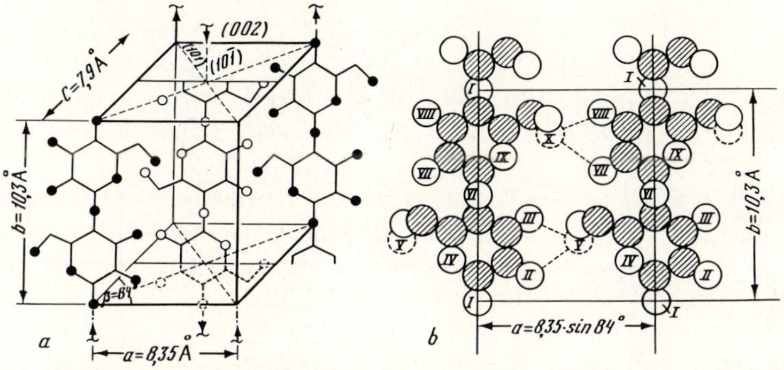

Fig. II.88. Meyer and Misch model of crystalline lattice of natural cellulose
a—spatial model; **b**—projection on the plane (the formation of intermolecular hydrogen bonds in the plane 002 is shown)

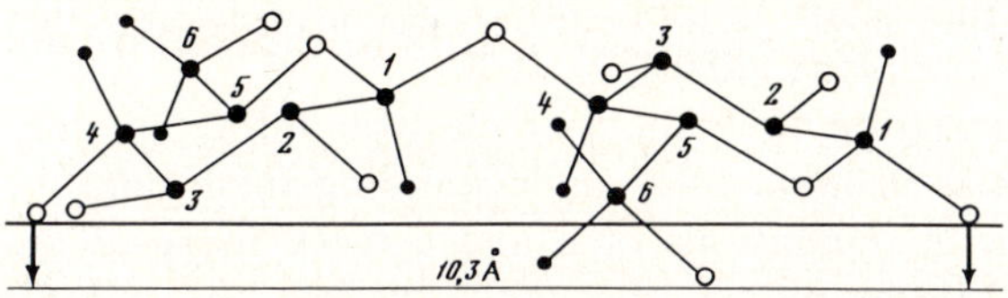

Fig. II.89. Conformation of cellobiose fragment of cellulose chain postulated by Meyer and Misch

units of the chain are alternately turned up and down with respect to the main axis. Thus the identity length corresponds to two glucopyranose units forming a cellobiose fragment. If the identical position of the oxygen atoms at C_6 in relation to the plane of the pyranose cycle is retained and the chain is turned 180° around its axis and it is shifted half of identity period in the direction of the main axis, the chain will accept the initial conformation (Fig. II.89).

The above model of the unit cell possess some evident defects:

1. The distance between C_1 and C_4 positioned on both sides of β-glucoside oxygen bridge is only 1.7 Å which is markedly less than the sum of the Van der Waals radii of two hydrogen atoms.
2. The distance between the oxygen atoms at C_2 of one cycle and C_6 of the adjacent cycle and the distance between the oxygen atom at C_2 and the hydrogen atoms at C_6 are markedly less than the sum of the corresponding Van der Waals radii.
3. The model does not demonstrate how oxygen atoms of sequential units form the system of hydrogen bondings.

The Meyer–Misch model of the crystalline lattice has been slightly modified by Frey–Wyssling [163] into the structure with hydrogen bonds in the plane

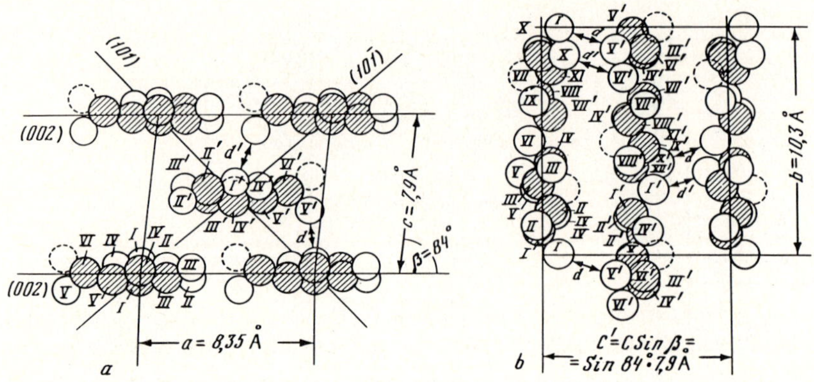

Fig. II.90. The projections of Meyer and Misch model of cellulose crystalline lattice modified by Frey-Wyssling

(101) rather than (002) since this plane is more important morphologically, according to his opinion, than the plane (002) (Fig. II.90).

These results exhibit no correlation, however, with electron diffraction data [170] for Valonia cellulose. It is necessary, for understanding these data, to accept the crystalline lattice with dimensions of *a* and *c* twice those of the relative parameters of Meyer–Misch unit cell, that is the triclinic unit cell (*p*I) with the following parameters: $a = 16.4$ Å, $b = 10.3$ Å, $c = 15.5$ Å, $\beta = 82°$, $\gamma = 90°$.

It is important that similar conclusions about parameters of Meyer–Misch unit cell follow the analysis of this model on the basis of efficient packing scheme of folding chains of natural cellulose (Nissan model analyzed by Jones [174]). These data also require the doubling of parameters *a* and *c* of the Meyer–Misch unit cell.

It was shown a bit later [214] that if one accepts the unit cell with parameters $a = 10.8$ Å, $b = 10.3$ Å, $c = 12.1$ Å, $\alpha = 90°$, $\beta = 90°$, $\gamma = 93.2°$ (triclinic cell also) it allows one to explain the experimental results with fewer assumptions than are required for the Meyer–Misch model. The comparison of these models of unit cells of natural cellulose is presented in Fig. II.91.

It should be stressed that the results of the Valonia cellulose study by the electron diffraction method [170] were confirmed by the Fischer and Mann paper [215]. It follows from it that the unit cell of natural cellulose is related to the *P*1 symmetry group; parameters *a* and *c* of this cell are twice as much as those accepted according to Meyer–Misch model. The results obtained from the investigation of bacterial cellulose by both X-ray diffraction and IR-spectroscopy [172] demonstrated that the dimensions and symmetry group of its unit cell are very similar to the previously obtained parameters of the Valonia cellulose unit cell.

Nevertheless the problem of the structure of the natural cellulose unit cell can hardly be considered solved. It is quite possible that the various pre-

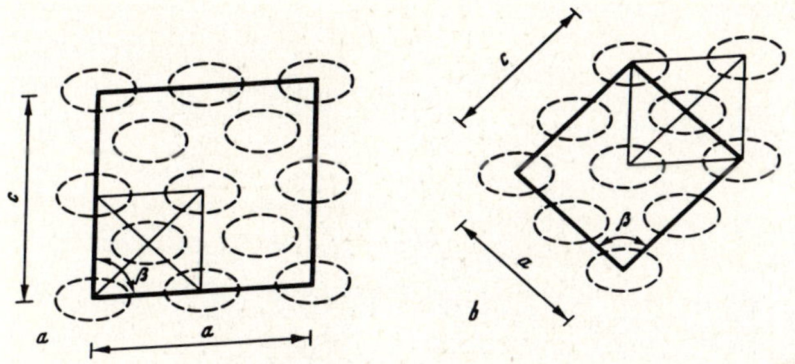

Fig. II.91. The comparison of unit cell models of natural celluloses proposed by **a** Meyer-Misch and Honjo-Watanabe and **b** Meyer-Misch and Ellis-Warwicker

parations of natural cellulose are distinguished by mutual disposition of the chains in the crystalline lattice taking into account the presence of several (see Chap. 4) stable conformations of the isolated cellulose helix. In this connection the unit cells of the crystalline regions of these preparations cannot be completely identical.

It is also possible that a definite variability of unit cell parameters can be an inherent property of natural cellulose caused by a certain lability of the system of hydrogen bonds arising from the aggregation of the isolated chains in the crystalline lattice. Some results supporting this possibility have been obtained in recent papers [64, 65] and they were considered in Chap. 7.

8.5 The Role of the Morphological Structure of Cellulose in Forming the Mechanical Properties of its Systems

The analysis of morphological forms characterizing the hypomolecular organization of cellulose macromolecules allows us to make a conclusion about the rules governing the appearance of highly ordered structures in particular on each of hypomolecular organization levels. It is interesting to note the similarity of the models, reflecting the structure of the elementary fibril (see Fig. II.82), the microfibril (see Fig. II.83) of the real cellulose fiber [217] to their actual analogs (see Fig. II.83, 92). It appears to be not by chance since these structures perform a general target function.

Actually both the elementary fibril (see Fig. II.82) surrounded by markedly less ordered (including non-glucosydic) macromolecules; and the microfibril (see Fig. II.83, 84) in the amorphous matrix of the cell wall (see also [218]); and the microfibril in the matrix from hemicellulose and lignin within cellulose fibers [217]; and thread from the modern composition materials like carton (see Fig. II.92) are known to have quite definite (often very high) strength properties due to some of the morphological forms of the hypomolecular organization

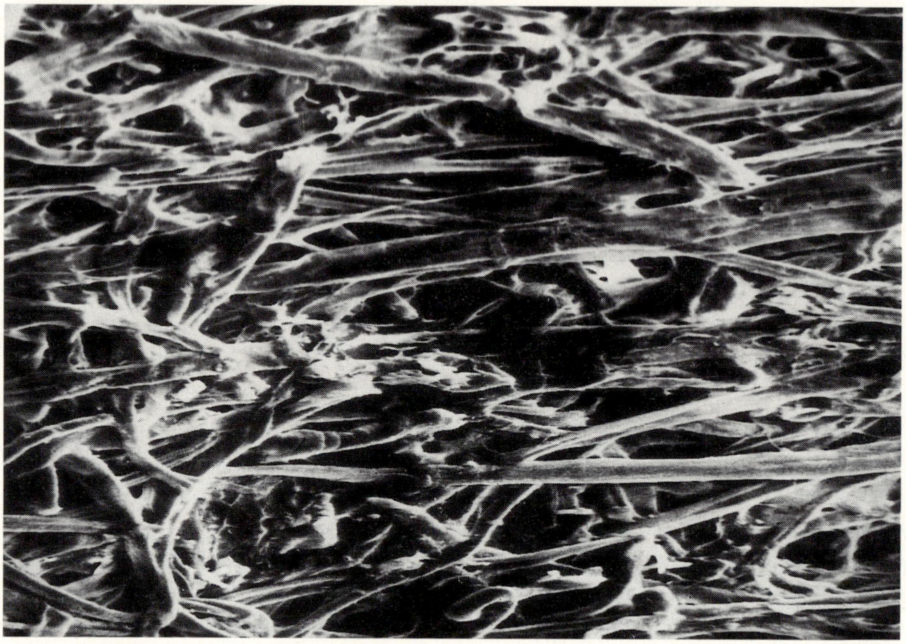

Fig. II.92. Arrangement of cellulose fibers in a composite material of the carton type

of cellulose macromolecules and the corresponding macroscopic species. Similarity of target function (and structural organization) of each morphological form makes possible the analysis of specific features of how they affect some types of deformation on the basis of the general methodological positions. It should be mentioned that this question was considered for the system microfibril—cell wall matrix by Frey–Wyssling [219] who found a certain similarity between the functions of microfibril in the cell matrix and a pile in soft soil. In this case the microfibril like a pile is considered as a reinforcing element of the system. According to this concept the destruction of the matrix upon extension of a single microfibril proceeds along the surface of so-called friction cone (Fig. II.93) not along the surface of its contact with matrix. The magnitude of this cone and thus the strength of fiber separation depends on the friction coefficient between fibril and matrix as well as on the length of fibril. According to Frey–Wyssling model the force required for displacement of the central rod is defined as:

$$P_z = \pi \mu^2 z^2 (P + vz/3 + R/\mu) \qquad (8.1)$$

where $\mu = \mathrm{tg}\, Q$ (coefficient of friction), z is the cone height equal to the length of matrix-involved rod (microfibril), μz is the radius of the cone base, P is the load on the cone base. According to this expression the specific property of the fibril on the basis of its effect on strength properties of the system is its length. Neither diameter nor the strength of the fibril itself comes into consideration.

In spite of its clarity, the above model seems to be over-simplified and it defines the behaviour of the actual cellulose thread upon deformation

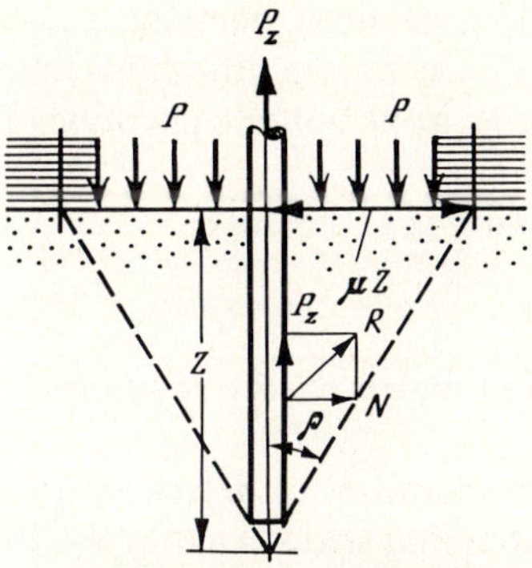

Fig. II.93. Friction cone of the rod extended vertically from the sheet substrate [219]

inadequately. It is related to the following facts.

1. The destruction of the system along the surface of the above friction cone may proceed only when the pressure on the fibril from the matrix side and the normal reaction from the fiber take place. In our opinion there are no reasons to suggest such a pressure in the above system.
2. The load in the real system is applied as a rule at a certain angle to the direction of the main axis of fibril.
3. It is hard to agree with the opinion that in the case of matrix reinforced with thin rods the final properties of the system are independent of the force properties of the reinforced elements. It is worth mentioning that the description of wear-resistant and deformation properties of the system-type cell wall and cellulose thread as it is, as well as the real system based on them is worthwhile within the scope of well-developed models and methods of the mechanics of composite materials. According to these ideas, the cellulose thread and perhaps the cell wall can be considered as single-directioned composite material reinforced by an infinite fibrous filler, where rigid microfibrils distributed within the corresponding matrices are used as the filler.

This concept was applied [220, 221] to the explanation of dependence of the strength of cellulose fibers on the slope of the extending force to the axis of microfibril packing. The satisfactory correlation was obtained between experimental results and theoretical deformation curves calculated with the use of the Hill criterion [222].

The relation between elastic constants of the components involved and the whole system in direction of packing of reinforced elements is governed by the additivity principle, which allows us to express the elastic constants of the cellulose thread (and apparently of the cell wall) along the packing of microfibrils via elastic constants of microfibrils and matrices by the following method:

$$E_{\parallel} = E_f V_f + E_m(1 + V_f) \tag{8.2}$$

$$\nu_{xy} = \nu_f V_f + \nu_m(1 + V_f) \tag{8.3}$$

where $E_{||}$ is Young's modulus of thread in direction of fibril packing, ν_{xy} is Poisson coefficient in tangential plane; E_f, E_m, ν_f, ν_m are Young's modulus and Poisson coefficients of fibril and matrix respectively; V_f is the volume portions of fibril in the thread.

Similarly the strength in direction of fibril packing is expressed:

$$\sigma_{b||} = V_f \sigma_{bf} + \sigma'_M(1 - V_f) \tag{8.4}$$

where σ_{bf} is the strength of fibrils, σ_M is the stress in matrix in the moment of fibril rupture.

According to traditional concepts [223] three mechanisms of destruction of composite material are possible with the effect of extending stresses at angle θ to the direction of the packing of reinforcing elements as a function of θ value.

1. At small θ the material is destroyed as a result of rupture of the reinforcing elements from normal stresses due to flowing of the matrix parallel to the reinforced elements. The material strength is determined in this case according to formula:

$$\sigma_{b\theta} = \sigma_{b||}/\cos^2\theta \tag{8.5}$$

where $\sigma_{b||}$ is defined by Eq. (8.4).

2. At a certain crytical value $\theta_{\text{cr.1}}$ the strength of the composite begins to be controlled by the second mechanism—destruction of the matrix or the interface thread-matrix as a result of shift at planes parallel to fibers. The strength of the material is defined in this case by the formula

$$\sigma_{b\theta} = \tau_n/\sin\theta \times \cos\theta \tag{8.6}$$

where τ_n is the limit of matrix strength or shear interface.

3. At large θ values the material strength is controlled by the normal strength of the matrix or interface in a direction perpendicular to the fibers. The strength in this case is expressed by equation

$$\sigma_\theta = (\sigma_b)_{M_i}/\sin^2\theta \tag{8.7}$$

where $(\sigma_b)_{M_i}$ is the strength limit of the matrix in conditions of flat deformation.

The values of angles related to exchange of destruction mechanism are defined by the following equations:

$$\operatorname{tg}\theta_{\text{cr.1}} = \tau/\sigma_{b||}; \quad \operatorname{tg}\theta_{\text{cr.2}} = (\sigma_b)_M/\tau_n \tag{8.8}$$

Thus the strength of composite for a given angle θ is defined by one of the Eqs. (8.5)–(8.7) depending on which one affords the smallest strength value. The above formulated condition is called the Kelley–Devies criterion. For evaluation of the dependence of fiber strength on the angle of fibril packing together with the Kelley–Devies criterion in the paper [220] the flow criterion of Hill was considered [224]. It was first proposed for undefined orthotropic material and as was demonstrated [217] this criterion was applicable to unidirectional composite materials. According to this criterion the strength conditions can be written as follows:

$$F(\sigma_{22} - \sigma_{33})^2 + G(\sigma_{33} - \sigma_{11})^2 + H(\sigma_{11} - \sigma_{22})^2 + 2L\sigma_{33}^2 + 2M\sigma_{31}^2 + 2N\sigma_{12}^2 = 1 \tag{8.9}$$

where σ_{ij} are the stresses denoted for symmetry axes of the material; F, G, H, L, M, N are some constants of the material expressed through strength of the material in various directions.

Equation (8.9) can be transposed to the final form:

$$\sigma_{b\theta} = \frac{1}{\dfrac{\alpha + \gamma}{(\sigma_b)_M^2} + \dfrac{\beta}{(\sigma_b)^2} + \dfrac{\delta}{\tau_n^2}} \tag{8.10}$$

where

$$\alpha = [\sin^4\theta + GA\sin^3\theta \times \cos\theta + (8A^2 - 1)\sin^2\theta \times \cos^2\theta - 2A\sin\theta \times \cos^3\theta],$$

$$\beta = [\cos^4\theta - GA\cos^3\theta \times \sin\theta + (8A^2 - 1)\sin^2\theta \times \cos^2\theta + 2A\cos\theta \times \sin^3\theta],$$

$$= [2A \times \cos^3\theta \times \sin\theta + (1 - 4A^2)\sin^2\theta \times \cos^2\theta - 2A\sin^3\theta \times \cos\theta],$$

$$\delta = [A^2\sin^4\theta - 2A \times \sin^3\theta \times \cos\theta + (1 - 2A^2)\sin^2\theta \times \cos^2\theta + 2A\sin\theta \times \cos^3\theta + A^2\cos^4\theta]$$

$A = S_{26}^*/S_{66}^*$, S_{26}^*, S_{66}^* are the elements of ductility matrix, written in the symmetry axes of the fiber [220, 221].

As has been already mentioned, the comparison of experimental data with theoretical dependences determined according to Kelley-Devies and Hill criteria demonstrated that that Hill criterion describes satisfactorily the dependence of the strength of cellulose fiber on the slope of the fibrils.

9 Experimental Methods for Studying Equilibrium and Kinetic Rigidity of Polymers in Solution

Extensive literature is devoted to structural studies of organic compounds, of polysaccharides in particular. A broad spectrum of the relevant methods involves the traditional, chemical ones which include identification and protection of functional groups, selective bond scission, total and partial hydrolysis, etc., as well as modern physical methods, viz. UV-, IR-, and Raman spectroscopy, NMR-spectroscopy, crystallography and electronography, optical rotatory dispersion, etc. Reference can be made to "Elucidation of Structures by Physical and Chemical Methods" (ed. K. W. Bentley, Interscience) and to "Cellulose and its Derivatives" (eds. N. Bikales, L. Segal).

At the same time books of this type do not consider practically the methods which are extremely important for the conformational characterization of polymers in solution, such as sedimentation and advancing diffusion, flow birefringence, electric field birefringence, etc., although their application is often highly valuable in the determination of the equilibrium and kinetic rigidity of polymers, this being of especial significance for cellulose and similar polymers. It is for this reason that it was thought reasonable to present here the basic principles of the most important methods.

9.1 Viscosity

Viscosity of the solution of a polymer is determined by an additional energy loss [69]. The viscosity as a parameter is manifested in a flow with a velocity gradient differing from zero. For laminar flow the velocity is determined as follows:

$$U = U_x = gy; \quad U_y = U_z = 0,$$

where $g = \mathrm{d}U_x/\mathrm{d}y$ is a velocity gradient directed orthogonally to the flow direction. This is tied to the shear stress τ by the Newton equation:

$$\tau = \tau_x = \eta \mathrm{d}U_x/\mathrm{d}y \tag{9.1}$$

and this is a definition of the viscosity coefficient. The work for overcoming internal friction in a unit volume can be expressed as:

$$A = \tau g t \tag{9.2}$$

Taking into consideration Eq. (9.1), the energy loss due to friction is thus expressed as:

$$E = \mathrm{d}A/\mathrm{d}t = \eta g^2 \tag{9.3}$$

If E_0 and E are energy losses due to the friction in a given solvent and in a solution, the additional frictional losses, Eq. (9.3) being taken into account, are equal to:

$$E - E_0 = \eta g^2 - \eta_0 g^2 \tag{9.4}$$

The ratio $(\eta - \eta_0)/\eta_0 = \eta_{sp}$ is usually called the specific viscosity. Intrinsic viscosity is determined as:

$$[\eta] = \lim_{C \to 0} (\eta_{sp}/C) \tag{9.5}$$

Intrinsic viscosities are usually determined with the use of capillary viscosimeters. Provided only the relative viscosity, η/η_0, is to be found, the variable pressure due to lowering liquid column can be excluded. The most popular instrument is the Ostwald viscosimeter which is widely used. The pressure which creates the flow is produced by a middle column of liquid "h" which remains constant in each measurement because the volume of the liquid introduced into the apparatus is constant as well. Thus the average pressure is equal to $hg\rho$ and, due to constant h and g values, $\eta = c\rho T$. Having determined time T of a shift within a given interval of level of a liquid with the known ρ (density) and η (viscosity), the viscosity of some other liquid with the density can be determined from respective T'-value:

$$\eta' = \eta(\rho' T'/\rho T) \tag{9.6}$$

Thus η' is the relative viscosity of $\eta = 1$, and absolute one provided the numerical value for η is used in Eq. (9.6). For viscosimeters of this type

$$\eta'_{sp} = (\eta - \eta_0)/\eta_0 = (T_i - T_0)/T_0 = T_i/T_0 - 1 \tag{9.7}$$

where T_0 and T_i are flow times for a solvent and a solution of concentration "i". Thus η_{sp} is determined by a T_i/T_0 ratio.

Using the Huggins equation

$$\eta_{sp}/C = [\eta] + K'C[\eta]^2 + \ldots \tag{9.8}$$

one can extrapolate η_{sp} upon successive dilution of solutions to zero concentration ($C \to 0$).

To study the gradient dependence of viscosity, other viscosimeters, e.g. the Eigner viscosimeter (Fig. II.94b), are used. This differs from the Ostwald viscosimeter (Fig. II.94a) in design and length of the capillary tube. This permits us to conduct measurements at low gradients of flow rate, g-values being an order of magnitude less than in the Ostwald viscosimeter. This permits us to extrapolate T_i/T_0 for the zero rate gradient 'g'. The g-value is then calculated

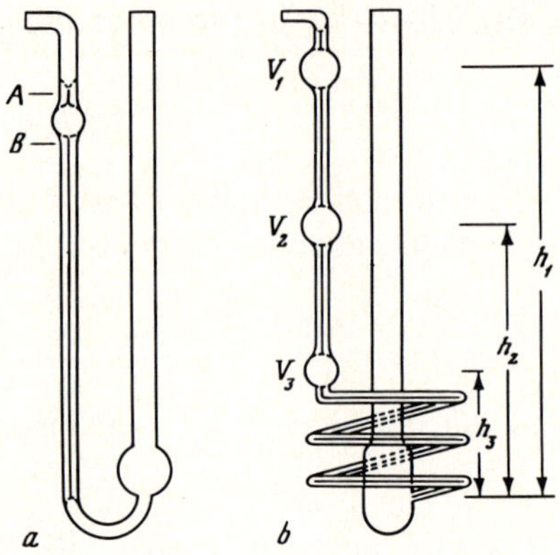

Fig. II.94. Design of **a** Ostwald and **b** Eigner capillary viscosimeters

from the equation:

$$g \cong 2/3 \times \Omega/R^3 \tag{9.9}$$

where R is the radius of a capillary tube, Ω is a discharge of liquid per time unit and is expressed by pressure (P) and length of capillary tube (L):

$$\Omega = \pi R^4/(8\eta) \times P/L \tag{9.10}$$

It is obvious that the smaller the column height the less the gradient value and functional dependences $\eta = f(g)$ and $T_i/T_0 = f(g)$ can be determined.

9.2 Translation Diffusion

The macroscopic direction of diffusion in a solution appears when a gradient ($\mathrm{d}c/\mathrm{d}x$) exists of a solute concentration (c) in direction (x). In stationary conditions the diffusion process is expressed as the Fick's first law:

$$L = -D(\mathrm{d}c/\mathrm{d}x) \tag{9.11}$$

which defines the coefficient of the translation diffusion D (D is equal to the flow at $\mathrm{d}c/\mathrm{d}x = 1\ \mathrm{g/cm^3}$).

According to Einstein, the diffusion coefficient (D) for the stationary diffusion process in an ideal solution with nonparticipating particles depends on the particle's friction coefficient (f) which, in turn, is directly related to the particle's size and shape:

$$D = kT/f \tag{9.12}$$

Under real conditions polymer solutions are not ideal and for determination of D it is necessary to take into account the intermolecular interaction of macromolecules in solution which is practically given by a concentration dependence:

$$D = (kT/f)(1 + 2A_2Mc + 3A_3M^2 + \dots) \tag{9.13}$$

9.3 A Procedure for Measurement of Translation Diffusion Coefficients

Translation diffusion coefficients of polymeric molecules in solution are usually determined with the use of special diffusometers [69], and the Tsvetkov diffusometer [225] is one of the most perfect and convenient instruments. Its main advantage is the uniquely high sensitivity of its optical registration system which permits us to measure the diffusion coefficients of polymer solutions with concentrations as low as $c = (3 \text{ to } 7) \times 10^{-4}$ g/cm^3 which is beyond the sensitivity level of other diffusometers. Studies of solutions which are so highly diluted makes it unnecessary, in certain cases, to extrapolate the data obtained to zero concentration due to the insignificant influence of intermolecular interactions. The design of the diffusometer is given in Fig. II.95.

The quartz wedge (Babin wedge) and the polaroid D_2 crossed relative to D_1 permit the interference of two polarized beams 1 and 2 to be observed on a screen D, the film plane optically conjugated to the middle part of a cell A. Optical axis and wedge ribs K are parallel to the axes of the Iceland spar crystals P_1 and P_2. When interfering beams 1 and 2 pass through an optically uniform medium the resulting interference pattern is displayed as a series of vertical lines. Provided beams 1 and 2 cross the layers in the cell with differing refractive indices, n_1 and n_2, the resulting path difference δ is manifested as nonuniform line shift in the compensator in a horizontal direction, parallel to the interface AB between two liquids (Fig. II.95). The path difference δ is equal to:

$$\delta = \frac{h}{\lambda}\frac{\mathrm{d}n}{\mathrm{d}c} \int_{\chi_0 - \frac{a}{2}}^{\chi_0 + \frac{a}{2}} \frac{\mathrm{d}c}{\mathrm{d}x}\,\mathrm{d}x \tag{9.14}$$

where (h) is the cell's height along the beam path, (λ) is the light wavelength, (a) is doubling of spars, ($\mathrm{d}n/\mathrm{d}c$) is a refractive index increment, (χ_0) is a distance from AB to a line equidistant from the beams 1 and 2. It thus follows that the shape of a diffusion curve is determined by the ratio of (a) to the width of a concentration boundary (the variable concentration region). Development of the theory of the polarization diffusometer [69, 225] has led to the following dependence of (δ) on standard deviation (σ) of $\mathrm{d}c/\mathrm{d}x$ distribution:

$$\delta = \frac{\delta_m^0}{2}\left[\Phi\left(\frac{\chi_0 + a/2}{\sigma 2^{1/2}}\right) - \Phi\left(\frac{\chi_0 - a/2}{\sigma 2^{1/2}}\right)\right] \tag{9.15}$$

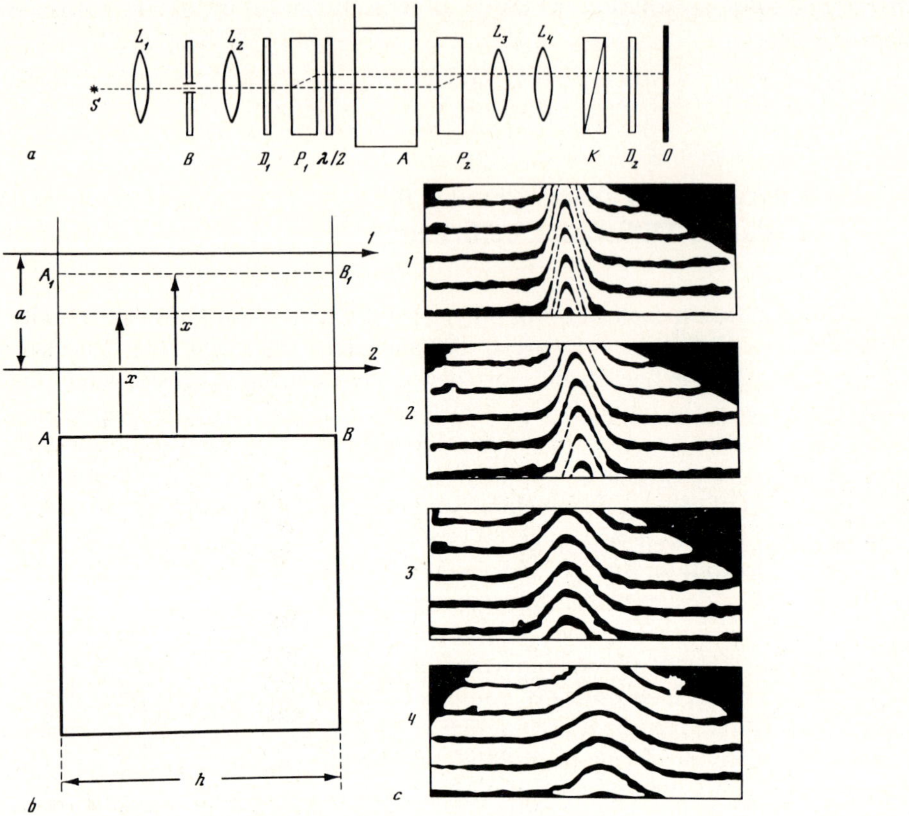

Fig. II.95. Advancing diffusion.

a—optical scheme of the Tsvetkov diffusometer. S is the light source, the lens L_1 projects light onto a horizontal slit in the plane of which the focus of the second lens L_2 is placed. The parallel beam is then polarized with a polaroid D_1 and split into two coherent beams *1* and *2* by a crystal of iceland spar P_1 whose main section is vertical and forms an angle of 45° with the plane of incident light. Beams *1* and *2* are polarized in mutually perpendicular directions. The plate $\lambda/2$ rotates polarization planes 1 and 2 by 90°. Having passed through the cell A the beams are combined by a spar crystal P_2 identical to P_L; **b**—calculation of difference of beams paths δ; **c**—interferogrammes recorded for diffusion of cellulose nitrate in ethyl acetate after (*1*) 1 h, (*2*) 2 h, (*3*) 4 h, and (*4*) 7 h

where $\delta_m^0 = (n - n_0)h/\lambda$ is the value of a function at $t = 0$; Φ is the probability integral whose values are tabulated. Thus to determine δ it is necessary to calculate the difference for two values of the probability integral Φ. From the interference pattern shown in Fig. II.95 it is possible to determine $\delta(\chi)$ and from that, the diffusion coefficient D. The latter is calculated from the area Q and the maximum ordinate δ_{max}. Obviously,

$$Q = h\Delta n a/\lambda = \delta_{m_0} a \tag{9.16}$$

It is time-independent and is equal to area of a rectangle at $t = 0$. From

$$\delta_m = \delta_{m_0}\Phi[(k_a)^{-1/2}/2] \tag{9.17}$$

where $k = (4Dt)^{-1} = (2\sigma^2)^{-1}$, and taking Eq. (9.16) into account it follows that:

$$\Phi = [(k_a)^{-1/2}/2] = \delta_m a/Q \tag{9.18}$$

From this equation k and D can be found. These methods are valid at $\sigma/(a/2) \geq 0.5$.

9.4 Sedimentation

Sedimentation studies are usually carried out with analytical ultracentrifuges equipped with polarization-interferometer accessories of the type shown in Fig. II.96 [69, 226–228]. With the use of a polarization interferometer, the sharper contour of the sedimentation boundary at lower solute concentrations can be achieved as well as differential and integral radial distribution curves for the refractive index of a column of liquid in a cell (Fig. II.96a, b).

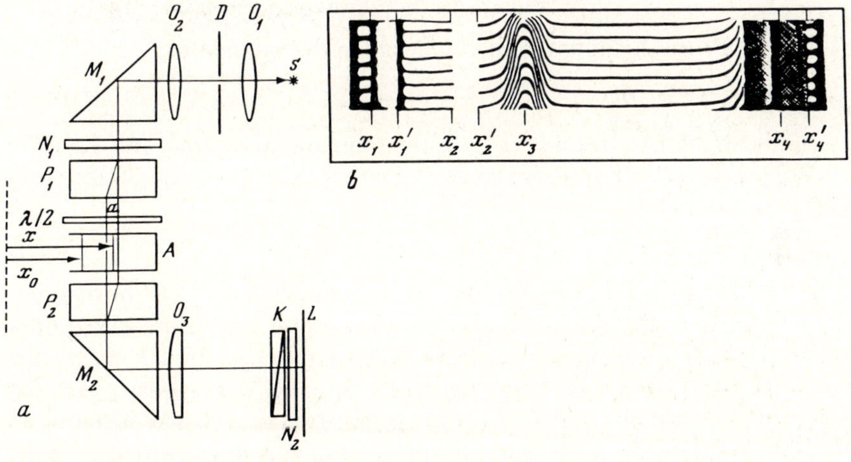

Fig. II.96. Polarization interferometer (scheme).

a—O_1, O_2 are objectives and a slit which form a light beam which is then polarized with a polaroid N_1 and split with a spar crystal P_1 into two beam systems shifted one from another by a distance a in radial direction X from the rotation axes; $\lambda/2$ is a plate of half-wave-length to rotate planes of the split and polarized beams by $\pi/2$; P_2 is a spar crystal to bring the beams together which having passed through a quartz wedge and a polaroid N_2 produce an interference pattern on a plate L; O_3 and O_4 are a telescopic system of lenses (cofocal) which forms a projection of a plane across the centre of the cell A onto the plane L; **b**—Sedimentation diagram

It is essential that the trajectory of the movement of each point of a rotating cell forms an arc of a circle and is projected onto a plate undistorted (unlike Philpot–Svensson and Rayleigh optics [228]) because the optical scheme of the interferometer is free from astigmatic elements. Therefore all distances on the sedimentation diagram are to be measured in a radial direction. To this end the vertical line in the cross hairs of the measuring microscope ocular is adjusted tangentially at every point of the index curve image. Then the horizontal shift of the microscope corresponds to the radial shift on a sedimentation diagram.

The interference contour (δ_x) shape is determined by the concentration gradient dc/dx in a cell:

$$\delta_x = \beta \int_{\chi - a/2}^{\chi + a/2} \frac{dc}{dx} \times \frac{1}{dx}, \quad \beta = (h/\lambda)\, dn/dc \tag{9.19}$$

where h is a cell height along the beam path, λ is a wavelength, dn/dc is a refractive index increment for a polymer-solvent system. To get both differential and integral curves simultaneously for distribution of a polymer in a cell it is necessary to have two pairs of polarizing crystals, with doubling of $a_1 \cong 0.1$ to 0.3 mm and $a_2 \cong 1.2$ mm. When sedimentograms obtained with the polarization interferometer are to be deciphered, one should take into account the double image of the meniscus indices lines having a step equal to the doubling character of a crystal.

For a polymer the sedimentation coefficient is determined as:

$$S = (1/\omega^2)(d \ln x_m/dt) \tag{9.20}$$

where ω is the angular rotation speed, t is the sedimentation time, and x_m is the distance from the axis to the maximum on the gradient curve. This is found from Eq. (9.21):

$$x_m = A + y_1 \tag{9.21}$$

where A is the (known) distance from the axis to an index line, y_1 is the distance from the index line to the sedimentation peak maximum and is measured on a sedimentation diagram using a measuring microscope. For the left index line image $y_1 = (x_3 - x_1) - a/2$, for the right one $y_1 = (x_3 - x_1) + a/2$, and the $\ln x_m$ vs t plot is a straight line with a slope proportional to S. The distance x_0 from the axis to the meniscus is equal to the sum $x_0 = A + y_2$, where y_2 is the distance from the index line to meniscus, $y_2 = x_2 - x_1 = x'_2 - x'_1$.

Calculation of x_m and x_0 requires magnification due to optical system of the interferometer having to be taken into account. The distance between the counterweight apertures is usually 16 mm. Then the magnification coefficient K can be determined from the sedimentation pattern (Fig. II.96b) as follows:

$$K = (x_4 - x_1)/16 \equiv (x'_4 - x'_1)/16$$

If the temperature varies in different runs the S values should be calculated for a single temperature, e.g. 21 °C. This requires temperature coefficients of solvent viscosity and density be estimated and that the correction factor be calculated:

$$S_{21} = S + \eta_0/\eta_{021} \times (1 - \bar{v}\rho_{021})/(1 - \bar{v}\rho_{0+})$$

where ρ and η are the solvent density and viscosity and $\bar{v}$ is a specific partial volume of a polymer.

To calculate molecular mass values from sedimentation and diffusion data the values obtained have to be extrapolated to zero concentrations, i.e. to exclude the effect of intermolecular interaction on the values under estimation. To this end the concentration dependencies of S-values for fractions are analysed and the coefficients S_0 and K_s are calculated from the linear approximation:

$$1/S = (1 + K_sC)/S_0 \tag{9.22}$$

To calculate molecular masses from the data of diffusion–sedimentation analysis the specific partial volume of the polymer should be determined, e.g. by picnometry using a picnometer of a volume V, for a series of successively diluted solutions:

$$\bar{v} = [1 - (P - P_0)/\omega(P_0)]/\rho_0 \tag{9.23}$$

where ρ_0 is the solvent density, P_0 is its weight for a volume V, P is the weight of a solution (for a volume V) with a weight fraction of the polymer.

And now, from the S_0, D_0, and $\bar{v}$ values found, the molecular masses can be calculated using the Svedberg equation:

$$M_{SD} = [RT/(1 - \bar{v}\rho_0)] \times S_0/D_0 \tag{9.24}$$

9.5 The Kerr Effect

One of the main methods for the determination of characteristics is the study of the electric birefringence (EBR, Kerr effect) of polymer solutions [70, 93]. Difficulties in studies of polymers like cellulose and its derivatives arise from a limited set of suitable solvents, many of which exhibit considerable conductivity. This hampers the measurement of the EBR of the solvent themselves and, even more so, of solutions in a constant and alternating sinusoidal electric field since optical nonuniformities which arise due to the electric conductivity of solutions mask the Kerr effect.

An instrument free of these drawbacks and eliminating the interference of electric conductivity is shown in Fig. II.97. Electric pulses of direct current are created by a gating circuit with a partial discharge of a capacitor. The gate is composed of three triodes (GI-41) in parallel, which are switched on with positive pulses from a generator (G5-6A) and bring about a discharge of the capacitor C.

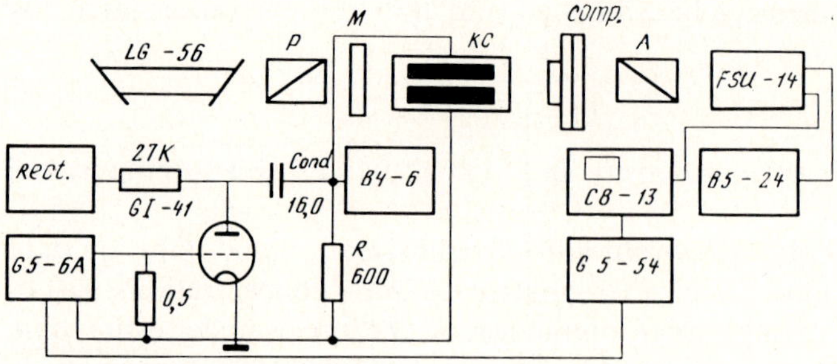

Fig. II.97. An apparatus to measure EBR in pulse field (scheme). *P*—polarizer; *M*—modulator; *KC*—Kerr cell; *comp.*—elliptic compensator with a graduated circle; A—analyser

Thus the rectangular pulses are produced at a load resistance *R* and the Kerr cell *C*, their amplitude is regulated by changing the voltage at the capacitor *C* and measured by a voltmeter (V4-7). The Kerr cell consists of two half-cylinder titanium electrodes ($l = 6$ cm) with a clearance of 0.042 cm fused through molibdenum plugs into a glass tube to whose ends optical glasses are glued, preferably tension-free.

The optical part of the apparatus comprises a high-intensity light source Ω (He–Ne laser, $\lambda = 632.8$ nm), crossed polarizer *P* and analyzer *A*, an elliptic compensator with a pace difference of 0.001, an elliptic modulator *M*, and a light recorder (photomultiplier of the type PEM-14). The light passed through the analyzer is transformed by the photomultiplier into a photocurrent, the permanent component of which and pulse changes are displayed on an oscillograph of type S8-13.

The light intensity after the analyzer (I) with the modulator switched off depends on the phase difference for normal and abnormal light waves following the anisotropic layer (δ) in the Kerr cell with the optical axis azimuth of $\pi/4$ relative to the polarizer axis and the compensator (δ_0) with an azimuth of φ_0. For two small anisotropies ($\delta, \delta_0 \ll 1$) between crossed polarizer and analyzer this dependence is expressed as follows:

$$I = I_0 \times (\delta^2 + \delta_0^2 \sin^2 2\varphi + 2\delta\delta_0 \sin 2\varphi)/8 = I_0 \times (\delta + \delta_0 \sin 2\varphi)^2/8 \quad (9.25)$$

By rotating the compensator to the cut out position ($I = 0$) the compensation of anisotropy under study takes place, its value and sign can be determined from the equation

$$\delta = (2\pi/\lambda) \times \Delta n l = -\delta_0 \sin 2\varphi \quad (9.26)$$

where $n = n_l - n_0$ the birefringence that is measured.

When the EBR is measured in a pulse field the method based on equalizing light intensities during pulses and in intervals between them is more sensitive. In

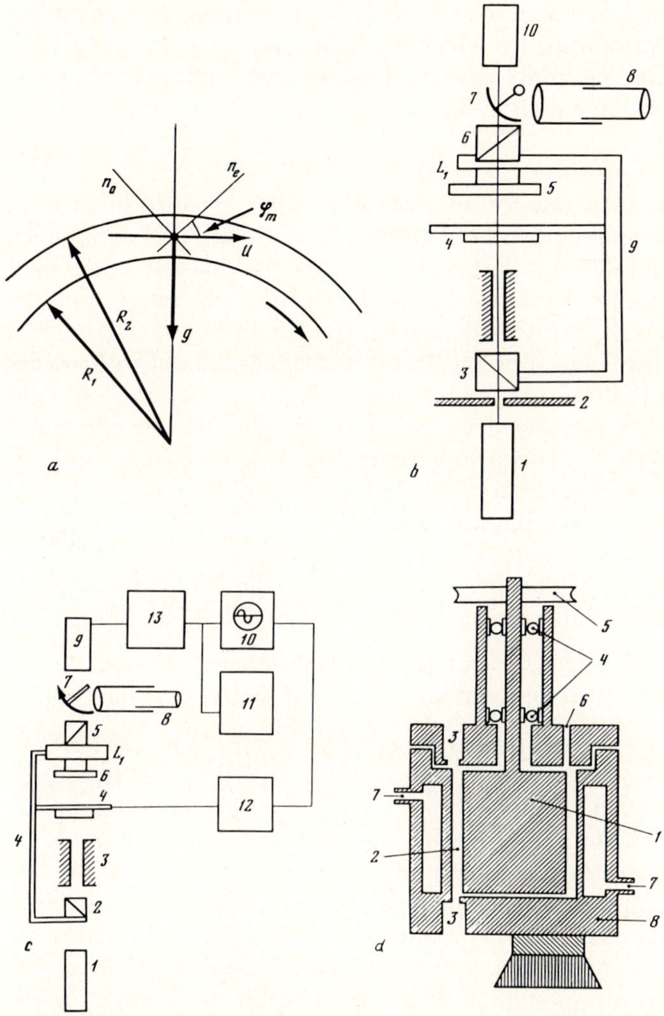

Fig. II.98. Flow birefringence.

a—Flow velocity gradient in a liquid layer within an annular space between stationary (1) and rotating (2) cylinders; U is the flow velocity, g is the direction of the velocity gradient, φ_m is the extinction angle (see also text); **b**—Optical scheme of an instrument to study flow birefringence by a photoelectric method: *1*—light source (laser LG-75), *2*—slit, *3*—polarizer, *4*—modulator, *5*—compensator, *6*—analyser, *7*—rotating mirror, *8*—telescopic tube, *9*—lever, *10*—photoreceiver; **c**—A principal outline of an apparatus to study flow birefringence with the use of photoelectric compensation: *1*—light source (laser LG-75), *2* and *5*—crossed polarizer and analyser, *3*—annular space in a dynamoptimeter, *4*—modulator, *6*—compensator, *7*—rotating mirror, *8*—telescopic tube, *9*—photoreceiver, *10*—oscillograph, *11*—synchronous detector, *12*—audiooscillator, *13*—selective amplifier, *14*—lever; **d**—Dynamoptimeter with an internal rotor: *1*—rotor, *2*—annular space, *3*—sight holes, *4*—bearing, *5*—pulley, *6*—feeding aperture, *7*—jacket tubes, *8*—stator this case the compensator is rotated so as to diminish the amplitude of an optical pulse. It is stopped at position φ for which the light intensity is the same both upon electric field action and in its absence. Under these conditions of partial compensation the following equation holds:

$$\delta = -2\delta_0 \sin 2\varphi \tag{9.27}$$

This method cannot be applied to determine EBR of nitrocellulose in acetone or cyclohexanone because of the amplitude modulation of light which occurs in solutions at a pulse duration as small as 10^{-3} s. Changes in the light intensity passed through the Kerr cell during the electric field action excludes the use of the method of light intensity equalizing.

Measurement of EBR under these conditions can be carried out with the use of a method based on the sinusoid modulation of elliptic polarization of light [229]. The modulation is performed by means of the photoelastic effect in a plane–parallel glass plate fixed to a longitudinally oscillating ferrite rod. Its resonance is induced by a magnetostrictive effect in an alternating magnetic field produced by a solenoid surrounding the rod.

Shown in Fig. II.98, the modulator produces an oscillating birefringence with an optical axis azimuth of $\pi/4$. The light passed through the modulator acquires elliptic polarization with an alternating phase difference $\delta \cong \delta_m \sin \omega t$ ($\delta_m \ll 1$), its intensity is

$$I = (I_0/8)[\delta_m^2 \sin^2 \omega t + (\delta + \delta_0 \sin 2\varphi)^2 + 2(\delta + \delta_0 \sin 2\varphi)\delta_m \sin \omega t] \tag{9.28}$$

from which it can be seen that in the absence of an electric pulse at the zero position of the compensator ($\varphi = 0$), the light intensity is $I = (I_0/16)\delta_m^{16} \times (1 - \cos 2\omega)$ and sinusoidal oscillations with a doubled modulation frequency 2ω are displayed. During the electric pulse, i.e. at $\delta \neq 0$, oscillations are displayed with the frequency and amplitude proportional to $(\delta + \delta_0 \sin 2\varphi)$. Thus the position of compensation of anisotropy to be measured corresponds to such an azimuth that the light intensity modulations during the pulse action are only of the frequency 2ω. Modulation of the elliptic light polarization permits EBR to be measured by a zero method which is independent of changes in the light intensity due to light scattering or absorption within a Kerr cell in the presence of electric field.

9.6 The Maxwell Effect

The dynamooptic Maxwell effect is the appearance of optical anisotropy due to shear stress in a laminar flow. The forced anisotropy appears as a double refraction in the solution. For a light beam normal to the directions of a gradient g and flow velocity U (i.e. parallel to a cylinder axis) (Fig. II.98a) the properties of an anisotropic layer are similar to those of a crystal plate with an angle φ_m

between its main section and the flow direction U. This angle is referred to as the extinction or orientation angle.

The orientation angle is one of two parameters measured directly when the Maxwell effect is studied. The second principal parameter is the difference in the indices of refraction $\Delta n = n_e - n_0$ for a liquid with the electric vector of a ray n_e parallel and n_0 normal to the main section (axis) of an anisotropic layer. Thus Δn characterises birefringence and it can be positive or negative. Study of the Maxwell effect of dilute solutions of polymers can be carried out with apparatus designed by Tsvetkov [69].

Measurement of the flow birefringence is best performed in a cylindrical apparatus, the so-called dynamooptimeter. The liquid under study is placed in the annular space between two coaxial cylinders, one of which is stationary and the other rotates at a constant velocity. The liquid layer is looked through along the rotation axis. A laminar flow produced in the annular space can be considered practically uniform within the small area investigated by an observer, it possesses a constant velocity gradient $g = \mathrm{d}U/\mathrm{d}x = \mathrm{d}U/\mathrm{d}R$ along the radius R and is equal to $g = 2\pi R/\Delta R$, where v is the angular velocity in revolutions per second, $R = (R_1 + R_2)/2$, is the mean radius and $\Delta R = R_1 - R_2$ is the gap width. The prerequisite for quantitative studies of the Maxwell effect is the maintenance of laminar flow.

When dynamooptic properties of polymers like cellulose nitrate are to be studied instruments with a photoelectric compensation of birefringence are recommended [229]. Such instruments permit us to achieve sensitivity higher by at least one order of magnitude compared to visual detection procedures. What is especially important is that orientation angles are determined reliably at low concentrations and velocity gradients.

Shown in Fig. II.98b is a scheme of an optical device for studying the small effects observed in dilute solutions of polymers. A very sensitive Brace system is used as a compensator, comprising a compensating thin mica plate 5 (of several thousandths of a wavelength, $\delta_0 = 0.006 \times 2\pi$) fixed on a graduated circle. The slit 2 is illuminated with a laser lightsource 1. The essential components are the elliptic light modulator 4 which produces an alternating phase difference $\delta = \delta_0 \sin \omega t$ and a photoreceiver 10 which detects light signals. Crossed polarizer 3 and analyzer 6 rotate together with the modulator 4 and the mica plate 5 by a lever 9 on the graduated circle L_1. Measurements may visually be checked by using a rotating mirror 7 and a telescopic tube 8. A schematic of an instrument used is shown in Fig. II.98c. The intensity of light coming out of the analyzer is determined first by a phase difference of normal and abnormal rays passing through the anisotropic solution δ with an optical axis azimuth of $\pi/4$ relative to the polarizer and secondly by a phase difference of the compensator δ_0 with an azimuth of φ. For small anisotropies ($\delta \ll 1$, $\delta_0 \ll 1$) and for modulator switched off $\delta \cong 0$, this dependence is expressed as follows:

$$I = I_0(\delta^2 + \delta_0^2 \sin^2 2\varphi + 2\delta\delta_0 \sin 2\varphi)/8 = I_0(\delta + \delta_0 \sin 2\varphi)^2/8 \tag{9.29}$$

Rotation of the compensator to the extinction position results obviously in compensation of anisotropy to be measured whose magnitude and sign are determined from the equation:

$$\delta = 2\pi \Delta n \times (l/\lambda) = -\delta_0 \sin 2\varphi_k \tag{9.30}$$

where Δn is the difference in the principal refractive indices for abnormal and normal light beams, λ is the wavelength, and l is the path length through a cell.

The use of photoelectric registration permits us to increase the instrument sensitivity by isolation from the constant intensity I (Eq. 9.29) of a component proportional to $\sin 2\varphi_k$. This can be done by inserting between polarizer and analyzer, along with the cell and the compensator, a modulator which creates an oscillating anisotropy with an optical axis azimuth of $\pi/4$.

After passing through the modulator, the light beam acquires elliptic polarization with an alternating phase difference: $\tilde{\delta} \cong \tilde{\delta}_0 \sin \omega t$, the intensity of light after the analyzer is the sum of one constant and two variable components and is expressed as:

$$\begin{aligned} I = I_0(\delta + \delta_0 \sin 2\varphi + \tilde{\delta}_0 \sin \omega t)^2/8 &= I_0[(\delta + \delta_0 \sin 2\varphi)^2 + (\tilde{\delta}_0)^2 \sin^2 \omega t \\ &+ 2(\delta + \delta_0 \sin 2\varphi)\tilde{\delta}_0 \sin \omega t] \end{aligned} \tag{9.31}$$

The amplitude of the latter component, $2(\delta + \delta_0 \sin 2\varphi)\tilde{\delta}_0 \sin \omega t$, is proportional to $\sin 2\varphi$ and equal to zero at the compensation position $\delta = \delta_0 \sin 2\varphi$. To register this component the light flow is converted into an electric signal which enters a selective amplifier tuned to the modulation frequency ω. Proportionality of the alteration of the signal isolated to changes in the compensator azimuth φ when the anisotropy under study is compensated as well as higher signal-to-noise ratio characteristic for selective devices permit us to increase sensitivity of the δ measurement to 10^{-5} radian, which is equivalent to $10^{-6}\lambda$.

The basis of the elliptic light modulator fixed to a lever is a ferrite rod in which mechanical vibration at its resonance frequency is stimulated by an alternating magnetic field. An oscillating optical anisotropy $\tilde{\delta} = \delta_0 \sin t$ (photoelastic effect) whose axis forms an azimuth of $\pi/4$ with the principal sections of the crossed polarizer and analyzer is created in a glass plate 5 glued to an end of the rod. The output of the selective amplifier tuned to a resonance frequency is connected to the oscillograph plates. The amplitude of a harmonic signal is displayed and the use of a synchronous detector favours a decrease in errors due to dust particles present in a solution.

The velocity of rotation of the dynamooptimeter rotor can be measured with the use of a frequency meter FZ-33. A radially perforated disk is attached to a rotor and this is used as a light-breaker. Rotation of the rotor produces at the output of a photodiode an alternating voltage $\underset{\sim}{U}$ whose frequency depends on velocity of the disk rotation. The voltage $\underset{\sim}{U}$ is amplified and fed to the input of the frequencymeter and rotation velocities from 0.05 to 50 revolutions per second can be reliably measured.

Studies on flow birefringence of cellulose nitrate solutions were carried out with a titanium dynamooptimeter with an internal rotor (Fig. II.98d). The critical (maximum) velocity gradient below which the flow is laminar is equal to

$$g_k = \pi R^{1/2} \eta / (A^{1/2} \rho \Delta R^{5/2}) \tag{9.32}$$

where ρ and η are the density and viscosity of a liquid. A proportionality constant A for almost every type of apparatus employed can be taken as being equal to 0.0571. Birefringence $\Delta n = n_l - n_0$ is found from the equation

$$\Delta n = \delta_0 \lambda \sin 2\varphi_k / 2\pi l \tag{9.33}$$

where φ_k is an angle for compensator rotation to compensate birefringence in solution; l is the path length through a cell. Polarizer and analyzer are made of Iceland spar prisms of the Frank–Retter normal field.

10 Conclusion

As can be seen from a rather short survey of the literature and the author's data on structural chemistry of cellulose and its derivatives, this branch of physics and chemistry of polymers is in a state of continuing and intensive development. It should be emphasized that the current status is characterized by application of both modern methods of investigation including computer-assisted modelling and the search for models of optimum structures, computational methods for prognostic evaluation of properties of cellulose, etc., and traditional ones which are widely used in conformation analysis of cellulose esters in dilute solutions (sedimentation and viscosity, advancing diffusion, Kerr and Maxwell effects, light-scattering). The latter methods have permitted us to accumulate comprehensive data on the conformational characteristics of different cellulose esters in dilute solutions. The importance of this lies in the fact that it permits rather exact estimation of the computational methods used. It is even more important that cellulose does not produce real solutions and experimental data on cellulose structure, in the broad sense, are as a rule indirect ones and characterize any particular, soluble derivative.

Specific at present is the essentially deeper insight that has been gained into conformation of an isolated cellulose helix and into packing of macromolecular cellulose chains in an elementary crystalline cell. The most important finding seems to be the proof of the relatively high lability of hydrogen bonds which fix individual chains one each other in the elementary cell. This fact determines the lability of the molecular and crystalline structures of cellulose as such. Obviously this is not random and reflects the environmental peculiarities under which formation of cellulose macromolecule in a plant cell wall has occurred to suit specific conditions. Changes in environmental conditions (temperature, humidity, solar radiation level, salt concentration in nutrients, etc.) in the course of plant evolution have extended and extend now over an extremely broad range. It seems quite natural that, during evolution, the optimization of cellulose structure, the specificity of the skeletal function cellulose microfibrils in the cell wall being taken into account, should be directed towards the formation of rather labile structures which can undergo easy reconstruction under the influence of permanently or periodically changing environmental conditions. It is presumably this property that forms the basis for the polymorphism typical of cellulose and that can account for the known fact of some deviations in crystalline unit parameters for native cellulose from various sources as determined by different authors.

It can thus be concluded, on the basis of data available, that the specificity of cellulose structure is primarily determined by specific conditions under which selection of optimum structures occurred in the course of evolution of a cellulose macromolecule as a component of the cell wall of a plant regarded as a biological object. This conclusion is important for specialists both in the field of biochemistry and physics and chemistry of polymers and in the field of chemistry and technology of cellulose processing because it can certainly be claimed that the structure of macromolecular cellulose chain has been optimized to a large extent during its prolonged evolution. And hence isolation of cellulose from plant sources and the conditions of its processing should be mild wherever possible and not affect the original level of structural organization of the cellulose which is specific for the native material.

Certainly, not all the questions discussed are actually elaborated to the extent necessary to meet the importance of the problem. But the fact that they have been precisely named predetermines a great interest in more profound studies in the extremely complex and at the same time attractive area of knowledge of both purely scientific and industrial value which has been used as the title for this book.

11 References

1. Terentiev AP, Potapov VM (1964) Fundamentals of stereochemistry, Khimia, Moscow
2. Eliel E (1965) Stereochemistry of carbon compounds, Mir, Moscow
3. Kochetkov NK, Bochkov AF, Dmytriev BA (1967) Chemistry of carbohydrates, Khimia, Moscow
4. Klenkova NI (1976) Structure and reactivity of cellulose, Nauka, Leningrad
5. Tudze R, Kawai T (1977) Physical chemistry of polymers, Moscow, Khimia
6. Meyer KH, Misch L (1937) Helv Chim Acta, 20: 232
7. Frenkel SY (1968) In: Jale FH (ed) Polymeric monocrystals, Khimia, Leningrad, pp 478, 535
8. Jale FH (1968) Polymeric monocrystals, Khimia, Leningrad
9. Fetzer WR, Grosby EK, Endel CE, Kierst LC (1953) Industr and Eng Chem 45: 1075
10. Gess K (1934) Chemistry of cellulose and its satellites, Goskhimizdat, Leningrad
11. Jones KN, Perry MB (1967) Elucidation of structure of organic compounds by physical and chemical methods, vol 2, Mir, Moscow, p 291
12. Marx–Figini M, Penzel E (1965) Makromol Chem 87: 307
13. Glegg PE, Tang LJ, Brewer RJD (1974) In: Bikales H, Segal L (eds) Cellulose and its derivatives, vol 1, Mir, Moscow, p 382
14. Eliel E, Allinger H, Engial S, Morrison G (1969) Conformational analysis, Mir, Moscow
15. Stoddart J (1975) Stereochemistry of carbohydrates, Mir, Moscow
16. Panov VP, Zhbankov RG (1975) Conformations of carbohydrates, Nauka i Tekhnika, Minsk
17. Zhbankov RG, Marupov RM, Ivanova NV, Shishko AM (1976) Spectroskopy of cotton, Nauka, Moscow
18. Zhbankov RG (1964) Infrared spectra of cellulose and its derivatives, Nauka i Tekhnika, Minsk
19. Barker SA, Bourne EG, Staceg M, Whiffen DN (1954) J Chem Soc 6: 171
20. Segal L (1967) In: Fettes E (ed) Chemical reactions of polymers, vol 1, Mir, Moscow, p 49
21. Garbisch EW, Griffith MG (1968) J Amer Chem Soc 90: 6543
22. Hendrikson GB (1961) J Amer Chem Soc 83: 4537
23. Coxon B (1965) Tetrahedron 21: 3481
24. Hansen FG, Craine EM (1954) J Biol Chem 208: 293
25. Eliel EL, Giza CA (1968) J Org Chem 33: 3754
26. Lugovsky AA, Dashevsky VG (1972) Zhur Struct Khim (English transl.: Sov J of Struct Chem) 13: 122
27. Dashevsky VG (1974) Conformations of organic molecules, Khimia, Moscow

28. Kozlov PV, Zhbankov RG (1975) In: I Conference on chemistry and physics of cellulose. Section 2. Tezysy dokl., Riga, p 5
29. Dunfield LJ, Whittington SG (1977) J Chem Soc Perkin Trans, Part II 5: 654
30. Kildeby K, Melberg S, Rasmussen K (1977) Acta Chem Scand, A31: 1
31. Melberg S, Rasmussen K (1979) Carbohydrate Res 69: 27
32. Neely WB (1969) J Med Chem 12: 16
33. Giacomini M, Pullman B, Maigret B (1970) Theor Chim Acta 19: 347
34. Saran A, Govil G (1971) Ind J Chem 9: 1095
35. Saran A, Govil G (1972) Ind J Chem 10: 547
36. Khodyreva NV, Stolyarova TV, Ivanov MA (1981) Khimia Drevesin (English transl.: Sov Chemistry of Wood) 6: 56
37. Bandura AV, Novoselov NP, Evarestov RA (1976) Theoret. i Experiment Khimia (English transl.: Sov Theoret and Experim Chem) 12: 598
38. Pople JA, Segal GA (1966) J Chem Phys 44: 3289
39. Komornick A, McIver JW (1973) J Amer Chem Soc 95: 4512
40. Bingham RC, Dewar MJS, Lo DH (1975) J Amer Chem 97: 1285
41. Ovchinnikov AA, Shamov AG, Tikhomirov VA (1981) Zhur Organ Khim (English transl.: Sov J Org Chem) 17: 298
42. Boyd R, Whitehead MA (1972) J Chem Soc Dalton Trans 1: 73
43. Boyd R, Whitehead MA (1972) J Chem Soc Dalton Trans 1: 78
44. Boyd R, Whitehead MA (1972) J Chem Soc Dalton Trans 1: 81
45. Burstein KY, Khurgin YI (1974) Izvest Akad Nauk, Ser Khim (English transl.: Sov Report of Academy Science, Ser Chem) 2: 294
46. Zielinski TJ, Breen DL, Rein RA (1978) J Amer Chem Soc 100: 6266
47. Klopman G, Andreozzi P, Hopfinger AJ (1978) J Amer Chem Soc 100: 6167
48. Hermans PH (1949) Physics and chemistry of cellulose fibers. Elsevier, New York
49. Sarko A, Marchessault RH (1967) J Amer Chem Soc 89: 6454
50. Sarko A, Muggli R (1974) Macromolecules 7: 486
51. Kolpak FG, Wein H, Blackwell J (1976) Macromolecules 9: 273
52. Sarko A, Southwick G, Hayashi G (1976) Macromolecules 5: 341
53. Kolpak FG, Wein H, Blackwell G (1978) Polymer 19: 123
54. Hunter R, Dweltz E (1979) J Appl Polym Sci 23: 249
55. Kitaigorodsky AI (1972) Molecular crystals, Nauka, Moscow
56. Pertsin AI, Nugmanov OK, Sopin VF, Marchenko GN, Kitaigorodsky AI (1981) Vysokomol Soed (English transl.: Sov Macromol Comp) A23: 2147
57. Nugmanov OK, Marchenko GN, Karasev YA (1989) Khimia Drevesin (English transl.: Sov Chem of Wood) 1: 88
58. Arnott S, Scott WE (1972) J Chem Soc Perkin Trans Part II 3: 324
59. Arnott S, Wonacott AJ (1966) Polymer 7: 157
60. Tumanyan VG, Manaeva OK, Bocharov AL (1974) Doklady Akad Nauk SSSR (English transl.: Sov Report of Acad Sci.) 214: 1452
61. Jones DW (1958) J Polym Sci 32: 371
62. Fletchir R (1970) Trans Amer Crystallogr Assoc 6: 57
63. Lafuma F, Quivoran C (1975) J Polym Sci Polym Symp 52: 261
64. Nugmanov OK, Pertsin AI, Marchenko GN (1983) Doklady Akad Nauk SSSR (English transl.: Sov Report of Acad Sci.) 273: 140
65. Pertsin AI, Nugmanov OK, Marchenko GN, Kitaigorodsky AI (1984) Polymer 25: 107

66. Newton MD, Jeffrey GA, Takagi S (1979) J Amer Chem Soc 101: 1997
67. Jeffrey GA, Mitra I (1983) Acta Crystallogr B39: 469
68. Benoit H (1948) J Polym Sci 3: 376
69. Tsvetkov VN, Eskin VE, Frenkel SY (1964) Structure of macromolecules in solutions, Nauka, Moscow
70. Tsvetkov VN (1979) Vysokomol Soed (English transl.: Sov Macromol Comp) 11: 2606
71. Kuhn H, Kuhn W, Silberber A (1954) J Polymer 14: 183
72. Hearst J, Stockmeyer W (1962) J Chem Phys 37: 1425
73. Jayme GA (1974) In: Bikales H, Segal L (eds) Cellulose and its derivatives, vol 1, Mir, Moscow, p 279
74. Pogodina NV, Pozhivilko KS, Evlampieva NI (1981) Vysokomol Soed (English transl.: Sov Macromol Comp) A23: 1252
75. Pogodina NV, Pozhivilko KS, Mel'nikov AB (1981) Vysokomol Soed (English transl.: Sov Macromol Comp) A23: 2454
76. Rumtsev EI, Andreeva LN, Pogodina NV (1975) Vysokomol Soed (English transl.: Sov Macromol Comp) A17: 61
77. Meyerhoff G (1958) J Polym Sci 29: 399
78. Jamakawa H, Fujii M (1973) Macromolecules 6: 407
79. Jamakawa H, Fujii M (1974) Macromolecules 7: 128
80. Korneev EV, Ryumtsev EI, Andreeva LN (1979) Vysokomol Soed (English transl.: Sov Macromol Comp) A21: 1547
81. Marx–Figini M, Schulz GV (1962) Makromol Chem 54: 102
82. Tsvetkov VN (1963) Vysokomol Soed (English transl.: Sov Macromol Comp) A5: 740
83. Pogodina NV, Mel'nikov AV, Evlampieva NP (1984) Khimia Drevesin (English transl.: Sov Chemistry of Wood) 6: 3
84. Pavlov CM, Kozlov AN, Marchenko GN, Tsvetkov VN (1982) Vysokomol Soed (English transl.: Sov Macromol Comp) B24: 284
85. Tsvetkov VN (1983) Vysokomol Soed (English transl.: Sov Macromol Comp) A25: 1571
86. Bushin SV, Lusenko EB, Cheraskov VA, Smirnov KP, Didenko SA, Marchenko GN, Tsvetkov VN (1983) Vysokomol Soed (English transl.: Sov Macromol Comp) A25: 1899
87. Bushin SV, Tsvetkov VN, Lusenko EB, Emeljanov VN (1981) Vysokomol Soed (English transl.: Sov Macromol Comp) A23: 2494
88. Tsvetkov VN, Stennikova JN, Mezheritskaya NA, Bolotnikova LS (1963) Cellulose and its derivatives, Izdat Akad Nauk SSSR, Moscow
89. Zwetkoff WN (1942) Uber die Molekularordnung in der Anisotropflussigen Phase; Acta Physiochimica (USSR) 16: 132
90. Pogodina NV, Tarabukina LB, Starchenko LV (1980) Vysokomol Soed (English transl.: Sov Macromol Comp) A22: 2219
91. Tsvetkov VN (1965) Doklady Akad Nauk SSSR (English transl.: Sov Report of Akad Sci.) 165: 360
92. Tsvetkov VN (1962) Vysokomol Soed (English transl.: Sov Macromol Comp) A4: 894
93. Tsvetkov VN, Rjumtsev EI, Pogodina NV (1977) Vysokomol Soed (English transl.: Sov Macromol Comp) A19: 2141
94. Noda G, Hearst GE (1971) J Chem Phys 54: 2342

95. Tsvetkov VN (1978) Vysokomol Soed (English transl.: Sov Macromol Comp) A20: 2066
96. Peterlin A (1963) J Chem Phys 39: 224
97. Daum U (1964) J Polym Sci A2: 141
98. Tsvetkov VN, Stennikova JN, Ryumtsev EI, Skazka VS (1965) Vysokomol Soed (English transl.: Sov Macromol Comp) A7: 1111
99. Tsvetkov VN, Kolomiets JP, Lezov AV, Marchenko GN (1982) Doklady Acad Nauk SSSR (English transl.: Sov Report of Acad Sci.) 265: 1202
100. Tsvetkov VN, Stennikova JN, Rjumtsev EI (1972) Doklady Akad Nauk SSSR (English transl.: Sov Report of Acad Sci.) 207: 1173
101. Zimm B (1956) J Chem Phys 24: 269
102. Zimm B, Roc G, Epstein L (1956) J Chem Phys 24: 279
103. Eskin VE (1964) Uspekhi Phis Nauk (English transl.: Sov Success of Phys Sciences) 82: 649
104. Debue P (1944) J Appl Phys 15: 338
105. Sundelof LO (1974) In: Bikales H, Segal L (eds) Cellulose and its derivatives, Mir, Moscow, 1: 308
106. Tsepelevich SO, Marchenko GN, Tsvetkov VN (1981) Vysokomol Soed (English transl.: Sov Macromol Comp) B23: 773
107. Benoit H, Doty P (1953) J Phys Chem 57: 958
108. Schultz GV, Penzel E (1968) Makromol Chem 112: 260
109. Jamakawa H, Stockmayer WH (1972) J Chem Phys 57: 2843
110. Tsvetkov VN, Mitin YV, Stennikova JN (1965) Vysokomol Soed (English transl.: Sov Macromol Comp) A7: 1098
111. Pavlinov LI, Korshak VV, Rusanov AL (1973) Doklady Akad Nauk (English transl.: Sov Report of Acad Sci.) 211: 385
112. Pavlinov LI, Papulova LV, Miroshnichenko EA (1975) Doklady Akad Nauk SSSR (English transl.: Sov Report of Acad Sci.) 224: 143
113. Matushin YN, Korchatova LI, Sopin VF (1980) In: 1st Conference on cellulose synthesis and its regulation. Tezysi dokladov, Kazan, p 36
114. Ranby B, Rabek Y (1978) Photodestruction, photooxidation, photostabilisation of polymers, Mir, Moscow
115. Heisenberg E (1931) Cellulosechemie 12: 159
116. Karrer P, Floroni W (1923) Helv Chim Acta 6: 396
117. Kharasch MS (1929) J Res Nat Bur Stand 2: 359
118. Vestrum E, McCallaf J (1967) Physics and chemistry of solid state of organic compounds, Mir, Moscow
119. Stull DR, Westrum EF, Sinke G (1971) Chemical thermodynamics of organic compounds, Mir, Moscow
120. Cox GD, Pilcher G (1970) Thermochemistry of organic and organometallic compounds, vol 11, Academic, New York, p 646
121. Brauning BL (1967) In: Brauning BL (ed) Chemistry of wood. Lesnaya Prom, Moscow, p 27
122. Rogovin ZA, Shorygina NN (1953) Chemistry of cellulose and its satellites, Goskhimizdat, Moscow
123. Immergut EH (1967) In: Brauning BL (ed) Chemistry of wood. Lesnaya Prom, Moscow, p 114
124. Sarkanen KV (1967) In: Brauning BL (ed) Chemistry of wood. Lesnaya Prom, Moscow, p 184

125. Rogovin ZA (1964) The basic chemistry and technology of production of chemical fibers, Khimia, Moscow, 1
126. Khalatur PG, Papulov YG (1982) "Computer experiment" in conformational analysis of polymers, University Press, Kaliningrad
127. Khalatur PG, Marchenko GN, Pletneva SG, Khrapkovsky GM (1986) Doklady Akad Nauk SSSR (English transl.: Sov Report of Acad Sci.) 291: 157
128. Khalatur PG (1978) Zhur Phys Khim (English transl.: Sov J Phys Chem) 52: 805
129. Khalatur PG (1978) Vysokomol Soed (English transl.: Sov Macromol Comp) B20: 305
130. Khalatur PG (1979) Vysokomol Soed (English transl.: Sov Macromol Comp) A21: 2190
131. Khalatur PG (1979) Vysokomol Soed (English transl.: Sov Macromol Comp) B21: 605
132. Khalatur PG (1979) Vysokomol Soed (English transl.: Sov Macromol Comp) A21: 2687
133. Khalatur PG, Pletneva SG, Papulov YG (1981) Vysokomol Soed (English transl.: Sov Macromol Comp) B23: 752
134. Papulov YG, Khalatur PG (1980) Conformational calculations, University Press, Kaliningrad
135. Belnikovich NG, Bolotnikova LS, Edilyan ES (1976) Vysokomol Soed (English transl.: Macromol Comp) B18: 485
136. Flory PJ (1956) Proc Roy Soc London A, 234: 60
137. Grosberg AY, Khokhlov AR (1980) Prepr. 10/80. 71. Intern Centre Theor Phys p 60
138. Pletneva SG, Marchenko GN, Pavlov AS (1982) Doklady Akad Nauk SSSR (English transl.: Sov Report of Acad Sci.) 264: 109
139. Kirste RC (1970) Discuss. Faradey Soc 49: 51
140. Rosenbluth AN, Rosenbluth AW (1955) Chem Phys 23: 356
141. Papkov OG, Kulichikhin VG (1977) Liquid crystalline state of polymers, Khimia, Moscow
142. Tripp VW (1974) In: Bikales H, Segal L (eds) Cellulose and its derivatives, vol 1, Mir, Moscow, p 214
143. Flory P (1971) Statistical mechanics of chain molecules, Mir, Moscow
144. De Jeune P (1977) Physics of crystals, Mir, Moscow
145. Solc R (1971) J Chem Phys 55: 335
146. Brown GM, Lewy HA (1965) Science 147: 1038
147. Chy SSC, Jeffrey GA (1968) Acta Crystallogr B24: 830
148. Dashevsky VG (1975) In: Org. Chem. VINITI, Moscow, p 99
149. Khalatur PG (1980) Vysokomol Soed (English transl.: Sov Macromol Comp) A22: 2226
150. Zimin PA, Papulov YG, Seregin EA (1985) Thermodynamical calculations, University Press, Kalinin, p 57
151. Nugmanov OK (1985) Determination of the structure of crystalline regions of cellulose. Ph.D. Thesis, Kazan, KHTI
152. Nugmanov OK, Pertsin AI, Zabelin LV, Marchenko GN (1987) Uspekhi Khimii (English transl.: Sov Success of Chem) 1: 1339
153. Dugley RL, Fufe CA, Stephenson PJ, Deslandes J, Namer FK, Marchessault RH (1983) J Amer Chem Soc 105: 2468
154. Stipanovic AG, Sarko A (1976) Macromolecules 9: 851
155. Bernstein J, Hagler AT (1978) J Amer Chem Soc 100: 673

156. Pertsin AI, Nugmanov OK, Marchenko GN (1986) Polymer 27: 597
157. Gardner KN, Blackwell J (1974) Biopolymers 13: 1957; Polymer 27: 597
158. French AD (1978) Carbohydr Res 61: 67
159. Pizzi A, Eaton N (1984) South Aft Forestry J 130: 26
160. Pizzi A, Eaton N (1984) South Aft Forestry J 131: 53
161. Pizzi A, Eaton N (1987) J Macromol Sci A24: 1065
162. Bailey RW (1973) Chem and Biochem Herbage 1: 157
163. Frey–Wyssling A (1954) Science 119: 80
164. Mukherjee SM, Woods HJ (1953) Biochim et Biophys Acta 10: 499
165. Kargin VA (1960) Vysokomol Soed (English transl.: Sov Macromol Comp) 2: 466
166. Rogovin ZA (1960) Vysokomol Soed (English transl.: Sov Macromol Comp) 2: 1588
167. Battista OA (1974) In: Bikales H, Segal L (eds) Cellulose and its derivatives, vol 2, Mir, Moscow, p 412
168. Petropavlovsky GA, Kotel'nikova NE (1979) Khimia Drevesin (English transl.: Chemistry of Wood) 6: 3
169. Kotel'nikova NE, Petropavlovsky GA, Pogodina TE (1982) Cellulose Chem and Technol 16: 303
170. Honjo G, Watanabe M (1958) Nature 181: 326
171. Sharkov VI (1967) Uspekhi khimii (English transl: Sov Success of Chem) 36: 312
172. Ellefsen O, Tonnesen BA (1974) In: Bikales H, Segal L (eds) Cellulose and its derivatives, vol 1, Mir, Moscow, p 154
173. Usmanov KhU (1975) In: 1st Conference on chemistry and physics of cellulose Abstracts, Riga, p 30
174. Jones DV (1974) In: Bikales H, Segal L (eds) Cellulose and its derivatives, vol 1, Mir, Moscow, p 119
175. Wolfrom ML, Dacons JS (1952) J Amer Chem Soc 74: 5331
176. Wellard HG (1954) Nature 174: 135
177. Kerr RW (1940) J Amer Chem Soc 62: 2735
178. Manley RSJ (1964) J Polym Sci A2: 4503
179. Yamashita Y (1965) J Polym Sci A3: 3251
180. Yundt AP (1951) Tappi 34: 89
181. Horio M, Imamura R (1964) J Polym Sci A2: 638
182. Whistler RR, Chen-Ghuan-Tu (1953) J Amer Chem Soc 75: 647
183. Manley RSJ (1960) J Polym Sci 47: 509
184. Manley RSJ (1961) J Polym Sci 189: 390
185. Manley RSJ (1963) J Polym Sci A1: 1893
186. Maeda BH, Kawada H, Kawai T (1970) Makromol Chem 131: 169
187. Manley RSJ (1963) J Polym Sci A1: 1875
188. Ranby B, Gand Noe (1961) J Polym Sci 51: 337
189. Kargin VA, Slonimsky GL (1967) In: Short review of physical chemistry of polymers, MGU, Moscow, p 44
190. Bonart R, Hoseman R (1960) Makromol Chem 39: 105
191. Holland VF, Lindenmeyer PH (1965) J Appl Phys 36: 3049
192. Koutsky IA, Walton AG, Baer E (1967) J Polym Sci B5: 177
193. Bassett GA, Blundell DC (1967) J Macromol Sci B1: 161
194. McCall DM, Anderson EW, Ernest W (1959) J Polym Sci 37: 385
195. Fischer EW, Peterlin A (1964) Makromol Chem 74: 1
196. Okui N, Kawai T (1972) Makromol Chem 154: 161
197. Grubb DT, Keller A (1978) Colloid and Polym Sci 256: 218

198. Kawai T, Ujihara K (1970) Makromol Chem 132: 87
199. Kawai T, Keller A (1964) J Polym Sci B2: 333
200. Kawai T (1966) Makromol Chem 90: 288
201. Kawai T (1965) Makromol Chem 84: 290
202. Keller A (1977) J Polym Sci: Polym Symp 59: 1
203. Keller A (1977) Philos Mag 2: 1117
204. Kawai T, Hama T, Ehara K (1968) Makromol Chem 113: 282
205. Anderson FR (1963) J Polym Sci C3: 123
206. Arlie JP, Spegt P, Sconlios A (1967) Makromol Chem 104: 212
207. Lindenmeyer PH (1967) J Polym Sci C20: 145
208. Bassett DA, Keller A, Mitsuhashi S (1963) J Polym Sci A2: 763
209. Jaccodine R (1955) Nature 176: 4476
210. Till PH (1957) J Polym Sci 24: 301
211. Fischer EW (1957) Ztschr Naturforsch A12: 753
212. Flory PJ (1961) J Polym Sci 49: 151
213. Wellard HG (1954) J Polym Sci 13: 71
214. Ellis KC, Warwicker GO (1962) J Polym Sci 56: 339
215. Fischer DG, Mann G (1960) J Polym Sci 42: 139
216. Nugmanov OK, Pertsin AI, Marchenko GN (1984) Doklady Akad Nauk SSSR (English transl.: Sov Report of Acad Sci.) 277: 128
217. Kerr T, Bailey GW (1934) J Arnold Arboretum 15: 327
218. Jayme G, Fengel D (1962) Papier (BRB) 16: 519
219. Frey–Wyssling A (1969) Forschr Chem Org Naturst 27: 1
220. El-Hosseini F, Page DH (1975) Fibre Sci and Technol 8: 21
221. Page DH, El-Hosseini F, Winkler K, Lankaster APS (1977) Tappi 4: 114
222. Azzi VD, Tsai SW (1965) Exp Mech 5: 283
223. Kelley A, Devies CJ (1965) Metall Revs 10: 1
224. Hill R (1950) Mathematical theory of plasticity, Oxford Univ. Press, Oxford
225. Tsvetkov VN (1951) Zhur Experim. i Theoret Physiki (English transl.: Sov J Exper and Theor Phys) 21: 701
226. Tsvetkov VN (1962) Vysokomol Soed (English transl.: Sov Macromol Comp) 4: 1575
227. Tsvetkov VN (1967) Vysokomol Soed (English transl.: Sov Macromol Comp) A9: 1249
228. Tsvetkov VN, Skazka VS, Lavrenko PN (1971) Vysokomol Soed (English transl.: Sov Macromol Comp) A13: 2251
229. Pen'kov SN, Stepanenko NZ (1963) Optica i Spectroscopia (English transl.: Sov. Optics and Spectroscopy) 14: 156
230. Prettre M (1930) Mem Poudres, 24: 232
231. Jessup RS, Prosen EJ (1950) J Res Nat Bur Standards 44: 387
232. Woodcock C, Sarko A (1980) Macromolecules 13: 1183
233. Panov VP, Zhbankov RG, Evdakov VN (1973) Vysokomol Soed (English transl.: Sov Macromol Comp) A15: 1468
234. Liang CY, Marchessault RH (1959) J Polym Sci B37: 385
235. Usmanov KU, Razikov KK (1974) Light electronic microscopy of structural transformations of cotton, FAN, Tashkent
236. Usmanov KU, Nikonovitch GV (1962) Electronic microscopy of cellulose, FAN, Tashkent
237. Fischer EW, Schmidt GE (1962) Angew Chem 74: 15
238. Jackson EJ, Flory PJ, Chang R (1963) Trans Faradey Soc 59: 1906

Index of Scientific Names

Subject Index